Industrial and Applied Mathematics

Industrial and Applied Mathematics, a Scopus-indexed book series, publishes high-quality research-level monographs, lecture notes and contributed volumes focusing on areas where mathematics is used in a fundamental way, such as industrial mathematics, bio-mathematics, financial mathematics, applied statistics, operations research and computer science.

Abdesslam Boutayeb · Mohamed E. N. Lamlili
Editors

Computational Mathematics and Modelling for Diabetes

Springer

Editors
Abdesslam Boutayeb
Department of Mathematics
Faculty of Sciences
University Mohamed Premier
Oujda, Morocco

Mohamed E. N. Lamlili
Department of Mathematics
Faculty of Sciences
University Mohamed Premier
Oujda, Morocco

ISSN 2364-6837 ISSN 2364-6845 (electronic)
Industrial and Applied Mathematics
ISBN 978-981-96-1924-5 ISBN 978-981-96-1925-2 (eBook)
https://doi.org/10.1007/978-981-96-1925-2

Mathematics Subject Classification: 92-10, 92B99

This Springer imprint is published by the registered company Springer Nature Singapore Pte Ltd.
The registered company address is: 152 Beach Road, #21-01/04 Gateway East, Singapore 189721, Singapore

This book is dedicated to all children suffering from diabetes and especially to Anouar Boutayeb and Wiam Boutayeb. Anouar was diagnosed with type 1 diabetes at the age of three years and he is now 12 years old. Wiam has been living with type I diabetes since the age of one year. She now has a Ph.D. in the mathematical modelling of diabetes and has contributed as a co-author of five of the 10 chapters included in this book. This is the best evidence proving that a child or adult with diabetes can have a quality of life like any child or adult without diabetes, provided good understanding and managing of diabetes in all its aspects are given.

Foreword

Diabetes is a complex disease that continues to pose challenges to healthcare systems around the world. This book, Computational Mathematics and Modelling for Diabetes, presents a comprehensive collection of reviews and research chapters at the intersection of mathematical modelling, data science, and medical research. The chapters address a broad spectrum of mathematical models, ranging from traditional to state-of-the-art, designed to understand various aspects of diabetes management.

The first part of the book consists of review chapters that provide a detailed overview of the most recent developments and challenges, including analyses of compartmental models adapted to understand diabetes progression and complications. A key focus is the shift to data-driven approaches, particularly machine learning, which represents a significant tool in managing glucose homeostasis and understanding diabetes pathways. These models make use of large datasets and complex algorithms to develop personalised treatment strategies.

The second part of the book contains research chapters that add depth to this collection by addressing specific issues, such as the impact of misinformation driven by social media and proposing mitigation strategies. Other contributions include compartmental models examining health state transitions and conditions like metabolic syndrome and gestational diabetes. These models provide a more comprehensive understanding of diabetes and its related conditions. Predictive modelling is used to identify risk factors and early complications, demonstrating the practical applications of advanced techniques in understanding diabetes care.

The growing use of machine learning models is key for researchers to make better predictions and interventions, ultimately improving patient outcomes and reducing healthcare costs. This book highlights the future of medical research, where interdisciplinary mathematical techniques provide more effective treatments for diabetes while advancing scientific understanding and addressing practical issues in the field.

July 2024

Mukandavire Zindoga
Institute of Applied Research
and Technology
Emirates Aviation University
Dubai, United Arab Emirates

Preface

Diabetes is a silent epidemic affecting about 550 million people worldwide, with more than 75% of them living in low- or middle-income countries. Beside killing some 7 million people every year, this disease is also causing a high rate of morbidity due to complications of diabetes such as kidney failure, blindness, foot ulcers and cardiovascular diseases. Due to its global socioeconomic burden, diabetes care is challenging health decision makers in developed and developing countries.

Although diabetes is not curable, its burden can be relatively controlled by acting regularly on some parameters such as a healthy diet, physical exercise, weight control and healthy lifestyle in general. Consequently, it is necessary to understand the pathways of diabetes, its different aspects and especially, how it evolves from different stages (pre-diabetes, undiagnosed diabetes, known treated or not treated diabetes, and diabetes with complications).

In this direction, the present book is proposed to show how mathematical modelling and data science can be used as an obvious scientific tool that helps predict diabetes, explains the requirement and essential role of continuous monitoring in order to achieve well-controlled diabetes through glucose homeostasis, and stresses the vital need to seek a sustainable healthy life by avoiding as far as possible complications of diabetes.

As a collection of seven chapters (five reviews and two original research chapters), this book is a valuable useful source for a large range of users, including advanced researchers, academic teachers, Ph.D. and M.Sc. students, scientists willing to start modelling as well as high pharmaceutical industry executives. It can also guide health decision makers seeking cost-efficient strategies to halt the highly increasing trend of diabetes and its disastrous consequences in terms of mortality, morbidity and socioeconomic burden.

Oujda, Morocco
July 2024

Abdesslam Boutayeb
Mohamed E. N. Lamlili

Acknowledgements

Professor E. H. Twizell was supposed to be a co-editor of this book but, for family reasons, he was unable to contribute to the project as he wished. He remains an honorary editor, however, and the editors are very grateful for his useful and constructive advice.

Sincere and cordial gratitude is addressed to the reviewers who constructively reviewed the seven chapters of this book, and provided comments, remarks and suggestions to improve the content of the first version submitted by the authors. Unfortunately, two other chapters were rejected based on the comments and decision of the two experts who reviewed each chapter.

List of Reviewers

Hannah Al Ali Department of Mathematics, Emirates Aviation University, Dubai, UAE. e-mail: hannah.alali@emirates.com
Houria Ali Merina Département de Mathématiques et Informatique, Faculté des Sciences Exactes et de l'Informatique, Université Abdelhamid Ben Badis Mostaganem, Algeria. e-mail: alimerina.houria@univ-mosta.dz
Mohamed Amnai Comput. Science Research Lab, Ibn Tofail University, Kenitra, Morocco. e-mail: mohamed.amnai@uit.ac.ma
Si Lhoussain Aouragh ENSIAS, University Mohamed V, Rabat, Morocco. e-mail: l.aouragh@um5r.ac.ma
Nabil Azouagh FSTMohammedia, Université Hassan II, Casablanca, Morocco. e-mail: azouaghnabil@gmail.com
Mohammed Badaoui High School of Technology. BP. 473, 60000 Oujda, Morocco. e-mail: med.badaoui@gmail.com or med.badaoui@ump.ac.ma
Omar Balatif Faculty of Sciences Ben M'Sik, Hassan II University Mohammedia, Casablanca, Morocco. e-mail: balatif.maths@gmail.com
Omar Belhamiti Faculté des Sciences Exactes et de l'Informatique, Université Abdelhamid Ben Badis Mostaganem, Algeria. e-mail: omar.belhamiti@univ-mosta.dz
Omar Chakrone Département of Mathematics, University Mohamed Premier, Oujda, Morocco. e-mail: chakrone@yahoo.fr
Abdelrachid El Amrouss Département of Mathematics, University Mohamed Premier, Oujda, Morocco. e-mail: elamrouss@hotmail.com
Hammou El Hachemi Higher School of Technology SALE Mohamed V University, Rabat, Morocco. e-mail: elhachmi.hammou@est.um5.ac.ma
Abdellatif El Ouissari ENSA, University Sidi Mohamed Ben Abdellah, Fez, Morocco. e-mail: abdellatif.elouissari@usmba.ac.ma
Abdelghani Ghazdali ENSA, Sultan Moulay Slimane University, Khouribga, Morocco. e-mail: a.ghazdali@gmail.com
Abdelfattah Kouidere University Hassan II, Casablanca, Morocco. e-mail: kouidere89@gmail.com

Abdelilah Monir Applied Mathematical Lab. University Moulay Ismail, Meknes, Morocco. e-mail: abdelilah.monir@gmail.com
Zindoga Mukandavire IART, Emirates Aviation University, Dubai, UAE. e-mail: zindoga.mukandavire@emirates.com
Mohammed Nasri ENSA, Sultan Moulay Slimane University, Khouribga, Morocco. e-mail: mohammed.nasri@gmail.com
Silvere Paul Nuiro Département de Mathématiques et Informatique appliquées, Université des Antilles, Guadeloupe. e-mail: silverenuiro.pro@icloud.com
Aissa Sghir Département of Mathematics, University Mohammed Premier, Oujda, Morocco. e-mail: aissa.sghir@gmail.com
Mostafa Zahri Department of Mathematics. Sharjah University, Sharjah, UAE. e-mail: mzahri@sharjah.ac.ae

Contents

Contributors

Hannah Al Ali Faculty of Mathematics and Data Science, Emirates Aviation University, Dubai, United Arab Emirates

Mohammed Badaoui Higher School of Technologie, University Mohammed First, Oujda, Morocco

Fatima Zahra Benassila Endocrinologist, Endocrinology Unit, Al Farabi Hospital, Oujda, Morocco

Abdesslam Boutayeb Faculty of Mathematics and Data Science, Emirates Aviation University, Dubai, United Arab Emirates

Wiam Boutayeb Higher School of Education and Training, Oujda, Morocco;
University Mohamed Premier, Oujda, Morocco;
Mohammed First University, Oujda, Morocco

Alireza Daneshkhah Faculty of Mathematics and Data Science, Emirates Aviation University, Dubai, United Arab Emirates

Mohamed Derouich Mohammed First University, Oujda, Morocco

Maghnia Hamou-Maamar Faculty of Exact Sciences and Computer Science, University of Mostaganem, Mostaganem, Algeria

Aissa Kerkour Elmiad Faculty of Sciences, University Mohammed First, Oujda, Morocco

Abdelaadim Khriss Faculty of Sciences, University Mohammed First, Oujda, Morocco

Mohamed E. N. Lamlili Mohammed First University, Oujda, Morocco

Yasser Lamlili E. N. Mohammed First University, Oujda, Morocco

Nora Merabet Institute of Applied Research and Technology, Emirates Aviation University, Dubai, United Arab Emirates

Woody Merat Campus de Fouillole, French Universite des Antilles, Pointe-à-Pitre, France

Zindoga Mukandavire Institute of Applied Research and Technology, Emirates Aviation University, Dubai, United Arab Emirates

Silvere Paul Nuiro Campus de Fouillole, Universite des Antilles, Pointe-à-Pitre, France

Review Chapters

A Systematic Review of Mathematical Models Dealing with Diabetes Population Dynamics

Abdesslam Boutayeb, Mohamed E. N. Lamlili, and Wiam Boutayeb

Abstract The present chapter is a systematic review describing and analyzing 85 mathematical models used for diabetes population dynamics during the last 25 years. Although diabetes is not a transmitted disease, its evolution goes through different stages, including pre-diabetes, diagnosed an undiagnosed diabetes without complications, and diabetes with treated and not treated complications. The main purpose of modeling the diabetes population dynamics is to show how the prevalence of this disease can be controlled and the complications avoided or at least limited. The models selected in this systematic review used different mathematical tools such as delay, ordinary and partial differential equations, discrete and numerical approximations, fractional-order approaches, optimal control and stochastic methods. The quasi totality of authors dealing with this theme advise health decision-makers by suggesting efficient strategies susceptible to limit the highly increasing trend of diabetes incidence and to improve the well-being of people with diabetes, essentially by avoiding the occurrence of complications.

1 Introduction

Diabetes is defined as a chronic disease caused by a high level of blood glucose due to a disorder in the secretion and/or action of insulin. Over time, people with diabetes are often exposed to serious complications such as kidney failure, eye disease, foot ulcer and cardiovascular diseases in general. The main known types of diabetes are type1, type 2 and gestational diabetes [109]. A recent book edited by Boutayeb explains in details, types, aspects and risk factors of diabetes [41]. In particular, chapter 1 of this book deals with definition, diagnosis, classification, incidence and prevalence, while chapter 2 is devoted to prevention, monitoring and management of diabetes. According to the statistics released by the International Diabetes Federation (IDF)

A. Boutayeb (✉)
Emirates Aviation University, Dubai, UAE
e-mail: x.boutayeb@gmail.com

M. E. N. Lamlili · W. Boutayeb
Mohammed First University, Oujda, Morocco

A. Boutayeb and M. E. N. Lamlili (eds.), *Computational Mathematics and Modelling for Diabetes*, Industrial and Applied Mathematics,
https://doi.org/10.1007/978-981-96-1925-2_1

Table 1 Prevalence of diabetes in adults (20–79 years) in IDF Regions in 2021 and 2045

	2021			2045		
IDF Region	Number of people with diabetes (millions)	Diabetes prevalence (%)	Comparative diabetes prevalence[ii] (%)	Number of people with diabetes (millions)	Diabetes prevalence (%)	Comparative diabetes prevalence[ii] (%)
North America and Caribbean	50.5	14.0	11.9	62.8	15.2	14.2
Middle East and North Africa	72.7	16.2	18.1	135.7	19.3	20.4
Europe	61.4	9.2	7.0	69.2	10.4	8.7
South and Central America	32.5	9.5	8.2	48.9	11.9	9.8
Africa	23.6	4.5	5.3	54.9	5.2	5.6
South-East Asia	90.2	8.7	10.0	151.5	11.3	11.3
Western Pacific	205.6	11.9	9.9	260.2	14.4	11.5
World	536.6	10.5	9.8	783.2	12.2	11.2

Source International Diabetes Federation [66]
i Prevalence is standardized to each national population for the respective year
ii Prevalence is standardized to world population for the respective year

[66], it is estimated that, globally, nearly 540 million adults (20–79 years) were living with diabetes in 2021 and the number is expected to be over 780 million in 2045. Table 1 shows the number and prevalence of adults with diabetes in different regions. Globally, the number of adults with diabetes is expected to increase by 46% between 2021 and 2045 while the world diabetes prevalence is expected to increase from 10.5 to 12.2% during the same period. However, a great variation is seen between regions, indicating that diabetes prevalence in Middle East and North Africa (16.2%) is nearly the quadruple of diabetes prevalence in Africa (4.5%).

2 Method

In this chapter, a systematic review is conducted to select and present papers dealing with mathematical models devoted to population dynamics. PubMed, Web of Science, Scopus and Embase were the main databases searched electronically. However, other sources were also investigated, including Google search, research gate and

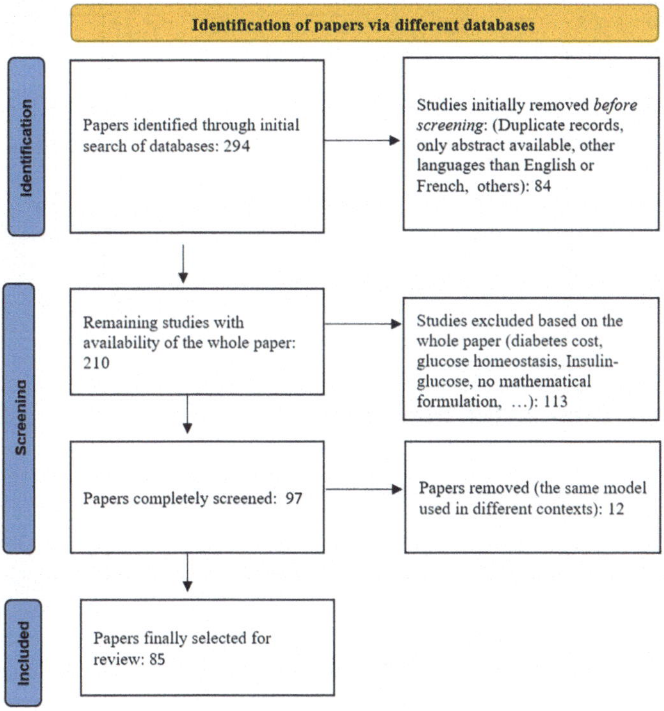

Fig. 1 Adapted from PRISMA 2020 flow diagram for new systematic review

academia. The following search terms were used: Mathematical models in diabetes, compartmental models in diabetes, diabetes population dynamics.

The search was limited to the last quarter of a century (January 2000–March 2024). Following the inclusion/exclusion criteria, all titles and abstracts obtained were screened independently by two reviewers, and a first selection was based on the availability of the whole text (in English or French). Diagram 1 indicates the selection process and the number of papers identified at each stage of the process, using PRISMA 2020 guidelines [100]. The 78 papers retained by this systematic review are presented in 8 sections, namely: discrete compartmental models, continuous compartmental models using partial differential equations, continuous compartmental models using ordinary differential equations, time-delay compartmental models, life style and genetic factors, compartmental modeling with optimal control, fractional-order models and modeling co-existence of diabetes and communicable diseases (Fig. 1).

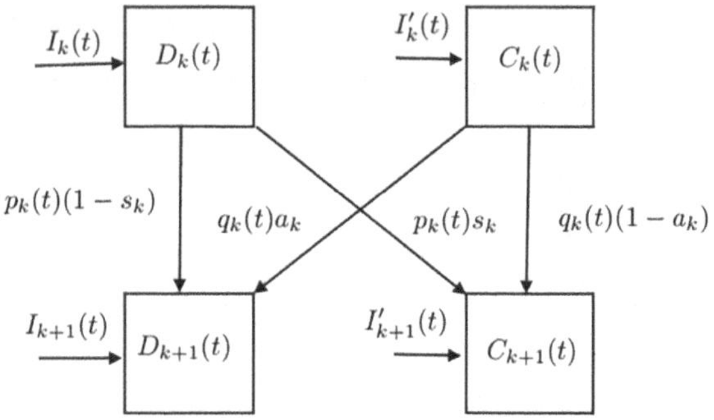

Fig. 2 Dynamics of a population of diabetes with and without complications [44]

3 Discrete Compartmental Models

Boutayeb and Derouich [44]

An early discrete mathematical model was proposed by Boutayeb and Derouich in 2002 [44]. They assumed an age structure with no complication of diabetes at birth. As indicated by the flow chart below, the total population of diabetics ($n_k(t)$) was subdivided into diabetes without complications ($D_k(t)$) and diabetes with complications ($C_k(t)$) where k stands for age and t for time (Fig. 2). The authors stressed that their model shows how the proportion of diabetics suffering from complications can be controlled by efficient strategies based on early diagnosis, regular control of blood sugar, healthy diet, physical exercise and diabetic education.

The discrete model is given by

$$\begin{aligned} D_0(t_{i+1}) = &\ I_0(t_{i+1}) + \sum_{k=0}^{m} \frac{\theta(t_i)}{2}\left(\frac{1}{\gamma_k(t_i)} - 1\right) b_k(t_i) n_k(t_i) + \frac{\theta'(t_i)}{2}(b'_k(t_i) D_k(t_i) \\ &+ b''_k(t_i) C_k(t_i)) \qquad (1) \\ C_0(t_{i+1}) = &\ 0 \qquad (2) \\ D_{k+1}(t_{i+1}) = &\ I_{k+1}(t_{i+1}) + p_k(t_i) D_k(t_i)(1 - s_k(t_{i+1})) + a_k(t_{i+1}) q_k(t_i) C_k(t_i) \qquad (3) \\ C_{k+1}(t_{i+1}) = &\ I'_{k+1}(t_{i+1}) + q_k(t_i) C_k(t_i)(1 - a_k(t_{i+1})) + p_k(t_i) s_k(t_{i+1}) D_k(t_i) \qquad (4) \end{aligned}$$

For $k = 0, 1, \ldots..m - 1$
with

$$I_{k+1}(t_{i+1}) = \frac{\varepsilon_k(t_i)}{\gamma'_k(t_i)} D_k(t_i) = \alpha_{k+1}(t_i) D_k(t_i) \qquad (5)$$

$$I'_{k+1}(t_{i+1}) = \frac{\varepsilon'_k(t_i)}{\gamma''_k(t_i)} C_k(t_i) = \alpha'_{k+1}(t_i) C_k(t_i) \qquad (6)$$

$$I_0(t_i) = \alpha_0(t_i) n_0(t_i) \qquad (7)$$

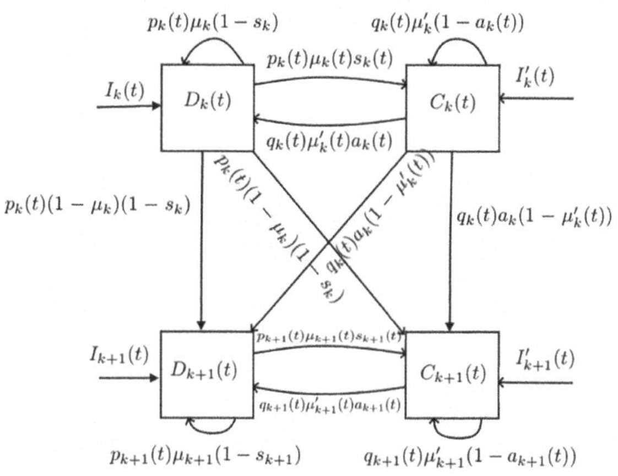

Fig. 3 Dynamics of a population of diabetes with and without complications, events being registered at the beginning of the year [44]

The following assumptions were made: $b_k(t_i) = 0, b'_k(t_i) = 0$ and $b''_k(t_i) = 0$ for $k < A_0$ and $k > A_1$.

Putting $n_{k+1}(t_{i+1}) = D_{k+1}(t_{i+1}) + C_{k+1}(t_{i+1})$ leads to the following simplified model:

$$
\begin{aligned}
n_0(t_{i+1}) &= I_0(t_{i+1}) + \sum_{k=0}^{m} \frac{\theta(t_i)}{2}(\frac{1}{\gamma_k(t_i)} - 1)b_k(t_i)n_k(t_i) + \frac{\theta'(t_i)}{2}(b'_k(t_i)D_k(t_i) \\
&\quad + b''_k(t_i)C_k(t_i)) && (8) \\
r_0(t_{i+1}) &= 0 && (9) \\
n_{k+1}(t_{i+1}) &= (p_k(t_i) + (q_k(t_i) - q_k(t_i))r_k(t_i)n_k(t_i) + \alpha_{k+1}(t_i)D_k(t_i) \\
&\quad + \alpha'_{k+1}(t_i)C_k(t_i) && (10)
\end{aligned}
$$

The rate $r_k(t_i)$ satisfies the recurrence relation:

$$
r_{k+1}(t_{i+1}) = \frac{s_k(t_{i+1})p_k(t_i)(1 - r_k(t_i)) + (\alpha'_{k+1}(t_i) + q_k(t_i)(1 - a_k(t_{i+1})r_k(t_i)}{\alpha_{k+1}(t_i) + p_k(t_i) + (q_k(t_i)) + \alpha'_{k+1}(t_i) - \alpha_{k+1}(t_i) - p_k(t_i))r_k(t_i)} \quad (11)
$$

A modified model was proposed in which migration and immigration were neglected, the number of males and females were the same and events (births, recovery, deaths...) were assumed to happen at the beginning of the year (Fig. 3).

The modified model was as follows

$$
D_0(t_{i+1}) = \mu_0(t_i)D_0(t_i) + I_0(t_{i+1}) + \sum_{k=0}^{m} \frac{\theta(t_i)}{2}(\frac{1}{\gamma_k(t_i)} - 1)b_k(t_i)n_k(t_i)
$$

$$+\frac{\theta'(t_i)}{2}(b'_k(t_i)D_k(t_i)+b''_k(t_i)C_k(t_i)) \tag{12}$$

$$C_0(t_{i+1})=0 \tag{13}$$

$$\begin{aligned}D_{k+1}(t_{i+1}) &= p_{k+1}(t_i)\mu_{k+1}(t_i)((1-s_{k+1}(t_{i+1}))D_{k+1}(t_i)\\ &+ q_{k+1}(t_i)\mu'_{k+1}(t_i)a_{k+1}(t_{i+1})C_{k+1}(t_i)+I_{k+1}(t_{i+1})\\ &+ p_k(t_i)(1-s_k(t_{i+1}))(1-\mu_k(t_i))D_k(t_i)\\ &+ a_k(t_{i+1})q_k(t_i)(1-\mu'_{k+1}(t_i))C_k(t_i)\end{aligned} \tag{14}$$

$$\begin{aligned}C_{k+1}(t_{i+1}) &= q_{k+1}(t_i)\mu'_{k+1}(t_i)(1-a_{k+1}(t_{i+1}))C_{k+1}(t_i)\\ &+ p_{k+1}(t_i)\mu_{k+1}(t_i)s_{k+1}(t_{i+1}))D_{k+1}(t_i)+I'_{k+1}(t_{i+1})\\ &+ q_k(t_i)(1-a_k(t_{i+1}))(1-\mu'_k(t_i))C_k(t_i)\\ &+ p_k(t_i)s_k(t_{i+1})(1-\mu_k(t_i))D_k(t_i)\end{aligned} \tag{15}$$

For $k=0,1,\ldots..m-1$.

More information on the model and its parameters can be found in the Ph.d thesis defended by Derouich in 2001 [56].

Kouidere et al. [73]

Following the mathematical model published by Boutayeb et al. in 2004 [46], Kouidere et al. [73] considered in 2019 a discrete mathematical model that describes the dynamics of a population of diabetics by highlighting the impact of the living environment, such as unhealthy food and health habits on diabetics without complications. The discrete compartmental model contained four compartments: a first compartment of potential diabetics (P) referring to people who are likely to have diabetes through genetics, a second compartment (D) of people who developed diabetes without complications, a third compartment (E), referring to people who are likely to have diabetes through the effect of living environment or psychological problems and finally a fourth compartment (C) containing diabetics with complications.
The parameters of the model are as follows:
μ: Natural death rate (Natural mortality is not due to diabetes and its complications).
Λ_1: Incidence of pre-diabetes.
Λ_2: Incidence due to effect of environment.
α: Probability of developing diabetes with complications because of the bad contact with th living environment.
β_1: Probability of developing diabetes without complications.
β_2: Probability for a diabetic person to develop a complication.
β_3: Probability of developing diabetes at an advanced stage of complications.
δ: Mortality rate due to complications.
γ: The rate of developing diabetes without complications due to the effect of the living environment.

$$\begin{cases} P_{k+1} = \Lambda_1 + (1 - \mu - \beta_1 - \beta_3)\, P_k, \\ D_{k+1} = \beta_1 P_k + \gamma E_k - \alpha \dfrac{D_k E_k}{N} + (1 - \mu - \beta_2)\, D_k, \\ E_{k+1} = \Lambda_2 + (1 - \mu - \gamma) E_k, \\ C_{k+1} = \beta_3 P_k + \beta_2 D_k + \alpha \dfrac{D_k E_k}{N} + (1 - \mu - \delta) C_k. \end{cases} \tag{16}$$

And T: final time with $P_0 \cdot D_0 \cdot E_0, C_0 \geq 0$.

The authors introduced optimal controls as a strategy aiming to reduce at the same time the probability of developing diabetes and its complications (Fig. 4).

The first control u_k can be interpreted as the proportion to be subjected to treatment. The second control $\mathrm{v_k}$ is related to the awareness program through media and education. The third control $\mathrm{w_k}$ aims to reduce the number of patients becoming diabetics without complications by acting on the effect of the living environment. Taking into account the three controls $\mathrm{u_k}$, $\mathrm{v_k}$ and $\mathrm{w_k}$, a mathematical model with optimal control is given by the following system.

$$\begin{cases} P_{k+1} = \Lambda_1 + (1 - \mu - \beta_1 - \beta_3)\, P_k \\ D_{k+1} = \beta_1 P_k + \gamma\,(1 - w_k)\, E_k - \alpha\,(1 - v_k)\, \dfrac{D_k E_k}{N} + (1 - \mu - \beta_2)\, D_k + u_k C_k \\ E_{k+1} = \Lambda_2 + (1 - \mu) E_k - \gamma\,(1 - w_k)\, E_k \\ C_{k+1} = \beta_3 P_k + \beta_2 D_k + \alpha\,(1 - v_k)\, \dfrac{D_k E_k}{N} + (1 - \mu - \delta) C_k - u_k C_k \end{cases} \tag{17}$$

The objective function to be minimized is given by

$$L(u, v, w) = C_T - D_T + \sum_{k=0}^{T-1} \left[C_k - D_k + \frac{A_k}{2} u_k^2 + \frac{B_k}{2} v_k^2 + \frac{G_k}{2} w_k^2 \right]$$

where $\mathrm{A_l} > 0$, $\mathrm{B_k} > 0$ and $\mathrm{G_k} > 0$ are the cost coefficients, they are selected to weigh the relative importance of $\mathrm{u_k}$, $\mathrm{v_k}$ and $\mathrm{w_k}$ at time t; T is the final time.

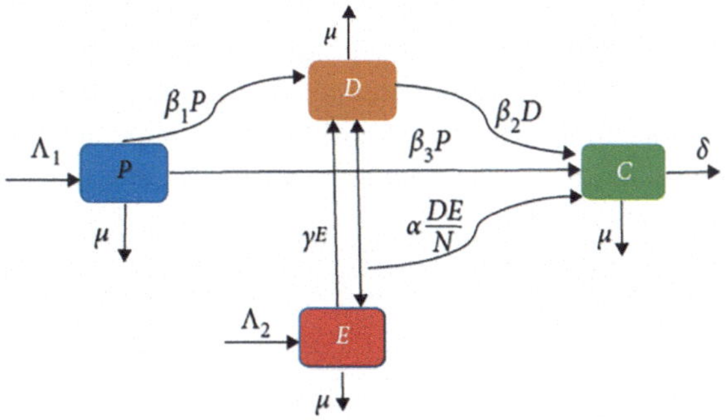

Fig. 4 The dynamics of a population of diabetics with and without complications *Source* Kouidere et al. [73], reproduced under the Creative Commons Attribution License

Different simulations were carried out using various values of parameters. Then the authors concluded that the numerical results showed the effectiveness of the proposed control strategies.

Kouidere et al. [77]

The previous discrete mathematical model was extended in 2020 by Kouidere et al. [77] in another paper introducing time delay in state and control of variables. The new model with time delay is given by the following system of differential equations.

$$\begin{cases} P_{k+1} = \Lambda_1 + (1 - \mu - \beta_1 - \beta_3) P_k \\ D_{k+1} = \beta_1 P_k + \gamma (1 - w_k) E_k - \alpha \dfrac{D_k E_k}{N} + \alpha v_{k-\tau_2} \dfrac{D_{k-\tau_2} E_{k-\tau_2}}{N} + (1 - \mu - \beta_2) D_k \\ \qquad + u_{k-\tau_1} C_{k-\tau_1} \\ E_{k+1} = \Lambda_2 + (1 - \mu) E_k - \gamma E_k + \gamma w_{k-\tau_3} E_{k-\tau_3} \\ C_{k+1} = \beta_3 P_k + \beta_2 D_k + \alpha (1 - v_k) \dfrac{D_k E_k}{N} + (1 - \mu - \delta) C_k - u_k C_k \end{cases} \tag{18}$$

In addition, for biological reasons, we assume, for $\varphi \in \{-\tau, \ldots, 0\}$, that P_φ, D_φ, E_φ and C_φ are non-negative continuous functions and $u_\varphi = 0$, $v_\varphi = 0$ and $w_\varphi = 0$.

The objective function to be minimized is given by

$$L(u, v, w) = C_T - D_T + \sum_{k=0}^{T-1} \left[C_k - D_k + \frac{A_k}{2} u_k^2 + \frac{B_k}{2} v_k^2 + \frac{G_k}{2} w_k^2 \right]$$

where $A_k > 0$, $B_k > 0$, $F > 0$ and $G_k > 0$ are the cost coefficients, they are selected to weigh the relative importance of u_k, v_k and w_k at time t; T is the final time.

The authors also carried out numerical simulations with different values of the parameters and stressed in their conclusion, the negative impact of delay in the awareness campaigns.

AlShurbaii et al. [13]

AlShurbaii et al. proposed a "Comprehensive Study of a Diabetes Mellitus Mathematical Model Using Numerical Methods with Stability and Parametric Analysis". Using the DC model of Boutayeb et al. [46], they focused on different numerical methods to monitor the size of the diabetic population with and without complications over time. They found the analytical solution (C(t), N(t)) given by equations (19) and (20) [13]:

$$C(t) = K_1 e^{-\eta_1 t} + K_2 e^{-\eta_2 t} + \frac{\alpha}{\beta} I \tag{19}$$

$$N(t) = K_1 e^{-\eta_1 t} + K_2 e^{-\eta_2 t} + \frac{\alpha}{\beta} I + \frac{\theta}{\lambda} K_1 e^{-\eta_1 t} + \frac{\theta}{\lambda} K_2 e^{-\eta_2 t} + \frac{\theta\alpha}{\lambda\beta} I - \frac{I}{\lambda} - \frac{1}{\lambda}\left(\eta_1 K_1 e^{-\eta_1 t} \eta_2 K_2 e^{-\eta_2 t}\right) \tag{20}$$

Then theoretical values were compared to numerical approximations obtained by using five numerical methods, namely: (1) explicit Euler method, (2) implicit Euler method, (3) Heun's method, (4) Runge-Kutta (RK) 4th order method and (5) Adam Bashforth-Moulton 4th order method.

Nasif et al. [87]

In the paper by Nasif et al., the authors added a control u to the model of Boutayeb et al. [46] to get the following system [87]:

$$\frac{dD}{dt} = I - \alpha(1-u)D - \mu D + \gamma C \tag{21}$$

$$\frac{dC}{dt} = \alpha(1-u)D - (\gamma + \mu + v + \delta)C \tag{22}$$

Then assuming that the probability of developing complications α is non-constant and given by $\alpha = \alpha$, (t) $= \beta$C(t)/N(t) where β is a constant, the authors considered the following optimal control model.

$$\frac{dC}{dt} = [\beta(1-u) - \theta]C - \beta\frac{C^2}{N}(1-u) \tag{23}$$

$$\frac{dN}{dt} = I - (v+\delta)C - \mu N \tag{24}$$

Applying Euler method with the step size to the previous system leads to the following non-linear discrete-time system:

$$C_{t+1} = C_t + m\left[[\beta(1-u) - \theta]C_t - \beta(1-u)\frac{C_t^2}{N_t}\right], \tag{25}$$

$$N_{t+1} = N_t + m\left[I - (v+\delta)C_t - \mu N_t\right], \tag{26}$$

This discrete system has two fixed points given by

$$I_1 = \left(0, \frac{I}{\mu}\right), \quad I_2 = \left(C^*, N^*\right),$$

where $C^* = \dfrac{kI}{k(v+\delta)+\mu\beta(1-u)}$, and $N^* = \dfrac{\beta(1-u)I}{k(v+\delta)+\mu\beta(1-u)}$, with $k = \beta(1-u)-\theta$ and $k > 0$.

The authors have also carried out numerical simulations and discussed how their model can be used to reduce and decrease the number of the population having diabetics with complications.

Akinsola et al. [7]

In 2015, Akinsola et al. [7] discussed a note on the divergence of the numerical solution of a mathematical model using the Euler method. They considered the model of Boutayeb et al. [46] approximated by the following discrete equations.

$$C_{n+1} = C_n + h(-(\theta+\lambda)C_n + \lambda C_n) \tag{27}$$

$$N_{n+1} = N_n + h(I - (v+\delta)C_n - \mu N_n) \tag{28}$$

Akinsola et al. [8]

Akinsola et al. [8] carried out numerical analysis using the solution of a previous model [6] using Eigenvalue Sensitivity Analysis (ESA) and Eigenvalue Elasticity Analysis(EEA), they showed that "The parameter with the greatest impact was determined to be the rate at which complications are controlled (γ)".

Cetinkaya [52]

Cetinkaya applied the non-standard finite difference method (NSFD) to solve equations of a mathematical model published by Aye et al. in 2021 and based on five compartments representing H(t), S(t), D(t), C(t) and T(t) denoting to the healthy (H(t)), susceptible (S(t)), diabetics without complication (D(t)), diabetics with complication (C(t)) and diabetics with complications receiving a cure (T(t)) [34]. The numerical scheme used by the author was as follows [52]:

$$H(n+1) = \frac{H(n) + (\beta\theta + \sigma S(n))\phi_1}{1 + (\mu+\tau)\phi_1} \tag{29}$$

$$S(n+1) = \frac{S(n) + (\beta(1-\theta) + \tau H(n))\phi_2}{1 + (\mu+\alpha+\sigma)\phi_2} \tag{30}$$

$$D(n+1) = \frac{D(n) + (\alpha S(n) + \omega T(n))\phi_3}{1 + (\mu+\lambda)\phi_3} \tag{31}$$

$$C(n+1) = \frac{C(n) + \lambda\phi_4 D(n)}{1 + (\mu+\delta+\gamma)\phi_4} \tag{32}$$

$$T(n+1) = \frac{T(n) + \gamma\phi_5 C(n)}{1 + (\mu+\omega)\phi_5} \tag{33}$$

where Φ_b, i = 1, 5 are denominator functions and determined as

$$\phi_1 = \frac{e^{h(\mu+\tau)} - 1}{\mu + \tau} \quad \phi_2 = \frac{e^{h(\mu+\alpha+\sigma)} - 1}{\mu + \alpha + \sigma} \quad \phi_3 = \frac{e^{h(\mu+\lambda)} - 1}{\mu + \lambda}$$
$$\phi_4 = \frac{e^{h(\mu+\delta+\gamma)} - 1}{\mu + \delta + \gamma} \phi_5 = \frac{e^{h(\mu+\omega)} - 1}{\mu + \omega}$$

The author concludes that "The accuracy of the NSFD scheme is supported by comparing the numerical results with the RKF45 method. The compatibility of the numerical results can be seen through graphics. One of the advantages of the NSFD method is to be convergent for the bigger step sizes".

Makanda [80]

In 2019, Makanda considered a mathematical model with the following four compartments: individuals diagnosed with diabetes (x_1), undiagnosed people (x_2), diabetics under treatment (x_3) and diabetic people not adhering to drugs (x_4). The flow between compartments is indicated by the chart flow given in Fig. 5 [80].

The proposed mathematical model is given by the following system of ordinary differential equations.

$$\begin{aligned} \dot{x_1} &= \Lambda S - (\mu + d_2 + 1)\, x_1 + \lambda x_2, \\ \dot{x_2} &= (1 - \Lambda)S - (\mu + d_1 + \lambda)\, x_2, \\ \dot{x_3} &= \tau x_1 - (\mu + \phi)x_3 + \nu x_4, \\ \dot{x_4} &= (1 - \tau)x_1 + \phi x_3 - (\mu + d_3 + \nu)\, x_4, \end{aligned} \tag{34}$$

with initial conditions

$$t > 0, x_1(0) = x_{10}, x_2(0) = x_{20}, x_3(0) = x_{30},$$
$$x_4(0) = x_{40}.$$

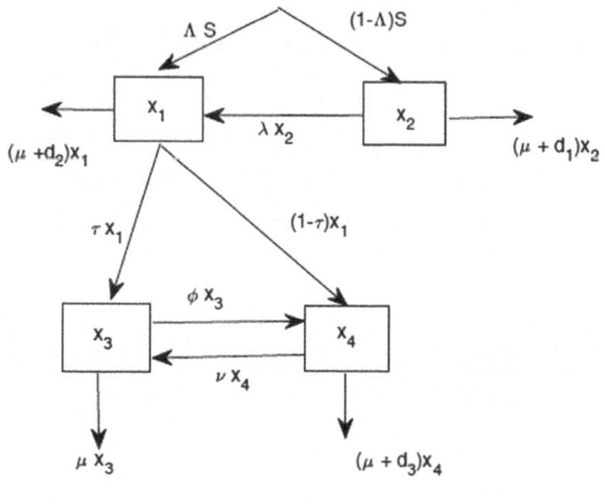

Fig. 5 Chart flow of the model *Source* Makanda [80] under the Creative Commons Attribution International License (CC BY 4.0). http://creativecommons.org/licenses/by/4.0/

The author carried out stability analysis and solved the system of ordinary differential equations (34) using two numerical methods, namely: The Jacobi method and the Gauss-Seidel method. Graphs were obtained for different values of the parameters Λ, λ, Φ and τ. As expected, the comparison between the two numerical methods showed that Gauss-Seidel method is more accurate than the Jacobi method.

Vanitha and Porchelvi [107]

Vanitha and Porchelvi solved numerical equations of the DC model of Boutayeb et al. [46]. They applied the numerical Euler-Cauchy method to find an approximation of the solution, using the following scheme [107]:

$$\begin{aligned} D_{n+1} &= D_n + \tfrac{1}{2}\,[k_{11} + k_{12}] \\ C_{n+1} &= C_n + \tfrac{f}{2}\,[k_{21} + k_{22}] \end{aligned} \tag{35}$$

where

$$\begin{aligned} k_{11} &= hF\,(t_0, D_0, C_0)\,; & k_{21} &= hG\,(t_0, D_0, C_0) \\ k_{12} &= hF\,(t_0 + h, D_0 + k_{11}, C_0 + k_{21})\,; & k_{22} &= hG\,(t_0 + h, D_0 + k_{11}, C_0 + k_{21}) \end{aligned}$$

By adding a Susceptible compartment to the DC model of Boutayeb et al. [46], Sisman and Merdan proposed in 2021, the following SDC model [68]:

$$\begin{aligned} \tfrac{dS}{dt} &= \alpha S + \alpha(1-\rho)(D+C) - \tfrac{\beta SD}{N} - \mu S \\ \tfrac{dD}{dt} &= \tfrac{\beta SD}{N} + \alpha\rho(D+C) - (\lambda+\mu)D + \gamma C \\ \tfrac{dC}{dt} &= \lambda D - (\gamma+\delta+\mu)C \end{aligned} \tag{36}$$

Then, using a numerical advanced difference method, they considered the following scheme:

$$\begin{aligned} S(n+1) &= S(n) + h(\alpha S(n) + \alpha(1-\rho)(D(n)+C(n)) \\ &\quad - \tfrac{\beta S(n)D(n)}{N} - \mu S(n)\Big) \\ D(n+1) &= D(n) + h\left(\tfrac{\beta S(n)D(n)}{N} + \alpha\rho(D(n)+C(n))\right. \\ &\quad -(\lambda+\mu)D(n) + \gamma C(n)) \\ C(n+1) &= C(n) + h(\lambda D(n) - (\gamma+\delta+\mu)C(n)) \end{aligned} \tag{37}$$

It should be stressed that, curiously, no indication was given by the authors on the meaning of the interaction $\beta SD/N$, usually explained by the rate of infection (in case of CDs)!

Side et al. [103]

Side et al. published a paper on "Numerical solution of diabetes mellitus model without genetic factors with treatment using Runge-Kutta method". Their $SEII_T$ mathematical model was based on four following compartments: a susceptible class including individuals who are not yet affected by diabetes (S), a class of exposed individuals who have bad habits, decreased insulin hormone and increased blood glucose (E) (latent population), and "infected population are classified into two classes, namely, infected class who gets treatment (I_T) and infected class without treatment (I)". The model is given by the following system: [103]

$$
\begin{aligned}
\frac{dS}{dt} &= A - \mu S - \beta S \frac{E}{N}, \\
\frac{dE}{dt} &= \beta S \frac{E}{N}, - \mu E - E, \\
\frac{dI}{dt} &= \alpha E - (\mu + \delta_1) I, \\
\frac{dI_T}{dt} &= (1 - \alpha) E - (\mu + \delta_2) I_T, \\
N &= S + E + I + I_T.
\end{aligned}
\tag{38}
$$

Important remark:
As highlighted for similar models, the authors treat diabetes as an infectious disease (I and IT). While the mathematical tools used are undoubtedly important, this approach is not biologically appropriate.

4 Continuous Compartmental Models Using Partial Differential Equations (Age Structure)

Boutayeb and Twizell [47]

An other age structured model was proposed by Boutayeb and Twizell [47] in 2004. They suggested a model using partial differential equations to illustrate the dynamics of a total population of diabetes $n(a, t)$ subdivided into a subpopulation of diabetics without complications $D(a, t)$ and a subpopulation of diabetics with complications $C(a, t)$ (Fig. 6).

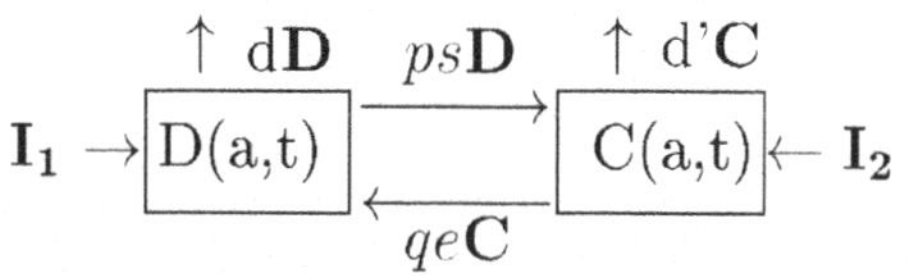

Fig. 6 A schematic representation of the model

The partial differential equation model proposed was as follows:

$$\frac{\partial D}{\partial t} + \frac{\partial D}{\partial a} = -dD - psD + qeC + I_1 \tag{39}$$

$$\frac{\partial C}{\partial t} + \frac{\partial C}{\partial a} = -d'D + psD - qeC + I_2 \tag{40}$$

$$\frac{\partial n}{\partial t} + \frac{\partial n}{\partial a} = -dD - d'D + I_1 + I_2 \tag{41}$$

Using the death rates d and $d + \delta$, a new equation is given for n by

$$\frac{\partial n}{\partial t} + \frac{\partial n}{\partial a} = -dn - \delta C + I_1 + I_2 \tag{42}$$

$$C(a, t) = r(a, t)n(a, t) \tag{43}$$

Then a differentiation with respect to a and t leads to the following equation

$$\frac{\partial r}{\partial t} + \frac{\partial r}{\partial a} = \delta r^2 - \varepsilon r + \omega \tag{44}$$

where
$\delta = \delta(a, t)$, $\varepsilon = \varepsilon(a, t) = [p(a, t)s(a, t) + e(a, t)q(a, t) + \delta(a, t) + \alpha(a, t)]$
and $\omega = \omega(a, t) = \alpha_2(a, t) + p(a, t)s(a, t)$

The b.c ($N(0, t)$ and $n(0, t)$) at zero ($a = 0$) are given respectively by

$$N(0, t) = \frac{1}{2}\int_0^A \beta_1(a, t)D(a, t) + \beta_2(a, t)C(a, t) + \beta_3(a, t)(N(a, t) - n(a, t))da \tag{45}$$

and

$$n(0, t)/N(0, t) = \tau(0, t) = \tau_1(0, t) + \tau_2(0, t) \tag{46}$$

consequently

$$r(0, t) = \frac{C(0, t)}{n(0, t)} = \frac{\tau_2(0, t)}{\tau(0, t)} \tag{47}$$

with i.c given by

$$r(a, 0) = r_0(a) = \frac{C(a, 0)}{n(a, 0)} \tag{48}$$

Table 2 Values of the parameters used in the model

Parameter	c_1	c_2	c_3	c_4	c_5	c_6	c_7	c_8
Value	0.001	0.009	0.00005	0.448	0.0001	0.00067	0.00005	0.00004
Parameter	c_9	c_{10}	c_{11}	c_{12}	c_{13}	c_{14}	c_{15}	c_{16}
Value	0.00001	0.0001	0.00001	0.00001	0.006	0.00001	0.00001	0.001

The authors used various scenarios and numerical approximations to estimate the parameters related to natural death rate (d_m^n), death rate due to complications (δ_m^n), incidence of diabetes with and without complications (α_m^n), rate at which complications are developed (p_m^n) and rate at which complications are cured (q_m^n) (Table 2).

The following expressions are used:

- $d_m^n = \dfrac{N}{N+n}(c_1 m - c_2)$
- $\delta_m^n = c_3 \dfrac{N}{N+n} \exp(c_4 m)$
- $\alpha_m^n = c_5 m(1 - c_6 m)$
- – $p_m^n = c_7 + c_8 m - c_9 m$ (Linear form)
  – $p_m^n = c_{10} + c_{11} m^2 - c_{12} n$ (Quadratic form)
  – $p_m^n = (c_{13} + c_{14} m - c_{15} n) \exp(m)$ (Exponential form)
- $q_m^n = c_{16} \dfrac{n}{N+n} \dfrac{M}{M+m}$

5 Continuous Compartmental Models Using Ordinary Differential Equations

Boutayeb et al. [46]

A continuous model using ordinary differential equations was proposed by Boutayeb et al. [46] in 2004. The total population of diabetes ($N(t)$) is subdivided into populations of diabetes without complications ($D(t)$) and diabetes with complications ($C(t)$).

The dynamics is illustrated by the flow chart below (Fig. 7) and the definition of different parameters can be found in [45].

The model proposed is governed by the following system of differential equations:

$$\frac{dD}{dt} = I(t) - (\mu + \lambda)D(t) + \gamma C(t) \; t > 0, \;\; D(0) = D_0 \tag{49}$$

$$\frac{dC}{dt} = \lambda D(t) - \theta C(t) \quad t > 0, \;\; C(0) = C_0 \tag{50}$$

where C_0 , D_0 are the i.v of $C(t)$ and $D(t)$ and $\theta = \gamma + \mu + \nu + \delta$.

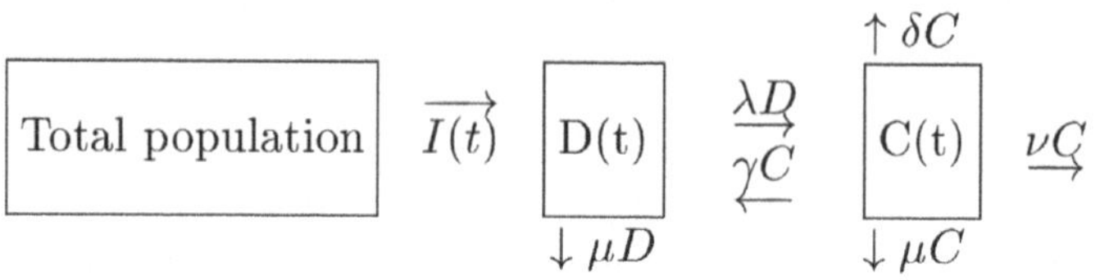

Fig. 7 Dynamics of populations with and without complications *Source* Boutayeb et al. [46]under the Creative Commons Attribution International License (CC BY 2.0). http://creativecommons.org/licenses/by/2.0/

Replacing $D(t)$ by $N(t) - C(t)$, the authors obtained the following initial-value problem (IVP):

$$\frac{dC}{dt} = -(\lambda + \theta)C(t) + \lambda N(t)\ t > 0, C(0) = C_0 \tag{51}$$

$$\frac{dN}{dt} = I(t) - (\nu + \delta)C(t) - \mu N(t)\ t > 0, N(0) = N_0 \tag{52}$$

Numerical approximations and numerical stability analysis were carried out and the authors concluded that investment in healthcare is a cost-effective strategy.

——— Boutayeb et al. [43] ———

The previous model {(51)–(52)} was modified by Boutayeb et al. [43] in 2006 by assuming that λ is given by the following expression:

$$\lambda = \lambda(t) = \beta \frac{C(t)}{N(t)}, \qquad \beta > 0 \tag{53}$$

Consequently the modified model is given by the following differential equations:

$$\frac{dC}{dt} = f_1(C, N) = (\beta - \theta)C(t) - \beta \frac{C^2}{N}\ t > 0,\ C(0) = C_0 \tag{54}$$

$$\frac{dN}{dt} = f_2(C, N) = I - (\nu + \delta)C(t) - \mu N(t)\ t > 0,\ N(0) = N_0 \tag{55}$$

The authors showed that the model has two critical points given by

1. $C^* = 0,\ N^* = \dfrac{I}{\mu}$ (an unstable trivial critical point)
2. $C^* = \dfrac{(\beta - \theta)I}{\mu\beta + (\nu + \delta)(\beta - \theta)},\ N^* = \dfrac{\beta I}{\mu\beta + (\nu + \delta)(\beta - \theta)}$ (a stable non-trivial critical point)

——— Rodrigues de Oliveira [95] ———

The previous model {(53)–(55)} was used Rodrigues de Oliveira et al. [95] in 2017 to show global stability. The author used the following Liapunov function

$$V(C, N) = [(C - C^*) - C^* ln(\frac{C}{C^*})] + W[(N - N^*) - N^* ln(\frac{N}{N^*})] \quad (56)$$

Boutayeb and Chetouani [42]

In 2007, a three-compartmental model using pre-diabetes (E), diabetes without complications (D) and diabetes with complications (C) was proposed by Boutayeb and Chetouani [42] (Fig. 8).

The three-compartmental model is given by the following system:

$$\frac{dE(t)}{dt} = I - (\mu + \lambda_3 + \lambda_1)E(t) \quad (57)$$

$$\frac{dD(t)}{dt} = \lambda_1 E(t) - (\mu + \lambda_2)D(t) + \gamma C(t) \quad (58)$$

$$\frac{dC(t)}{dt} = \lambda_3 E(t) + \lambda_2 D(t) - (\mu + \gamma + \nu + \delta)C(t) \quad (59)$$

The definition of parameters of this model can be found in [45].

Adamu et al. [1]

The model {(49)–(50)} was modified by Adamu et al. [1] in 2012 by introducing diabetic gene in susceptibility (Fig. 9).

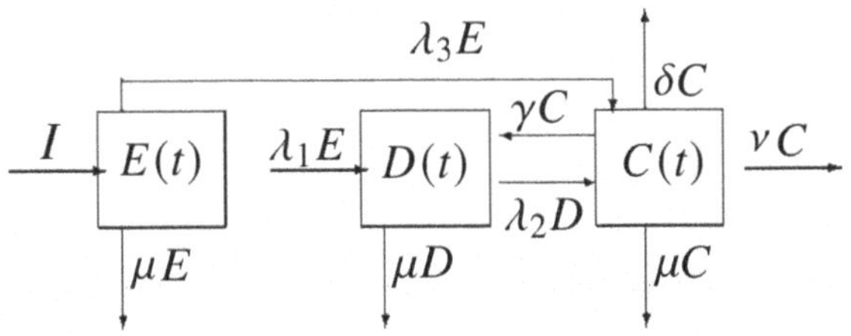

Fig. 8 Dynamics of subpopulations of pre-diabetes (E), diabetes without complications (D) and diabetes with complications (C)

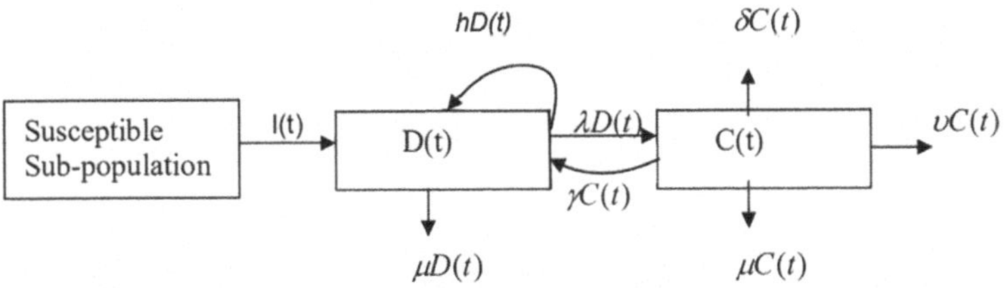

Fig. 9 Dynamics of subpopulations of diabetes without complications (D) and diabetes with complications (C)

The new model proposed is given by the following two differential equations:

$$\frac{dD}{dt} = I(t) - (\mu + \lambda - h)D(t) + \gamma C(t), \; D(0) = D_0 \quad (60)$$

$$\frac{dC}{dt} = \lambda(t))D(t) - (\gamma(t) + \mu(t) + \nu(t) + \delta(t))C(t), \; C(0) = 0 \quad (61)$$

The authors used the amount of calories burnt as a result of physical exercise "*ex*", and the level of dietary calorie intake "*d*" to estimate the parameter "*h*" by $h = \frac{ex}{d}$.

Considering the incidence as a constant ($I(t) = I$,), the following second-order derivative $\frac{d^2D}{dt^2}$ was used:

$$\frac{d^2D}{dt^2} = (\mu + \lambda - h)^2 D(t) - (\mu + \lambda - h)\gamma C(t) - (\mu + \lambda - h)I + \gamma(\lambda(t))D(t) - \theta C(t)) \quad (62)$$

This equation leads to

$$\frac{d^2D}{dt^2} = -\sigma\frac{dD}{dt} + \beta D + \theta I$$

or

$$\frac{d^2D}{dt^2} + \sigma\frac{dD}{dt} - \beta D = \theta I \quad (63)$$

where

$\theta = \gamma + \mu + \nu + \delta$

$\sigma = \theta + \mu + \lambda - h)$

$\beta = \gamma\lambda - \lambda\theta - h\theta - \theta\mu$

The authors found the auxiliary equation ($m^2 + \sigma m - \beta = 0$) and its roots $m_1 = \frac{-\sigma + \sqrt{\sigma^2 + 4\beta}}{2}$ and $m_2 = \frac{-\sigma - \sqrt{\sigma^2 + 4\beta}}{2}$

A particular solution is given $D_p(t) = -\frac{\theta I}{\beta}$ and the general solution of the quadratique equation in $D(t)$ is as follows

$$D(t) = C_1 \exp(m_1)t + C_2 \exp(m_2)t - \frac{\theta I}{\beta} \quad (64)$$

The solution of the quadratique in $C(t)$ is given by a similar form

$$C(t) = \frac{1}{\gamma}[C_1(-\eta_1 + \mu + \lambda - h)\exp(-\eta_1)t + C_2(-\eta_2 + \mu + \lambda - h)\exp(-\eta_2)t - \frac{(\mu + \lambda - h)\theta I}{\beta} - I] \quad (65)$$

where

$$\eta_1 = \frac{\sigma + \sqrt{\sigma^2 + 4\beta}}{2} \quad \text{and } \eta_2 = \frac{\sigma - \sqrt{\sigma^2 + 4\beta}}{2}$$

$$C_1 = \frac{\beta D_0(-\eta_2 + \mu + \lambda - h) - \theta I \eta_2 - I\beta}{(\eta_1 - \eta_2)\beta}$$

$$C_2 = \frac{-\beta D_0(-\eta_1 + \mu + \lambda - h) + \theta I \eta_1 + I\beta}{(\eta_1 - \eta_2)\beta}$$

Sensitivity analysis and limiting case behavior were carried out by the authors.

——— Adamu et al. [2] ———

Other modifications were introduced by Adamu et al. [2] on the previous model in 2016 ({60–61}) leading to the following model:

$$\frac{dDC(t)}{dt} = \rho(t) - \beta I(t) - (\beta + \mu + h)DC(t) + KDN(t) + \gamma C(t) \quad (66)$$

$$\frac{dDN(t)}{dt} = hDC(t) - \mu DN(t) - KDN(t) \quad (67)$$

$$\frac{dC}{dt} = \beta DC(t) - (\gamma + \mu + v + \delta)C(t),\ D(0) = 0 \quad (68)$$

$C(0) = 0$, $DC(0) = C_0$ and $DN(0) = D_0$

The reader can find more information about this model in [45].

Carrying out stability analysis, the authors found the following equilibrium points.

$$DC(t) = \frac{\rho(t) + \gamma C(t)}{(\beta + \mu + h) - \dfrac{hK}{\mu + K}} \quad (69)$$

$$DN(t) = \frac{h(\rho(t) + \gamma C(t))}{(\mu + K)[(\beta + \mu + h) - \dfrac{hK}{\mu + K}]} \quad (70)$$

$$C(t) = \frac{\beta(\rho(t) + \gamma C(t))}{(\gamma + v + \delta))[(\beta + \mu + h) - \dfrac{hK}{\mu + K}]} \quad (71)$$

The disease-free equilibrium state is given by

$$E_0 = \left(\frac{\rho(t)(\mu + K)}{[\beta + \mu + h)(\mu + K) - hK]}, \frac{h\rho(t)}{(\gamma + v + \delta))[(\beta + \mu + h) - \dfrac{hK}{\mu + K}]}, \frac{\beta\rho(t)}{[(\gamma + v + \delta))[(\beta + \mu + h) - \dfrac{hK}{\mu + K}] - \gamma]} \right)$$

Enagi et al. [63]

The model {(49)–(50)} was again used by Enagi et al. [63] in 2017 who applied the Homotopy Perturbation Method to show that the solution is given by

$$C(t) = C_0 + \alpha t + \frac{1}{2}(\lambda\beta - \alpha q)t^2 \tag{72}$$

$$N(t) = N_0 + \beta t + \frac{1}{2}(-m\alpha - \mu\beta)t^2 \tag{73}$$

where

$\alpha = (-qC_0 + \lambda N)$
$\beta = (I - mC_0 - \mu N_0)$
$\theta = \gamma + \mu + \nu + \delta$
$q = \lambda + \theta$
$m = \nu + \delta$

Simulations with different values of the parameters were also used by the authors who concluded that the parameters involved played a crucial role in the size of the population of diabetes.

Akinsola et al. [6, 9]

By introducing incidence I in both compartments D and C, Akinsola et al. [6] in 2014 modified the model of Boutayeb {(51)–(52)} to obtain the following system of equations:

$$\frac{dC}{dt} = I(t) - (\lambda + \theta)C(t) + \lambda N(t) \tag{74}$$

$$\frac{dN}{dt} = 2I(t) - (\nu + \delta)C(t) - \mu N(t) \tag{75}$$

This system has two critical points given by

$$C^*(t) = \frac{(2\lambda + \mu)I(t)}{\lambda(\nu + \delta + \mu) + \theta\mu} \tag{76}$$

$$N^*(t) = \frac{(2(\lambda + \theta) - (\mu + \delta))I(t)}{\lambda(\nu + \delta + \mu) + \theta\mu} \tag{77}$$

To find the exact solution the authors derived $\frac{dC(t)}{dt}$ and $\frac{dN(t)}{dt}$ with respect to t to obtain the second-order differential equations $\frac{d^2C(t)}{dt}$ and $\frac{d^2N(t)}{dt}$.

The second-order $\frac{d^2C(t)}{dt}$ was given by

$$\frac{d^2C(t)}{dt} + \sigma\frac{dC(t)}{dt} + \beta C(t) = \alpha I(t) \tag{78}$$

with
$\theta = \gamma + \mu + \nu + \delta$
$\sigma = \theta + \mu + \lambda$
$\beta = \lambda(\nu + \delta + \mu) + \theta\mu$
$\alpha = 2\lambda + \mu$
Finally the solution is given by

$$C(t) = K_1 \exp(m_1)t + K_2 \exp(m_2)t + \frac{\alpha}{\beta} I \tag{79}$$

$$\begin{aligned} N(t) = {} & K_1 \exp(-m_1)t + K_2 \exp(-m_2)t + \frac{\alpha}{\beta} I + \frac{\theta}{\lambda} K_1 \exp(-m_1)t + \frac{\theta}{\lambda} K_1 \exp(-m_2)t \\ & + \frac{\alpha\theta}{\lambda\beta} I - \frac{I}{\lambda} - \frac{I}{\lambda}(K_1 m_1 \exp(-m_1)t + K_2 m_2 \exp(-m_2)t) \end{aligned} \tag{80}$$

where
$$K_1 = \frac{\beta C_0(\lambda + \theta - m_2) + (\alpha m_2 - \beta)I - \lambda\beta N_0}{(\eta_1 - \eta_2)\beta}$$
$$K_2 = \frac{-\beta C_0(\lambda + \theta - m_1) - (\alpha m_1 - \beta)I + \lambda\beta N_0}{(\eta_1 - \eta_2)\beta}$$
$$m_1 = \frac{-\sigma + \sqrt{\sigma^2 + 4\beta}}{2} \text{ and } m_2 = \frac{-\sigma - \sqrt{\sigma^2 + 4\beta}}{2}$$

Akinsola and Oluyo [9] in 2020 found an analytic solution to the mathematical of their previous model using the fundamental matrix method.

Ali Merina [12]

In 2017, Ali Merina [12] assumed that the diabetic population is subdivided into two subpopulations (Fig. 10):

- The population of diabetics without complications noted D(t) dependent on time $t > 0$.
- The population of diabetics with at least one complication noted:
  - $C_a(t)$ the compartment of diabetics with acute complications at time t.
  - $C_c(t)$ the compartment of diabetics with chronic complications at time t.

with:

- $\lambda_a(t)$: probability that a diabetic has an acute complication at time t.
- $\lambda_c(t)$: probability that a diabetic has a chronic complication at time t.
- δ: rate of diabetics with acute complications who recover from their complications and return to class D.
- α: rate of diabetics with acute complications who develop chronic complications.
- μ_N: Natural mortality rate of the overall population assumed to be constant.
- υ: mortality rate due to the chronic complication.
- λ: Abrupt mortality rate due to an acute complication.

Fig. 10 Diagram of diabetic complications

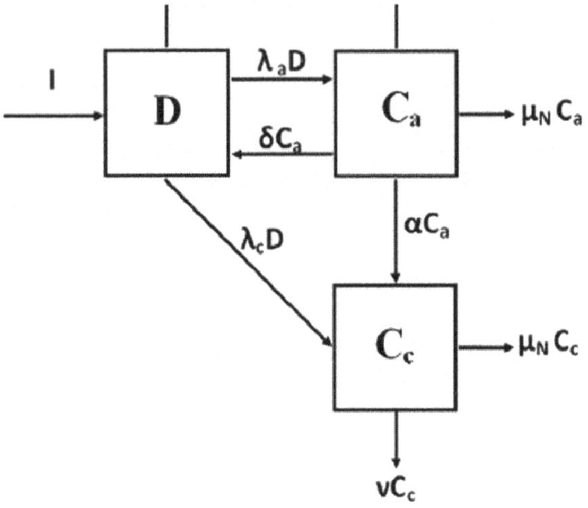

- I: The time-dependent incidence function of diabetes.

The mathematical model proposed by the author was as follows:

$$\begin{cases} \frac{d\mathrm{D}(t)}{dt} = \mathrm{I} - (\lambda_c + \lambda_a + \mu_{\mathrm{N}})\,\mathrm{D}(t) + \delta \mathrm{C}_a(t) \\ \frac{d\mathrm{C}_a(t)}{dt} = \lambda_a \mathrm{D}(t) - (\delta + \lambda + \alpha + \mu_{\mathrm{N}})\,\mathrm{C}_a(t) \\ \frac{d\mathrm{C}_c(t)}{dt} = \lambda_c \mathrm{D}(t) + \alpha \mathrm{C}_a(t) - (v + \mu_{\mathrm{N}})\,\mathrm{C}_c(t) \\ \mathrm{D}_0, \mathrm{C}_{a_0}, \mathrm{C}_{c_0} > 0, \\ \mathrm{C}(t) = \mathrm{C}_a(t) + \mathrm{C}_c(t), \\ \mathrm{N}(t) = \mathrm{C}(t) + \mathrm{D}(t). \end{cases} \tag{81}$$

Modu et al. [82]

Modu et al. [82] used an extension of the previous compartmental models used by the team of Boutayeb [46]. They subdivided the compartment of diabetics with complications $C(t)$ into two compartments: $C_1(t)$ with minor complications and $C_2(t)$ with major complications. In the first step, they considered the following linear model.

$$\begin{aligned} \frac{dD}{dt} &= -(\lambda_1 + \lambda_2 + \mu)\,D + \gamma_1 C_1 + \gamma_2 C_2 + I \\ \frac{dC_1}{dt} &= \lambda_1 D - (\delta_1 + \eta + \gamma_1 + \mu)\,C_1 \\ \frac{dC_2}{dt} &= \lambda_2 D + \eta C_1 - (\delta_2 + \gamma_2 + \mu + \nu)\,C_2 \end{aligned} \tag{82}$$

And since $N = D + C_1 + C_2$, the initial-value problems (IVP) in term of C_1, C_2 and N are

$$\begin{aligned}
\frac{dC_1}{dt} &= -(\xi + \lambda_1) C_1 - \lambda_1 C_2 + \lambda_1 N, \\
\frac{dC_2}{dt} &= (\eta + \lambda_1) C_1 - (\theta + \lambda_2) C_2 + \lambda_2 N, \\
\frac{dN}{dt} &= -\delta_1 C_1 - \Lambda C_2 - \mu N + I,
\end{aligned} \tag{83}$$

$C_1(0) = C_{10}, C_2(0) = C_{20}, N(0) = N_0, \xi = \delta_1 + \gamma_1 + \eta + \mu, \theta = \delta_2 + \gamma_2 + \mu + \nu, \Lambda = \delta_2 + \nu$, where C_{10}, C_{20} and N_0 are the initial conditions of C_1, C_2 and N respectively.

Then following Boutayeb et al. [43], they obtained the non-linear model by assuming that the probability of developing minor and major complications are respectively given by $\lambda_1 = \alpha\,(C_1/\text{N})$ and $\lambda_2 = \alpha\,(C_1/N)\,, \alpha \in [0, 1]$. Thus, by substituting $\lambda_1 = \alpha\,(C_1/N)$ and $\lambda_2 = \alpha\,(C_1/N)$, in the linear system (83) they obtained the non-linear mathematical model as follows:

$$\begin{aligned}
\frac{dC_1}{dt} &= (\alpha - \xi)C_1 - \alpha\frac{C_1C_2}{N} - \alpha\frac{C_1^2}{N}, \\
\frac{dC_2}{dt} &= \eta C_1 + (\alpha - \theta)C_2 - \alpha\frac{C_1C_2}{N} - \alpha\frac{C_2^2}{N}, \\
\frac{dN}{dt} &= -\delta C_1 - \Lambda C_2 - \mu N + I.
\end{aligned} \tag{84}$$

The authors showed that the linear system has one equilibrium which is globally asymptotically stable while the non-linear model has three equilibrium points but only one of them is globally asymptotically stable. The authors also carried out numerical simulations before concluding that their models have given insights into the various complications of diabetes. They also stressed that health decision-makers should invest in health sectors in order to minimize the high social and economic costs induced by uncontrolled diabetes and its delayed diagnosis [82].

——— Aye et al. [32–35] ———

The team of Aye and colleagues developed and published four-compartmental models dealing with the dynamics of the diabetes population [32–35].

In 2018, Aye et al. [34] were inspired by the models developed by Permatasari et al. [98] in 2018. They carried out global stability and analysis for a mathematical model of the dynamics of diabetes mellitus and its complications with five compartments representing the numbers of healthy people ($H(t)$), susceptible ($S(t)$) and diabetics without complications ($D(t)$), diabetics with complications ($C(t)$) and diabetics with complications that undergo treatment ($T(t)$). Their model was governed by the following differential equations:

$$
\begin{aligned}
\frac{dH}{dt} &= \beta - \mu H - \tau H + \sigma S \\
\frac{dS}{dt} &= \tau H - \mu S - \alpha S - \sigma S \\
\frac{dD}{dt} &= \alpha S - \mu D - \lambda D + \omega T \\
\frac{dC}{dt} &= \lambda D - \mu C - \nu C - \delta C - \gamma C \\
\frac{dT}{dt} &= \gamma C - \mu T - \omega T
\end{aligned} \tag{85}
$$

where β is the birth rate, α is the incidence of diabetes, μ is natural mortality deaths, parameter τ is the rate at which healthy individual become susceptible, σ is the rate at which susceptible individual becomes healthy, λ is the probability of a diabetic person developing a complication. Parameter γ is the rate at which diabetic with complication are treated and cured, υ is the rate at which diabetic patients with complication become severely disabled, while ω is the rate at which diabetic with complication that recovered from complications after treatment return to diabetic without complications, δ is the mortality rate due to complication.

In 2021, Aye et al extended the DC model of Boutayeb et al. [46] to a model based on the following five classes of population: Healthy (H(t)), Susceptible (S(t)), Diabetics without complications (D(t)), Diabetics with complications (C(t)) and Diabetics with complications undergoing treatment (T(t)). As indicated by the chart flow given below (Fig. 11), the definition and biological significance of the parameters used for this model are as follows:

Probability rate of incidence of diabetes (α), birth rate (β), natural mortality rate (μ), rate at which healthy individuals become susceptible (τ), rate at which susceptible individual becomes healthy (σ), rate at which individuals in $D(t)$ develop a complication (λ), rate at which individuals in $C(t)$ are treated (γ), rate at which individuals in $C(t)$ after treatment return to $D(t)$ (ω), mortality rate due to complications (δ). The population dynamics is explained by the chart flow below (Fig. 11) [33]:

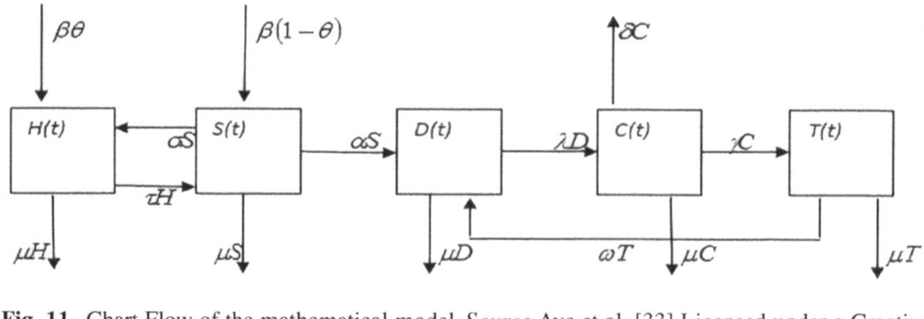

Fig. 11 Chart Flow of the mathematical model *Source* Aye et al. [33] Licensed under a Creative Commons Attribution-NonCommercial-ShareAlike 4.0 International License

The proposed mathematical model is governed by the following system of ordinary differential equations:

$$
\begin{aligned}
\frac{dH(t)}{dt} &= \beta\theta - \tau H(t) + \sigma S(t) - \mu H(t) \\
\frac{dS(t)}{dt} &= \beta(1-\theta) - \mu S(t) - \alpha S(t) - \sigma S(t) + \tau H(t) \\
\frac{dD(t)}{dt} &= \alpha S(t) - \mu D(t) - \lambda D(t) + \omega T(t) \\
\frac{dC(t)}{dt} &= \lambda D(t) - \gamma C(t) - \delta C(t) - \mu C(t) \\
\frac{dT(t)}{dt} &= \gamma C(t) - \omega T(t) - \mu T(t) \\
N(t) &= H(t) + S(t) + D(t) + C(t) + T(t)
\end{aligned}
\tag{86}
$$

The authors obtained the analytical solution of the model equations by the Homotopy Perturbation Method. Then variables and parameters values were estimated based on the available data from different international bodies like the United Nations (2019), International Diabetes Federation (IDF 2014; IDF 2018) and the World Health Organization (WHO 2016; WHO 2017). The authors also carried out numerical simulation and presented graphical profile of the system responses. Finally, they concluded that "The model described well the dynamics of diabetes mellitus and its complications and the results obtained showed that the parameters involved have an effect in determining the size of diabetic population in the country. However, an effective control measure is proposed to curtail the menace of diabetes thereby reducing the number of deaths attributable to complications of diabetes".

Aye modified the previous model (86) by incorporating two control measures, namely: the positive lifestyle (ϕ_1) and effective management of diabetes conditions (ϕ_2). Consequently, the new module illustrating the effect on control on the dynamics of diabetes mellitus is given by the following system of ordinary differential equations [32]:

$$
\begin{aligned}
\frac{dS(t)}{dt} &= \beta(1-\theta) - \mu S(t) + \tau H(t) - \alpha\,(1-\phi_1)\,S(t) - \sigma S(t), \\
\frac{dD(t)}{dt} &= \alpha\,(1-\phi_1)\,S(t) + \omega T(t) - \lambda\,(1-\phi_2)\,D(t) - \mu D(t), \\
\frac{dC(t)}{dt} &= \lambda\,(1-\phi_2)\,D(t) - \gamma C(t) - \delta C(t) - \mu C(t), \\
\frac{dT(t)}{dt} &= \gamma C(t) - \omega T(t) - \mu T(t).
\end{aligned}
\tag{87}
$$

The author solved the model equations analytically using the homotopy perturbation method and the conclusion was as follows: "The result showed that if control parameters rate could be increased, the transition rate from susceptible class to diabetics without complications class and transition rate from diabetics without complications to diabetics with complications would be drastically reduced. This will

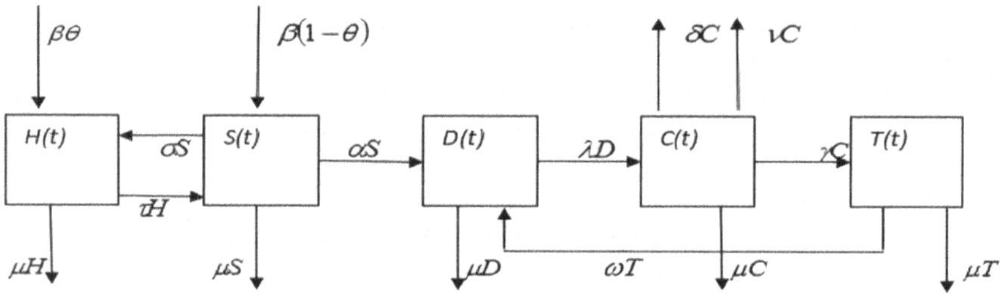

Fig. 12 Chart Flow of the mathematical model *Source* kindly reproduced from DAAM issues, University Malaysia Pahang Publishing, under CC- BY-NC-ND license

significantly reduce the incidence of diabetes and occurrence of complications in a population and the number of deaths attributable to diabetes and its complications would be minimized".

In 2022, building on the models previously published by Boutayeb et al. [46], Akinsola and Oluyo [6], Adamu et al. [2], Aye [35] presented a compartmental model dealing with a dynamics of a population subdivided into five compartments and taking into account lifestyle factors and genetic factors as causes influencing the incidence of diabetes mellitus (Fig. 12).

The author assumed that the incidence of diabetes is caused by two factors, namely: unhealthy lifestyle and genetic factors of parents who have a history of diabetes. The main unhealthy lifestyle factors are lower physical activity, irregular eating patterns and other unhealthy habits such as smoking and alcoholism. The main difference between this model and other similar published models consists in the assumption that healthy individuals will give birth to healthy children who will be born into the healthy compartment (H(t)) while parent who have diabetes or history of diabetes will give birth to children (with genetic factors) who will be born into the susceptible compartment (S(t)). Moreover, the proportion of children born in H(t) is denoted by θ while the proportion of children born in $S(t)$ is denoted (1- θ). Diabetics without complications (D(t)), diabetic with complications (C(t)) and diabetic with complications undergoing treatment (T(t)) are the other compartments of the model HSDCT which is governed by the following ordinary differential equations:

$$\frac{dH}{dt} = \beta\theta - (\mu + \tau)H + \sigma S$$

$$\frac{dS}{dt} = \beta(1 - \theta) + \tau H - (\mu + \alpha + \sigma)S$$

$$\frac{dD}{dt} = \alpha S - \mu D - \lambda D + \omega T \tag{88}$$

$$\frac{dC}{dt} = \lambda D - (\mu + \upsilon + \delta + \gamma)C$$

$$\frac{dT}{dt} = \gamma C - (\mu + \omega)T$$

Mat Daud et al. [81]

Mat Daud et al. considered the population dynamics of diabetes mellitus during pregnancy. They developed a mathematical model based on the compartments containing the following four subpopulations: non-diabetic non-pregnant women (V(t)), diabetic non-pregnant women (N(t)), diabetic pregnant women (K(t)) and pregnant women with the complication of diabetes (G(t)). Using 18 constant parameters for recruitment, flow between compartments and death rates, the proposed mathematical model is given by the following system of differential equations [81]:

$$\begin{aligned} \tfrac{dV}{dt} &= \psi - \phi_1 V(t) + \eta K(t) + \tau G(t), \\ \tfrac{dN}{dt} &= \varpi - \phi_2 N(t) + \sigma V(t) + \chi K(t) + \gamma G(t), \\ \tfrac{dK}{dt} &= \alpha V(t) + \Omega N(t) - \phi_3 K(t), \\ \tfrac{dG}{dt} &= \pi K(t) - \phi_4 G(t), \end{aligned} \tag{89}$$

where

$$\begin{aligned} \phi_1 &= \mu + \xi + \alpha + \sigma, \\ \phi_2 &= \varepsilon + \Omega + \mu + \kappa, \\ \phi_3 &= \eta + \chi + \pi + \upsilon + \delta, \\ \phi_4 &= \tau + \gamma + \theta + \beta. \end{aligned}$$

The authors showed that the solution set {V, N, K, G} of (89) with non-negative initial conditions remains non-negative for all time $t > 0$. In terms of stability analysis, they found an equilibrium point (V^*, N^*, K^*, G^*) and proved that it is asymptotically stable under suitable conditions. They also carried out numerical simulation and performed sensitivity analysis. Finally, they concluded that: "The results also show that the disease outcomes may depend on several important factors, which are represented by model parameters. If the dependence of disease outcomes on the factors are made easier to visualize by policymakers, it may assist the decision making".

Nasir and Mat Daud [93]

Following the DC model proposed by Boutayeb et al in 2004 [46], Nasir and Mat Daud suggested in 2021, a model with three compartments of the diabetes population, consisting of the diabetics without and never experienced any complication D(t), diabetics with complications $D_c(t)$ and post-treatment diabetics $D_R(t)$. The proposed model was governed by the following ordinary differential equations [93]:

$$\begin{aligned} \frac{dD(t)}{dt} &= I - \alpha D(t) - (\mu + \mu_1) D(t), \quad D(0) > 0 \\ \frac{dD_c(t)}{dt} &= \alpha D(t) + \eta D_p(t) - (\gamma + \mu + \mu_2) D_c(t), \quad D_c(0) > 0 \\ \frac{dD_p(t)}{dt} &= \gamma D_c(t) - \eta D_p(t) - (\mu + \mu_3) D_p(t), \quad D_p(0) > 0 \end{aligned} \tag{90}$$

Andima et al. [16]

In 2022, Andima et al. extended the DC model of Boutayeb et al. [46] by dividing the entire population into four compartments, including non-diabetic people who are susceptible to develop diabetes (S(t)), Diabetics (D(t)), Diabetics with complications (C(t)) and Hospitalized diabetics (H(t)). As indicated by the flow chart (Fig. 13), it was assumed that "the number of incidents that occur due to lifestyle factors is $(1-\varepsilon)$ SD/N" (Relation R). According to the authors, this assumption was proposed by Hill et al. [65]. However, although the paper published by Hill et al. in 2013 was dedicated to the relationship between social determinants of health and diabetes, no mathematical relation was given. Consequently, the use of the mixed product SD in the relation R is not justified but rather "falsely borrowed" from communicable diseases [16]:

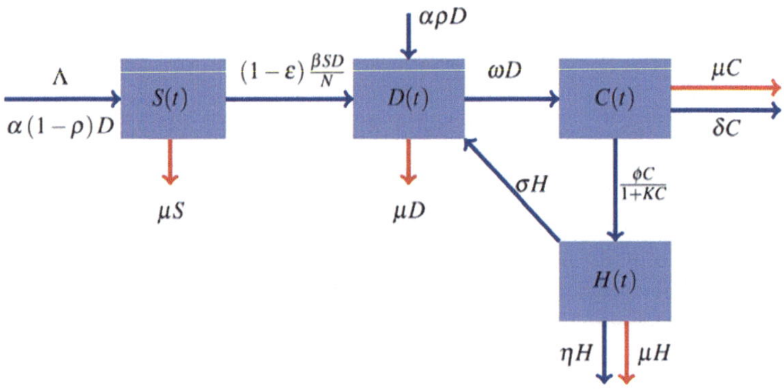

Fig. 13 Flowchart for the model considering intervention. *Source* From Andima et al. [16], under the Creative Commons Attribution International License (CC BY 4.0). http://creativecommons.org/licenses/by/4.0/

The mathematical model proposed was governed by the following differential equations:

$$
\begin{aligned}
\frac{dS}{dt} &= \Lambda + \alpha(1-\rho)D - (1-\varepsilon)\beta\frac{SD}{N} - \mu S\\
\frac{dD}{dt} &= \alpha\rho D + (1-\varepsilon)\beta\frac{SD}{N} + \sigma H - \omega D - \mu D,\\
\frac{dC}{dt} &= \omega D - \frac{\phi C}{1+KC} - \delta C - \mu C\\
\frac{dH}{dt} &= \frac{\phi C}{1+KC} - \sigma H - \eta H - \mu H.
\end{aligned}
\tag{91}
$$

Moya et al. [85]

Moya et al. proposed a mathematical dealing with overweight and obesity in a population and their contribution to new cases of diabetes. As indicated by the chart flow below, the model is based on the following four compartments: normal weight individuals (S), overweight individuals (O_w), obese individuals (O_b) and people with diabetes (D) [85] (Fig. 14).

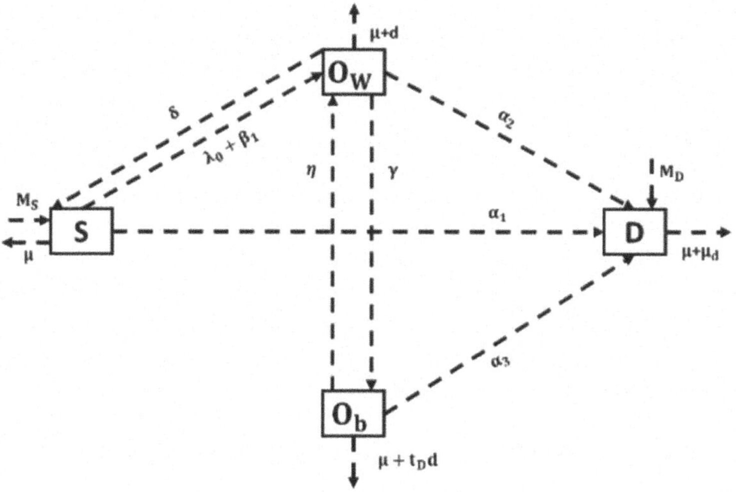

Fig. 14 Chart flow of the model. *Source* From Moya et al. [85], under the Creative Commons Attribution International License (CC BY 4.0). http://creativecommons.org/licenses/by/4.0/

Moya et al. [85] based their model on the fact that an overweight and/or obese person can have an influence on people who have a normal weight (S) and consequently, they assumed a transmission rate of the form: $\lambda_0 = \alpha * (O_w + \varepsilon O_b)/N$ where "the effective contact rate ($\alpha*$) is defined as the contact between an individual from the O_w and O_b compartments and an individual from S which will exert a negative effect by changing the lifestyle of the individual from S". The proposed mathematical model is governed by the following differential equations:

$$\begin{aligned}
\frac{dS}{dt} &= M_S + \delta O_w - (\lambda_o + \mu + \alpha_1 + \beta_1)\, S \\
\frac{dO_w}{dt} &= (\lambda_o + \beta_1)\, S + \eta O_b - (\gamma + \mu + d + \delta + \alpha_2)\, O_w \\
\frac{dO_b}{dt} &= \gamma O_w - (\eta + \mu + t_D d + \alpha_3)\, O_b \\
\frac{dD}{dt} &= M_D + \alpha_1 S + \alpha_2 O_w + \alpha_3 O_b - (\mu_d + \mu)\, D
\end{aligned} \tag{92}$$

with the initial conditions

$$S_0 > 0, \quad O_{w0} \geq 0, \quad O_{b0} \geq 0, \quad D_0 \geq 0.$$

The authors analyzed the system of differential equations (92), determined the basic reproductive number and then carried out numerical simulations.

Awad et al. [26]

In 2018, Awad and colleagues constructed a deterministic compartmental mathematical model describing the four dimensions of sex, age, risk factor status and T2DM status for a given population. The model incorporates the following four classes: healthy (H), obese (O), smoker (S) and physically inactive (F) where healthy stands for non-obese, non-smoker, physically active and non-diabetic. They applied their model to the Qatari population disaggregated into 20 five-year age bands ($0-4, 5-9\ldots.$... 95–99 years old) hence leading to a model with a set of 640 differential equations, including the 16 differential equations separately for men and women in each age category (a) and considering equations related to susceptible population and population with T2DM for up to one risk factor and for more than one risk factor, namely obese-smoker (OS), obese-physically inactive (OF), smoker-physically-inactive (SF) and obese-smoker-physically-inactive (OSF). The chart flow of the model is below [26] (Fig. 15):

The mathematical model is governed by the following system of differential equations:

$$a = 1:$$

$$\frac{dH_1}{dt} = bN(t) - (\mu_1 + \zeta)\, H_1(t)$$

Susceptible population with up to one risk factor (for > 4 years old)

$$a > 1:$$

$$\begin{aligned}
\frac{dH_a}{dt} &= \varsigma H_{a-1}(t) + \sigma_a O_a(t) + \delta_a S_a(t) + \varphi_a F_a(t) - (\lambda_a + \alpha_a + \beta_a + \Im_a + \mu_a + \varsigma)\, H_a(t) \\
\frac{dO_a}{dt} &= \varsigma O_{a-1}(t) + \alpha_a H_a(t) + \varepsilon_a OS_a(t) + \theta_a OF_a(t) - (\lambda_a RR_O + \upsilon_a + \sigma_a + \mu_a + \varsigma)\, O_a(t) \\
\frac{dS_a}{dt} &= \varsigma S_{a-1}(t) + \beta_a H_a(t) + \gamma_a OS_a(t) + \pi_a SF_a(t) - (\lambda_a RR_S + \chi_a + \omega_a + \delta_a + \mu_a + \varsigma)\, S_a(t) \\
\frac{dF_a}{dt} &= \varsigma F_{a-1}(t) + \Im_a H_a(t) + \rho_a SF_a(t) + \ni_a OF_a(t) - (\lambda_a RR_F + \xi_a + \psi_a + \mu_a + \varsigma)\, F_a(t)
\end{aligned} \tag{93}$$

Susceptible population with overlap of more than one risk factor (for > 4 years old)

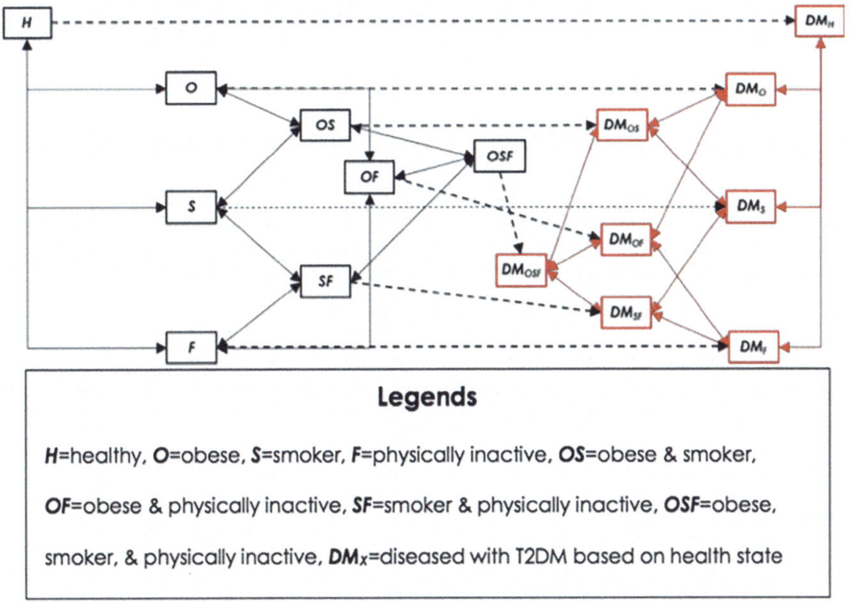

Fig. 15 Conceptual framework for the epidemiological dynamics of type 2 diabetes mellitus in the population, and the effects of T2DM-related risk factors on T2DM. *Source* From Awad et al. [26], under the Creative Commons Attribution International License (CC BY 4.0). http://creativecommons.org/licenses/by/4.0/

$$
\begin{aligned}
\frac{dOS_a}{dt} &= \varsigma OS_{a-1}(t) + \upsilon_a O_a(t) + \chi_a S_a(t) + \ddot{\imath}_a OSF_a(t) \\
&\quad - (\varepsilon_a + \gamma_a + \kappa_a + \lambda_a RR_{aS} + \mu_a + \varsigma)\, OS_a(t) \\
\frac{dOF_a}{dt} &= \varsigma OF_{a-1}(t) + \eta_a O_a(t) + \psi_a F_a(t) + \eta_a OSF_a(t) \\
&\quad - (\ni_a + \theta_a + \epsilon_a + \lambda_a RR_{OF} + \mu_a + \varsigma)\, OF_a(t) \\
\frac{dSF_a}{dt} &= \varsigma SF_{a-1}(t) + \omega_a S_a(t) + \xi_a F_a(t) + \upsilon_a OSF_a(t) \\
&\quad - (\pi_a + \rho_a + \Omega_a + \lambda_a RR_{SF} + \mu_a + \varsigma)\, SF_a(t) \\
\frac{dOSF_a}{dt} &= \varsigma\, \mathrm{OSF}_{a-1}(t) + \kappa_a OS_a(t) + \epsilon_a OF_a(t) + \Omega_a SF_a(t) \\
&\quad - (\ddot{a}_a + \eta_a + \upsilon_a + \lambda_a RR_{OSF} + \mu_a + \varsigma)\, OSF_a(t)
\end{aligned}
\tag{94}
$$

Populations with T2DM with up to one risk factor (for > 4 years old)

$$
\begin{aligned}
\frac{dD_{H,a}}{dt} &= \varsigma D_{H,a-1}(t) + \lambda_a H_a(t) + r_a D_{o,a}(t) + q_a D_{S,a}(t) + p_a D_{F,a}(t) \\
&\quad - (m_a + n_a + o_a + \mu_a + cf_a + \varsigma)\, D_{H,a}(t) \\
\frac{dD_{O,a}}{dt} &= \varsigma D_{O,a-1}(t) + \lambda_a RR_O O_a(t) + m_a D_{H,a}(t) + a_a D_{OS,a}(t) + b_a D_{OF,a}(t) \\
&\quad - (e_a + f_a + r_a + \mu_a + cf_a + \varsigma)\, D_{O,a}(t) \\
\frac{dD_{S,a}}{dt} &= \varsigma D_{S,a-1}(t) + \lambda_a RR_S S_a(t) + n_a D_{H,a}(t) + i_a D_{OS,a}(t) + c_a D_{SF,a}(t) \\
&\quad - (j_a + g_a + q_a + \mu_a + cf_a + \varsigma)\, D_{S,a}(t) \\
\frac{dD_{F,a}}{dt} &= \varsigma D_{F,a-1}(t) + \lambda_a RR_F F_a(t) + o_a D_{H,a}(t) + k_a D_{OF,a}(t) + d_a D_{SF,a}(t) \\
&\quad - (p_a + l_a + h_a + \mu_a + cf_a + \varsigma)\, D_{F,a}(t)
\end{aligned} \tag{95}
$$

Populations with T2DM with more than one risk factor (for > 4 years old)

$$
\begin{aligned}
\frac{dD_{OS,a}}{dt} &= \varsigma D_{OS,a-1}(t) + \lambda_a RR_{OS} OS_a(t) + s_a D_{OSF,a}(t) + e_a D_{O,a}(t) + j_1 D_{S,a}(t) \\
&\quad - (x_a + a_a + i_a + \mu_a + cf_a + \varsigma)\, D_{OS,a}(t) \\
\frac{dD_{OF,a}}{dt} &= \varsigma D_{OF,a-1}(t) + \lambda_a RR_{OF} OF_a(t) + t_a D_{OSF,a}(t) + f_a D_{O,a}(t) + l_a D_{F,a}(t) \\
&\quad - (w_a + b_a + k_a + \mu_a + cf_a + \varsigma)\, D_{OF,a}(t) \\
\frac{dD_{SF,a}}{dt} &= \varsigma D_{SF,a-1}(t) + \lambda_a RR_{SF} SF_a(t) + u_a D_{OSF,a}(t) + h_a D_{F,a}(t) + g_a D_{S,a}(t) \\
&\quad - (d_a + c_a + v_a + \mu_a + cf_a + \varsigma)\, D_{SF,a}(t) \\
\frac{dD_{OSF,a}}{dt} &= \varsigma D_{OSF,a-1}(t) + \lambda_a RR_{OSF} OSF_a(t) + x_a D_{OS,a}(t) + w_a D_{OF,a}(t) + v_a D_{SF,a}(t) \\
&\quad - (s_a + t_a + u_a + \mu_a + cf_a + \varsigma)\, D_{OSF,a}(t)
\end{aligned} \tag{96}
$$

Ayoade and Agboola [36]

Starting from the DC model of Boutayeb et al. [46], Ayoade and Agboola proposed a DCA model, claiming that their "analysis considered diabetes mellitus of type 2 and attributed its incidence to reckless lifestyles and heredity". Consequently, as indicated by the flow chart in Fig. 16, their modification introduced a compartment A(t) standing for "diabetics who have become critically disabled" [36]:

The proposed mathematical model is given by the following ordinary differential equations:

$$
\begin{aligned}
\frac{dD}{dt} &= \phi - (\lambda + \mu)D(t) + \gamma C(t), \\
\frac{dC}{dt} &= \lambda D(t) - (\gamma + \mu + \delta + \nu)C(t), \\
\frac{dA}{dt} &= \nu C(t) - (\mu + \delta)A(t).
\end{aligned} \tag{97}
$$

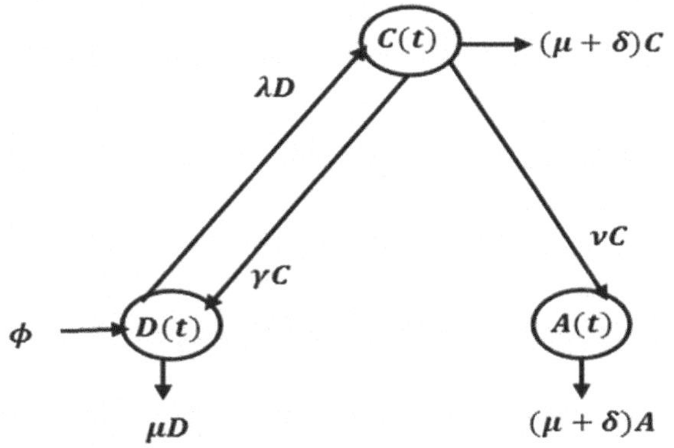

Fig. 16 Flow chart of the mathematical model *Source* From Ayoade and Agboola [36], under the Creative Commons Attribution International License (CC BY 4.0). http://creativecommons.org/licenses/by/4.0/

However, the authors stressed that the third equation $\left(\dot{\mathrm{A}}(\mathrm{t})\right)$ can be neglected from the analysis and hence, the model they really reconsidered is the DC model of Boutayeb et al given by

$$\begin{aligned}\frac{dD}{dt} &= \phi - (\lambda + \mu)D(t) + \gamma C(t),\\ \frac{dC}{dt} &= \lambda D(t) - \alpha C(t),\end{aligned} \tag{98}$$

where

$$\alpha = \gamma + \mu + \delta + v.$$

The authors pointed that this model "is developed by following the same assumptions and terminology as in [46]. The definitions for the model parameters also remain as in [46] except that (I) is changed to ϕ and it denotes the recruitment rate of diabetics due to heredity and unbalanced lifestyles".

——— Anusha and Athithan [18] ———

Anusha and Athithan developed a deterministic mathematical model of diabetes (type 2 diabetes) dealing with the diabetes population dynamics. The model proposed is based on the following four compartments: Susceptible (S), Imbalance Glucose Level (I), Treatment (T) and Restrain population (R). The model is governed by the following ordinary differential equations [18]:

$$\begin{aligned}\frac{dS}{dt} &= \Lambda - \sigma S - \nu S + lR,\\ \frac{dI}{dt} &= \sigma S - \beta I - \rho IT - \nu I,\\ \frac{dT}{dt} &= \beta I + \rho IT - \gamma T - v T,\\ \frac{dR}{dt} &= \gamma T - \nu R - lR.\end{aligned} \tag{99}$$

The authors showed that their model exhibits a Disease Free Equilibrium (DFE) and Endemic Equilibrium (EE). They also computed the expression for the Treatment reproduction number R_T and proved that the equilibrium points of the model are locally and globally asymptotically stable under suitable conditions. Finally, they performed numerical simulations and tried to fit their model with the data given by the International Diabetes Federation (IDF. Atlas 8th edition 2017).

6 Time Delay Compartmental Models

——— Nasir [90] ———

In a time-delay model, Nasir [90] investigated the stability and Hopf bifurcation with respect to a time delay τ using the following system of differential equations:

$$\begin{aligned} \frac{dD(t)}{dt} &= I - \gamma D(t-\tau) - (\mu + \mu_1)\, D(t) \\ \frac{dD_c(t)}{dt} &= \gamma D(t-\tau) + \sigma D_p(t) - (\kappa + \mu + \mu_2)\, D_c(t) \\ \frac{dD_p(t)}{dt} &= \kappa D_c(t) - (\sigma + \mu + \mu_3)\, D_p(t) \end{aligned} \tag{100}$$

where the variables $D(t)$, $D_C(t)$, and $D_P(t)$ are the people with diabetes who never presented any complications, the people with diabetes with complications and the people with diabetes with recovered complications at time t, respectively. The parameter I, which is constant over time, is the incidence rate of diabetes, γ is the rate of the first incidence of diabetes complications, τ is the time delay in developing the first complication, κ is the recovery rate of complications, σ is the recurrence rate of complications, μ is the death rate due to other causes than diabetes and $\mu_i (i = 1, 2, 3)$ are the diabetes-related death rates in the respective compartments. All the parameters in model (101) are assumed to be positive.

——— Nasir [88] ———

Stressing that in the previous model, the recovery rate from the stage of people with diabetes with complications to people with diabetes with recovered complications was of the form $h\,(D_C(t)) = \kappa D_C(t)$, Nasir [88] presented another two delays model in which the recovery rate has the unsaturated from

$$h\,(D_c(t)) = \frac{\kappa D_c(t)}{1 + \beta D_c(t)}$$

The new model is governed by the following differential equations:

$$\begin{aligned}
\frac{dP(t)}{dt} &= \Lambda - \alpha P(t-\tau_1) - \mu P(t) \\
\frac{dD(t)}{dt} &= \alpha P(t-\tau_1) - \gamma D(t-\tau_2) - (\mu+\mu_1) D(t) \\
\frac{dD_c(t)}{dt} &= \gamma D(t-\tau_2) - \frac{\kappa D_c(t)}{1+\beta D_c(t)} + \sigma D_p(t) - (\mu+\mu_2) D_c(t) \\
\frac{dD_p(t)}{dt} &= \frac{\kappa D_c(t)}{1+\beta D_c(t)} - (\sigma+\mu+\mu_3) D_p(t)
\end{aligned} \tag{101}$$

where $P(t)$ is the number of people without diabetes at time t, Λ is the influx of people without diabetes, α is the rate of developing diabetes, τ_1 is the time delay in developing diabetes, τ_2 is the time delay in developing the first complication after the onset of diabetes and β is the inhibition effect measuring the limited medical resources. The other symbols are de ned similarly as in model (101).

Inspired by his previous works [88, 90], Nasir indicated five types of the function h (Dc(t)) for the recovery rate of diabetes complications defined as follows:

1. The linear treatment rate,

$$h(D_c(t)) = \kappa D_c(t),$$

2. The saturated treatment rate,

$$h(D_c(t)) = \frac{\kappa D_c(t)}{1+\beta D_c(t)},$$

3. The treatment rate of the form,

$$h(D_c(t)) = \begin{cases} \kappa D_c(t), & \text{if } 0 \le D_c(t) \le D_c^{\max}, \\ \kappa D_c^{\max}, & \text{if } D_c(t) > D_c^{\max} \end{cases}$$

4. The saturated treatment rate of the form,

$$h(D_c(t)) = \frac{D_c^{\max} D_c(t)}{\omega + D_c(t)},$$

5. The Monod-Haldane type of treatment rate,

$$h(D_c(t)) = \frac{a D_c(t)}{b + D_c^2(t)}.$$

The author recalled that the linear treatment rate (1) used in model [90] and the saturated treatment rate (2) was used in model [88] and then proposed a new two delays model described by the following system differential equations:

$$\begin{aligned}
\frac{dP(t)}{dt} &= \Lambda - \alpha P(t-\tau_1) - \mu P(t) \\
\frac{dD(t)}{dt} &= \alpha P(t-\tau_1) - \gamma D(t-\tau_2) - (\mu+\mu_1) D(t) \\
\frac{dD_c(t)}{dt} &= \gamma D(t-\tau_2) - h(D_c(t)) + \sigma D_p(t) - (\mu+\mu_2) D_c(t) \\
\frac{dD_p(t)}{dt} &= h(D_c(t)) - (\sigma+\mu+\mu_3) D_p(t)
\end{aligned} \tag{102}$$

The state $P(t)$ is the number of non-diabetics at time t, $D(t)$ is the diabetics who have never presented any complications, $D_C(t)$ is the diabetics with complications, and $D_P(t)$ is the diabetics with recovered complications. The parameters in the model (102) are explained as follows. The positive constant Λ is the recruitment rate of non-diabetics, α is the incidence rate of diabetes, τ_1 is the delay in developing diabetes, γ is the rate of developing the first complication, τ_2 is the delay in developing the first complication, $h(D_C(t))$ is a function for the rate at which complications are recovered, σ is the rate of recurrence of complications, μ_1 is the diabetes-related mortality rate among diabetics without complications, μ_2 is the diabetes-related mortality rate among diabetics with complications, μ_3 is the diabetes-related mortality rate among diabetics with recovered complications, and μ is the mortality rate from other causes than diabetes.

The function h (D_C(t)) is assumed to satisfy the following conditions:

(HI) $h(D_c(t)) \geq 0$ for $D_c(t) \geq 0$ with equality if and only if $D_c(t) = 0$, and $\frac{dh(D_c(t))}{dD_c(t)} > 0$ for $D_c(t) \geq 0$.

7 Life Style and Genetic Factors

——— Widyaningsih et al. [110] ———

The model {(49)–(50)} was modified by Widyaningsih et al. [110] in 2018, suggesting that genetic factors and life style factors are causes of incidence. The suggested Susceptible Diabetes Complication (SDC) model is as follows:

$$\frac{dS}{dt} = \alpha S + \alpha(1-\rho)(D+C) - \beta SD/N - \mu S \tag{103}$$

$$\frac{dD}{dt} = \alpha\rho(D+C) - (\lambda+\mu)D + \gamma C \tag{104}$$

$$\frac{dC}{dt} = \lambda D - (\gamma+\delta+\mu)C \tag{105}$$

with initial conditions $S(0) > 0$, $D(0) > 0$, and $C(0) > 0$.

The parameters used in this model are: birth rate, interaction rate, recovery rate of complications, death rate due to complications, occurrence rate of complications, natural mortality rate and the proportion of genetic disorder's birth, denoted respectively by $\alpha, \beta, \gamma, \delta, \lambda, \mu, \rho$ with $0 \leq \rho \leq 1$.

——— Nasir [89] ———

Criticizing Widyaningsih et al. [110] for using a standard incidence rate of the form (βDS/N) (usually used for communicable diseases), Nasir [89] proposed a more general function of the form: $f(N) = \beta(1 + \eta N)$, leading to the following model.

$$\begin{aligned} \frac{dP}{dt} &= \Lambda - \beta(1 + \eta(P + D))P - \mu P, \\ \frac{dD}{dt} &= \beta(1 + \eta(P + D))P - (\mu + \delta)D. \end{aligned} \tag{106}$$

Then the authors dealt with the positivity of solutions, equilibrium points, local sensitivity analysis, existence and characterization of optimal control.

——— Kouidere et al. [75] ———

In 2021, highlighting the effect of lifestyle, Kouidere et al. [75] extended the mathematical model published by Boutayeb et al. [46]. The authors considered a mathematical model PEDC with four compartments in which P represents pre-diabetic people through genetic factors (Genetic predisposition), E stands for pre-diabetics due to the negative impact of socio-environmental factors on diabetics (without Genetic predisposition), D is the compartment of diabetics without complications and C that of diabetics with complications (cardiovascular disease, foot damage, neuropathy, nephropathy and retinopathy). The total population of individuals noted N, is given at a time t by $N(t) = P(t) + E(t) + D(t) + C(t)$. The model proposed is given by the following system of ordinary differential equations:

$$\begin{cases} \dfrac{dP(t)}{dt} = I_1 - (\mu + \beta_1 + \beta_3)\, P(t) \\ \dfrac{dE(t)}{dt} = I_2 - (\mu + \gamma)E(t) \\ \dfrac{dD(t)}{dt} = \beta_1 P(t) + \gamma E(t) - \alpha \dfrac{D(t)E(t)}{N} - (\mu + \beta_2)\, D(t) \\ \dfrac{dC(t)}{dt} = \beta_3 P(t) + \beta_2 D(t) + \alpha \dfrac{D(t)E(t)}{N} - (\mu + \delta)C(t) \end{cases} \tag{107}$$

with $P(0) \geq 0$, $D(0) \geq 0$, $E(0) \geq 0$, $C(0) \geq 0$ are the given initial states.
μ: Natural death rate (Natural mortality is not due to diabetes and its complications).
I_1: Denote the incidence of pre-diabetes due to genetic factors. I_1: Denote the incidence of pre-diabetes due to lifestyle (living environment).
α: The rate of negative effect of socio-environmental factors $E(t)$ on diabetics people D(t).
β_1: The rate of pre-diabetics who have been diabetics without complications.

β_2: rate of diabetics whose complications are cured.
β_3: The rate of diabetics becoming diabetics with complications through the sudden sock.
δ: Mortality rate due to complications.
γ: The rate of diabetics people probability of developing diabetes through the negative effect of socio-environmental factors on diabetics.

Hill et al. [65]

The authors stressed that, according to Hill et al. [65], the interaction between individuals without diabetes E(t) who have unhealthy lifestyles and diabetics without complications $D(t)$ leads to negative health outcomes for those with diabetes.

Then the authors adopted a strategy of control through an awareness program aiming to minimize the negative effect of behavioral factors on diabetics without complication. In order to minimize the number of people evolving from the stage of pre-diabetes to the stages of diabetes with and without complications, three controls $u(t)$, $v(t)$ and $w(t)$ were introduced. The first control $u(t)$ represents the treatment of complications of diabetes. The second control $v(t)$ enhances the awareness through media and education. Finally, the third control $w(t)$ concerns non-diabetics in terms of an awareness program about the negative impact of behavioral, economic and social factors that lead to diabetes at a given time t. Taking into account the three controls, the previous model becomes a model with optimal control given by the following system of differential equations.

$$\begin{cases} \dfrac{dP(t)}{dt} =I_1 - (\mu + \beta_1 + \beta_3)\, P(t) \\ \dfrac{dE(t)}{dt} =I_2 - \mu E(t) - \gamma(1 - w(t))E(t) \\ \dfrac{dD(t)}{dt} =\beta_1 P(t) + \gamma(1 - w(t))E(t) - \alpha(1 - v(t))\dfrac{D(t)E(t)}{N} - (\mu + \beta_2)\, D(t) \\ \qquad\qquad + u(t)C(t) \\ \dfrac{dC(t)}{dt} =\beta_3 P(t) + \beta_2 D(t) + \alpha(1 - v(t))\dfrac{D(t)E(t)}{N} - (\mu + \delta)C(t) - u(t)C(t) \end{cases} \tag{108}$$

with $P(0) \geq 0$, $D(0) \geq 0$, $E(0) \geq 0$ and $C(0) \geq 0$.

The authors also considered "the impact of behavioral factors that can result from the influence that non-diabetics wield over diabetics without complications by age groups. The negative effect of a group of children under the age of 20 on each other behaviorally, or a group between the age of 18 and 50 years on each other, as well as a group over the age of 50. It will also highlight how different age groups affect others, just like adults and young adults. As we divided these age groups k into three main groups: The first category $k = 1$ is less than 20 years old, the second category $k = 2$ is between 20 years and 50 years, and the third category $k = 3$ is over 50 years old".

The diabetic model with age groups is given by the following system of differential equations:

$$\begin{cases} \dfrac{dP_k(t)}{dt} =\Lambda_1 + (-\mu - \beta_1 - \beta_3)\, P_k(t) \\ \dfrac{dE_k(t)}{dt} =\Lambda_2 + (-\mu - \gamma)E_k(t) \\ \dfrac{dD_k(t)}{dt} =\beta_1 P_k(t) + \gamma E_k(t) - \sum_{j=1}^{3} \alpha_{k.j} \dfrac{D_{k,j}(t)E_{k,j}(t)}{N} + (-\mu - \beta_2)\, D_k(t) \\ \dfrac{dC_k(t)}{dt} =\beta_3 P_k(t) + \beta_2 D_k(t) + \sum_{j=1}^{3} \alpha_{k,j} \dfrac{D_{k,j}(t)E_{k,j}(t)}{N} + (-\mu - \delta)C_k(t) \end{cases} \quad (109)$$

with $P_k(0) \geq 0$, $D_k(0) \geq 0$, $E_k(0) \geq 0$, $C_k(0) \geq 0$.

Finally, the authors also proposed a strategy of optimal control to optimize mathematical modeling with multi-category of age. They included three controls $u(t)$, $v(t)$ and $w(t)$ respectively for awareness program, treatment for diabetics people and sensitization about the negative impact of behavioral, economic and social factors. Hence, a mathematical model with multi-category of age and optimal control was formulated as follows:

$$\begin{cases} \dfrac{dP_k(t)}{dt} =\Lambda_1 + (-\mu - \beta_1 - \beta_3)\, P_k(t) \\ \dfrac{dE_k(t)}{dt} =\Lambda_2 - \mu E_k(t) - \gamma\, (1 - w_k(t))\, E_k(t) \\ \dfrac{dD_k(t)}{dt} =\beta_1 P_k(t) + \gamma\, (1 - w_k(t))\, E_k(t) - \sum_{j=1}^{3} \alpha_{k,j} \left(1 - v_{k.j}(t)\right) \dfrac{D_{k,j}(t)E_{k,j}(t)}{N} \\ \qquad - (\mu + \beta_2)\, D_k(t) + u_k(t)C_k(t) \\ \dfrac{dC_k(t)}{dt} =\beta_3 P_k(t) + \beta_2 D_k(t) + \sum_{j=1}^{3} \alpha_{k,j} \left(1 - v_{k,j}(t)\right) \dfrac{D_{k,j}(t)E_{k,j}(t)}{N} - (\mu + \delta)C_k(t) \\ \qquad - u_k(t)C_k(t) \end{cases} \quad (110)$$

The problem is to minimize the objective functional

$$J(u, v, w) = C_k(T) - D_k(T) + \int_0^T \left[C_k(t) - D_k(t) + \frac{A}{2}u^2(t) + \frac{B}{2}v^2(t) + \frac{G}{2}w^2(t) \right] dt$$

where $A > 0$, $B > 0$ and $G > 0$ are the cost coefficients. They are selected to weigh the relative importance of $u(t)$, $v(t)$ and $w(t)$ at time t. T is the final time.

The authors concluded that the numerical simulations they carried out showed the effectiveness of the proposed control strategies based on the reduction of the negative impact of lifestyle on diabetic patients.

Kompas et al. [72]

Kompas et al. analyzed a model for diabetes transmission through social contact. Their $S_l S_h EIT$ model included the following five subpopulations: susceptible individuals with healthy lifestyles and low risk (S_l)), susceptible individuals with healthy lifestyles and high risk (S_h), susceptible individuals with unhealthy lifestyles (E), individuals infected with diabetes without treatment (I) and individuals infected with diabetes with treatment (T).

Figure 17 shows the transmission and transition process modeling the spread of diabetes transmission through social contact is as follows [72]:

$$
\begin{aligned}
\frac{dS_l}{dt} &= (1-p)A - \beta \frac{S_l}{S_l + S_h + E + I} E - \mu S_l, \\
\frac{dS_h}{dt} &= pA - k\beta \frac{S_h}{S_l + S_h + E + I} E - \mu S_h, \\
\frac{dE}{dt} &= \beta \frac{S_l}{S_l + S_h + E + I} E + k\beta \frac{S_h}{S_l + S_h + E + I} E - (\sigma + \mu) E, \\
\frac{dI}{dt} &= \sigma E - (\delta_1 + \mu + \gamma) I + T\varphi, \\
\frac{dT}{dt} &= I\gamma - T\varphi - (\mu + \delta_2) T.
\end{aligned}
\tag{111}
$$

As already stressed for similar models, the authors of this model are considering diabetes as an infectious disease and this is not biologically appropriate.

Asmaidi et al. [21]

Asmaidi et al. published a mathematical model called SEIIT, standing for Susceptible (S), Exposed (E), Infected population untreated (I) and Infected population treated

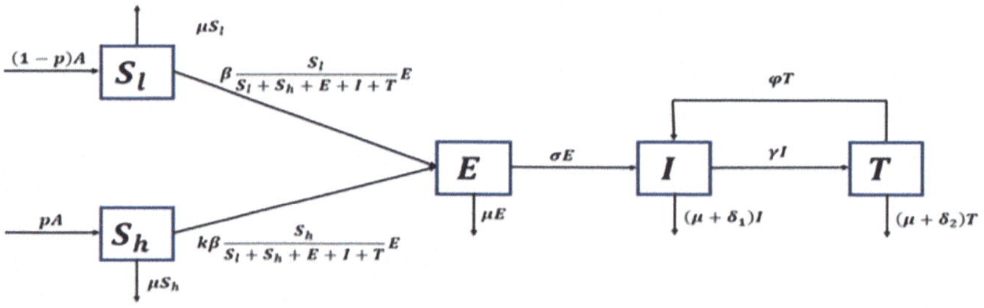

Fig. 17 The transmission and transition diagram for diabetes transmission through social contact. *Source* From Kompas et al. [72], under the Creative Commons Attribution International License (CC BY 4.0). http://creativecommons.org/licenses/by/4.0/

(IT) (Fig. 18). However, the model uses "infective contact among population" which is a biological error since diabetes is not a communicable disease [21]:

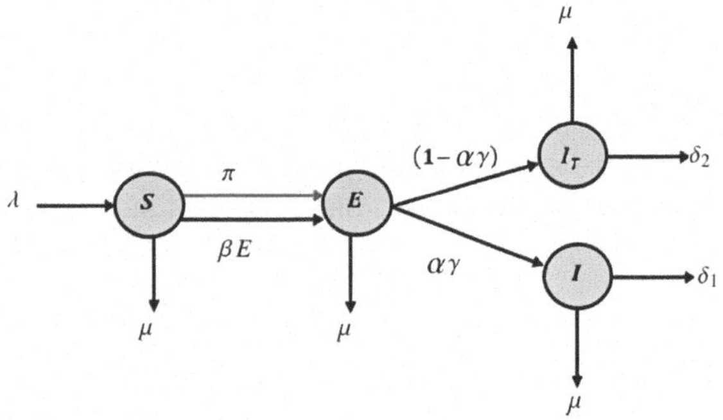

Fig. 18 Mathematical Modeling of Type SEI T_T with Treatment and Genetic Factors *Source* From Asmaidi and Suryanto [21], under the Creative Commons Attribution International License (CC BY 4.0). http://creativecommons.org/licenses/by/4.0/

λ	=	Birth rate
μ	=	Mortality rate by nature
δ_1	=	Mortality rate by illness to infected untreated population
δ_2	=	Mortality rate by illness to infected treated population
π	=	Population-shift rate susceptible to exposed by genetical factor
β	=	Population-shift rate susceptible to exposed by invective contact among population
α	=	Population-shift rate exposed to infected untreated

Ferdous [64]

In 2023, Ferdous [64] proposed an ordinary differential equation model to assess the impact of lifestyle intervention on the type 2 diabetes epidemic. She subdivided the total population N(t) of adults aged 20–79 years old into the following five compartments: Susceptible S(t), Affected A(t), Treated T(t), Healthy Lifestyle L(t) and Prevented P(t). The proposed model was governed by the following system of ordinary differential equations:

$$\begin{aligned}
\frac{dS}{dt} &= b - (\mu + \varepsilon + \alpha)S,\\
\frac{dA}{dt} &= \varepsilon S - (\tau + \mu + \delta_1)\,A + \gamma L,\\
\frac{dT}{dt} &= \tau A - (\mu + \delta_2)\,T,\\
\frac{dL}{dt} &= \alpha S - (\gamma + \mu + \beta)L,\\
\frac{dP}{dt} &= \beta L - \mu P,
\end{aligned} \tag{112}$$

with N(t) = S(t) + A(t) + T(t) + L(t) + P(t).

The parameters associated with this model are: adult birth rate (b), adult natural mortality rate (μ), rate at which adults become affected from susceptible class (ε), treatment rate (τ), rate of susceptible adults maintaining a healthy lifestyle (α), rate at which the healthy lifestyle population goes to the prevented class (β), rate at which adults become affected from healthy lifestyle class (γ), death rate in class A due to diabetes-related complications ($\delta 1$) and death rate in class T due to diabetes-related complications ($\delta 2$).

The author carried out stability analysis, performed sensitivity analysis in order to identify the most important parameters and used diabetes data set from Bangladesh for numerical simulations. She concluded that the results of her study showed that disease progression can be slowed by maintaining a healthy lifestyle.

Mollah and Biswas [83]

In 2023, Mollah and Biswas [83] incorporated the role of awareness programs by media in the dynamics of type 2 diabetes. They subdivided the total population into four classes: unaware susceptible (S_U), aware susceptible (S_A), type 2 diabetes patients with complications (X_C) and type 2 diabetes patients without complications (X_w) then proposed a deterministic model governed by the following four ordinary differential equations:

$$\begin{aligned}
\frac{dS_U}{dt} &= A - pS_U M - \beta S_U - dS_U \\
\frac{dS_A}{dt} &= pS_U M - \beta\beta_1 S_A - dS_A \\
\frac{dX_C}{dt} &= \beta S_U + \beta\beta_1 S_A - \frac{buX_C}{1+\gamma u X_C} - (d+e)X_C + \theta X_W \\
\frac{dX_W}{dt} &= \frac{buX_C}{1+\gamma u X_C} - \theta X_W - dX_W
\end{aligned} \tag{113}$$

In terms of parameters, the authors made the following assumptions:

- A is the constant rate of immigration at any time and all newly recruited individuals go to the unaware susceptible class,
- β is the incident rate of unaware susceptible to diabetes complication,
- M is the media effect while a portion pSS of unaware class directly joins the aware class with p being the awareness rate at which it is implemented,
- θX_W is a proportion of patients without complication of diabetes that joins X_C after becoming affected by complications (θ is a coefficient),
- d is the natural death rate and e is the additional death rate due to complication of diabetes.

Stressing that direct recovery from type 2 diabetes to susceptible is not possible and only remission is possible, they considered treatment effect (u) and treatment function f_u given by

$$f_u(X_C) = \frac{buX_C}{1 + \gamma u X_C}$$

Their analysis included positivity of solutions, boundedness, equilibrium and stability analysis. In order to determine the best cost-effective strategy with the limited resources for treatment, they assessed the efficacy and costs of several strategies. They found that the dynamics of type 2 diabetes was highly influenced by the parameters β, p, M and u. On the basis of their numerical simulations, they suggest that both awareness and treatment controls have a significant impact on the optimal system and are economically feasible to reduce the prevalence of type 2 diabetes.

8 Compartmental Modeling with Optimal Control

Bernard et al. [40]

The model {(49)–(50)} was also considered by Bernard et al. [40] et al. in 2015. Assuming that all the parameters are dependent of time, the authors studied the limit cycle of the model given by

$$\frac{dD}{dt} = I(t) - (\mu(t) + \lambda(t))D(t) + \gamma(t)C(t)\ t > 0,\ N(0) = N_0 \quad (114)$$

$$\frac{dC}{dt} = \lambda(t))D(t) - \theta(t)C(t)\ t > 0,\ C(0) = 0 \quad (115)$$

where $\theta = \gamma + \mu + \nu + \delta$.

The control $u(t) = \gamma(t)$ was chosen by the authors who stress other choices of control may be interesting.

The authors proved the following two theorems:

Theorem 1 *"For all fixed control $u = \gamma$, there exists one and only one maximal solution $([0, t_m(u)], D_u(.), C_u(.))$ of the Cauchy problem {(114)–(115)} with initial conditions $D(0) = D_0 \in IR$, $C(0) = C_0 \in IR$, and $t_m(u) \in IR_+ \bigcup\{+\infty\}$"*

Theorem 2 *"If $\mu = 0, \alpha < 1 - \sigma^{-1}I$ and $r > I(1 - \alpha < 1 - \sigma^{-1}I\ )^{-1}$, then there is an equilibrium state for the system"*

$$\frac{dD(t)}{dt} = I(t) - (\mu(t) + \lambda(t))D(t) + \gamma(t)C(t) \quad (116)$$

$$\frac{dC(t)}{dt} = \lambda(t)D(t) - \theta(t)C(t) \quad (117)$$

$$\frac{dP_D(t)}{dt} = (r + \mu + \lambda)P_D(t) - \lambda P_C(t) - 1 \quad (118)$$

$$\frac{dP_C(t)}{dt} = (r + \mu + \nu + \delta)P_C(t) - (C(t))^{-1} + \gamma(P_C(t) - P_D(t)) \quad (119)$$

Showing the existence of control and carrying out stability analysis, the authors concluded that the number of diabetics with complications which are cured and those without complications can be maximized.

——— Bernard et al. [39] ———

Bernard et al. [39] published another paper in 2018, studying the unexistence of limit cycle in an optimal control problem with reference to the model {(49)–(50)}.

Extending the result of their previous paper published in 2015, the authors stressed that: "Theoretically, any parameter of the model, that is $\lambda(t)$, $\mu(t)$, $\gamma(t)$, $\nu(t)$ or $\delta(t)$ can be chosen as control, but in practice, this is not possible since we cannot control the natural mortality rate for example. In this state of minds, we can control the probability of developing complications by informing the diabetic patients about a necessary good lifestyle with physical activity and healthy diet. We can also control the rate at which complications are healed or the rate at which patients with complications become severely disabled, with an increase in health expenditure. Consequently, in the following, u can be chosen as $\lambda(t)$, $\gamma(t)$, $\nu(t)$".

The author showed the following theorems:

Theorem 3 *"Whatever the control u and whatever the concave performance index chosen as previously specified, there is an equilibrium state for the system"*

$$\frac{dD(t)}{dt} = I(t) - (\mu(t) + \lambda(t))D(t) + \gamma(t)C(t) \tag{120}$$

$$\frac{dC(t)}{dt} = \lambda(t)D(t) - \theta(t)C(t) \tag{121}$$

$$\frac{dP_D(t)}{dt} = (r + \mu + \lambda)P_D(t) - \lambda P_C(t) - \frac{\partial F}{\partial D}(u, D, C) \tag{122}$$

$$\frac{dP_C(t)}{dt} = (r + \mu + \nu + \delta)P_C(t) - (C(t))^{-1} + \gamma(P_C(t) - \frac{\partial F}{\partial C}(u, D, C)) \tag{123}$$

Theorem 4 *"There is no limit cycle between the number of diabetics with complications, the one of diabetics without complications and the controlled parameter chosen as previously specified."*

——— Derouich et al. [57] ———

Starting from the model {(57)–(59)}, Derouich et al. [57] added Optimal Control in 2014 and analyzed the following model:

$$\frac{dE(t)}{dt} = I - (\mu + \beta_3 + \beta_1)(1 - u(t))E(t) \tag{124}$$

$$\frac{dD(t)}{dt} = \beta_1(1 - u(t))E(t) - (\mu + \beta_2)(1 - u(t))D(t) + \gamma C(t) \tag{125}$$

$$\frac{dC(t)}{dt} = \beta_3(1 - u(t))E(t) + \beta_2(1 - u(t))D(t) - (\mu + \gamma + \nu + \delta)C(t) \tag{126}$$

where u is a control.

The objective function considered and the results obtained in terms of existence, uniqueness, positive solutions and numerical simulations can be found in [45].

——— Boutayeb et al. [50] ———

Optimal control was also used by Boutayeb et al. [50] in 2015 on the model {(49)–(50)} leading to the following optimal control model:

$$\frac{dD}{dt} = I - (\beta(1-u) + \mu)D(t) + (\gamma + v)C(t)\ t > 0,\ N(0) = N_0 \tag{127}$$

$$\frac{dC}{dt} = \beta(1-u)D(t) - (\gamma + \mu + v + \delta)C(t)\ t > 0,\ C(0) = C_0 \tag{128}$$

where u and v are controls.

The objective function used in this model and the results obtained in terms of existence, uniqueness, positive solutions and numerical simulations can be found in [45].

——— Yusuf [111] ———

Optimal control was also introduced by Yusuf [111] in 2015 on the model {(49)–(50)} as follows:

$$\frac{dD}{dt} = \Lambda + u_2\gamma C - (1-u_1)\nu D - \mu D \tag{129}$$

$$\frac{dC}{dt} = (1-u_1)\nu D - u_2\gamma C - \mu C - \delta C \tag{130}$$

$$\frac{dN}{dt} = \Lambda - \mu N - \delta C \tag{131}$$

where u and v are controls.

The author proved the existence and uniqueness of the optimal control and characterized the controls based on Pontryagin's Maximum principle using the following objective function:

$$Z = \min_{u_1,u_2} \int_0^{tf} (k_1 C + \frac{1}{2}k_2 u_1^2(t) + \frac{1}{2}k_3 u_2^2(t))dt \tag{132}$$

——— Boutayeb et al. [51] ———

In 2016, Boutayeb et al. [51] proposed a four-compartmental model studying the dynamics between the subpopulations: P (Not diabetics), E (pre-diabetics), D (Diabetics without complications) and C (Diabetics with complications).

As indicated by the flow chart (Fig. 19), the model used the following parameters:

n (the incidence of adult population), I_1 (the rate of healthy persons to become prediabetic), I_2 (the rate of healthy persons to become diabetic), I_3 (the rate of healthy

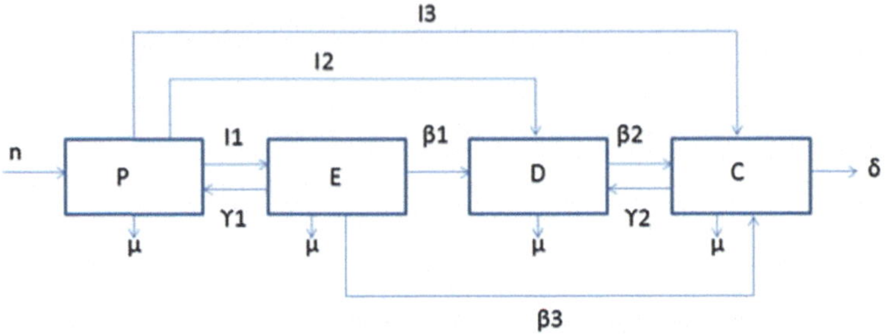

Fig. 19 Dynamics of diabetes from pre-diabetes to diabetes with complications

persons developing complications), μ (natural mortality rate), β_1 (the probability of a pre-diabetic person to become diabetic), β_2 (the probability of a diabetic person developing complications), β_3 (the probability of a pre-diabetic person developing complications), γ_1 (rate at which a pre-diabetic person becomes healthy), γ_2 (rate at which complications are cured), δ (mortality rate due to complications) and μ (natural mortality rate).

In order to prevent adults from diabetes and its complications, an optimal control approach was used to reduce the number of pre-diabetics.

The model with optimal control is given by

$$\frac{dP(t)}{dt} = n - I_1(1-u)P(t) - (I_2 + I_3 + \mu)P(t) + \gamma_1 E(t) \tag{133}$$

$$\frac{dE(t)}{dt} = I_1(1-u)P(t) - (\gamma_1 + \beta_1 + \beta_3 + \mu)E(t) \tag{134}$$

$$\frac{dD(t)}{dt} = I_2 P(t) + \beta_1 E(t) + \gamma_2 C - (\beta_2 + \mu)D(t) \tag{135}$$

$$\frac{dC(t)}{dt} = I_3 P(t) + \beta_2 D(t) + \beta_3 E(t) - (\gamma_2 + \mu + \delta)C(t) \tag{136}$$

where u is control.

Again different parameters, the objective function and the results obtained can be found in [45].

Kouidere et al. [76]

In 2020, Kouidere et al. [76] extended the mathematical model published by Boutayeb et al. [46]. They considered a compartmental model on the dynamics of a population of diabetics divided into six compartments, namely: Healthy people (H), pre-diabetics through genetic factors (P), pre-diabetics through the negative effect of behavioral factors (E), diabetics without complications (D) and diabetics with treatable complications (CT) and diabetics with serious complications (CS). The population dynamics is depicted in Fig. 20.

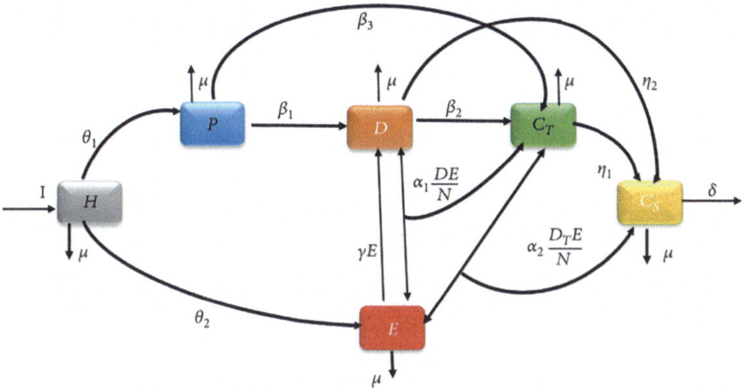

Fig. 20 The dynamics of a population of diabetics *Source* Kouidere et al. [77], reproduced under the Creative Commons Attribution License

The mathematical model proposed is the following:

$$\begin{cases} \dfrac{dH(t)}{dt} =I - (\mu + \theta_1 + \theta_2)\, H(t) \\ \dfrac{dP(t)}{dt} =\theta_1 H(t) - (\mu + \beta_1 + \beta_3)\, P(t) \\ \dfrac{dE(t)}{dt} =\theta_2 H(t) - (\mu + \gamma) E(t) \\ \dfrac{dD(t)}{dt} =\beta_1 P(t) + \gamma E(t) - \alpha_1 \dfrac{D(t)E(t)}{N} - (\mu + \beta_2 + \eta_2)\, D(t) \\ \dfrac{d\mathbf{C}_T(t)}{dt} =\beta_3 P(t) + \beta_2 D(t) + \alpha_1 \dfrac{D(t)E(t)}{N} - \alpha_2 \dfrac{\mathbf{C}_T(t)E(t)}{N} - (\mu + \eta_1)\, C_T(t) \\ \dfrac{d\mathbf{C}_S(t)}{dt} =\eta_2 D(t) + \eta_1 \mathbf{C}_T(t) + \alpha_2 \dfrac{\mathbf{C}_T(t)E(t)}{N} - (\mu + \delta) C_S(t) \end{cases} \tag{137}$$

Then optimal control was added in order to minimize the number of people evolving from pre-diabetes to the stages of diabetes without complications and with treatable complications. To this end, the authors introduced four controls $u_1(t), u_2(t), u_3(t)$ and $u_4(t)$ for $t \in [0, T_f]$.

The control $u_1(t)$ is for treating patients with treatable complications. The second control $u_2(t)$ acts on the awareness program through media and education. The third control u$_3$(t) deals with sensitivity of the negative effect of behavioral factors on diabetics with complications and finally, the fourth control $u_4(t)$ stands for awareness program for non-diabetics against the negative impact of behavioral economic and social factors that lead to develop diabetes. The controlled mathematical model is given by the following system of differential equations.

$$\begin{cases} \dfrac{dH(t)}{dt} = I - (\mu + \theta_1 + \theta_2)\, H(t), \\ \dfrac{dP(t)}{dt} = \theta_1 H(t) - (\mu + \beta_1 + \beta_3)\, P(t), \\ \dfrac{dE(t)}{dt} = \theta_2 H(t) - \mu E(t) - \gamma\,(1 - u_4(t))\, E(t), \\ \dfrac{dD(t)}{dt} = \beta_1 P(t) + \gamma\,(1 - u_4(t))\, E(t) - \alpha_1\,(1 - u_2(t))\, \dfrac{D(t)E(t)}{N} \\ \qquad - (\mu + \beta_2 + \eta_2)\, D(t) + u_1(t) C_T(t), \\ \dfrac{d\mathbf{C}_T(t)}{dt} = \beta_3 P(t) + \beta_2 D(t) + \alpha_1\,(1 - u_2(t))\, \dfrac{D(t)E(t)}{N} - \alpha_2\,(1 - u_3(t))\, \dfrac{\mathbf{C}_T(t)E(t)}{N} \\ \qquad - (\mu + \eta_1 + u_1(t))\, C_T(t), \\ \dfrac{dC_S(t)}{dt} = \eta_2 D(t) + \eta_1 \mathbf{C}_T(t) + \alpha_2\,(1 - u_3(t))\, \dfrac{\mathbf{C}_T(t)E(t)}{N} - (\mu + \delta) C_S(t). \end{cases} \tag{138}$$

The problem is to minimize the objective functional

$$J\,(u_1, u_2, u_3, u_4) = C_T\left(T_f\right) - D\left(T_f\right) - E\left(T_f\right) + \int_0^{T_f} \left[C_T(t) - D(t) \right.$$
$$\left. - E(t) + \frac{A}{2} u_1^2(t) + \frac{B}{2} u_2^2(t) + \frac{F}{2} u_3^2(t) + \frac{G}{2} u_4^2(t)\right] dt$$

where $A > 0,\ B > 0,\ F > 0$ and $G > 0$ are the cost coefficients, they are selected to weigh the relative importance of $u_1(t)$, $u_2(t)$, , $u_3(t)$ and $u_4(t)$ at time t; T is the final time. The authors concluded that numerical simulations showed the effectiveness of the control strategy proposed.

Permatasari et al. [98]

Following the mathematical models published by the team of Boutayeb in [48, 51, 57], Permatasari et al. [98] published a compartmental model with optimal control in 2018. In [98], the authors presented a dynamic model of diabetic population, illustrating the development and evolution of diabetes from healthy stage until the disabled stage due to complications. They considered models without control (called linear system) and with control (called non-linear system). The main purposes of the authors were to investigate the global stability (using the Lyapunov method) of linear system and controllability of the non-linear system (using Lie brackets).

The system of differential equations resulting from this populations dynamics is given by the following system (139):

$$\begin{aligned} \frac{dH}{dt} &= p - \sigma_1 H - (\sigma_2 + \sigma_3 + \mu)\, H + \gamma_1 E \\ \frac{dE}{dt} &= \sigma_1 H - (\gamma_1 + \mu + \beta_1 + \beta_3)\, E + \gamma_2 D \\ \frac{dD}{dt} &= \sigma_2 H + \beta_1 E - (\mu + \beta_2 + \gamma_2 + \nu_2)\, D + \gamma_3 C \\ \frac{dC}{dt} &= \sigma_3 H + \beta_3 E + \beta_2 D - (\mu + \mathcal{H} + \gamma_3 + \nu_1)\, C \\ \frac{dB}{dt} &= \nu_2 D + \nu_1 C - (\mu + \tau) B \end{aligned} \tag{139}$$

where ρ is the incidence of healthy adult population (H(t)), the rates σ_1, σ_2 and σ_3 represent the probabilities of becoming pre-diabetic (E(t)), diabetic without complications (D(t)) or diabetic with complications ($C(t)$) respectively. The rates β_1 and β_3 represent the probabilities of evolving from the pre-diabetes stage (E(t)) to the stage of diabetic without (D(t)) and with complications ($C(t)$) while β_2 is the probability of a diabetic person developing a complication. The rates γ_1, γ_2 and γ_3 represent respectively the probability of evolving back from the compartment E (pre-diabetes) to P, from the compartment $D(t)$ to $E(t)$ and from $C(t)$ to $D(t)$ while v_1 and v_2 represent respectively the rates of diabetics without and with complications who become disabled (B(t)). Finally, the rates of natural mortality, mortality due to disability and mortality due to complications are represented by μ, τ and δ respectively.

Then the authors considered the controlled model given by the following system (140):

$$\begin{aligned}
\frac{dH}{dt} &= \rho - \sigma_1(1-u)H - (\sigma_2+\sigma_3+\mu)\,H + \gamma_1 E \\
\frac{dE}{dt} &= \sigma_1(1-u)H - (\gamma_1+\mu+\beta_1+\beta_3)\,E + \gamma_2 D \\
\frac{dD}{dt} &= \sigma_2 H + \beta_1 E - (\mu+\beta_2+\gamma_2+v_2)\,D + \gamma_3 C \\
\frac{dC}{dt} &= \sigma_3 H + \beta_3 E + \beta_2 D - (\mu+\delta+\gamma_3+v_1)\,C \\
\frac{dB}{dt} &= v_2 D + v_1 C - (\mu+\tau)B
\end{aligned} \tag{140}$$

where u is a control.

The objective function to be minimized is defined by

$$J(u) = \int_0^t \left(E(t) + Au^2(t)\right) dt$$

The authors proved theoretical results concerning the global stability of the linear system and the controllability of the non-linear system. Precisely, they used the Lyapunov method to show that the linear system (140) is asymptotically stable while for controlled models, Lie brackets without linearization allowed them to show that the system (140) is locally accessible.

——— Permatasari et al. [97] ———

In 2018, Permatasari et al. [97] modified a model published by Boutayeb W et al. in 2016 [51]. They used optimal control in a model with five compartments representing respectively healthy people (P(t))), pre-diabetics (E(t)), diabetics without complications (D(t)), diabetics with complications ($C(t)$) and finally diabetics who became disabled due to complications ($B(t)$). The model proposed was governed by the following system of ordinary differential equations.

$$
\begin{aligned}
\frac{dP}{dt} &= \rho - (1-u)\sigma_1 P - (\sigma_2 + \sigma_3 + \mu)\, P + \gamma_1 E \\
\frac{dE}{dt} &= (1-u)\sigma_1 P - (\gamma_1 + \beta_1 + \beta_3 + \mu)\, E + \gamma_2 D, \\
\frac{dD}{dt} &= \sigma_2 P + \beta_1 E + \gamma_3 C - (\beta_2 + \gamma_2 + \nu_2 + \mu)\, D \\
\frac{dC}{dt} &= \sigma_3 P + \beta_2 D + \beta_3 E - (\delta + \gamma_3 + \mu_1 + \mu)\, C \\
\frac{dB}{dt} &= \mu_2 D + \mu_1 C - (\tau + \mu) Q
\end{aligned}
\tag{141}
$$

where ρ is the incidence of a healthy adult population while the rates σ_1, σ_2 and σ_3 represent the probabilities of becoming pre-diabetic, diabetic without complications or diabetic with complications respectively. The rates β_1 and β_3 represent the probabilities of evolving from the pre-diabetes stage to the stage of diabetes without and with complications while β_2 is the probability of a diabetic person developing a complication. The rates γ_1, γ_2 and γ_3 represent respectively the probability of evolving back from the compartment $E(t)$ (pre-diabetes) to P(t), from the compartment $D(t)$ to $E(t)$ and from $C(t)$ to $D(t)$ while μ_1 and μ_2 represent respectively the rates of diabetics without and with complications who become disabled. Finally, the rates of natural mortality, mortality due to disability and mortality due complications are represented by μ, τ and δ respectively.

In order to prevent adults from diabetes and its complications, the authors introduced an optimal control $u = u(t)$ (percentage of healthy people who are prevented from pre-diabetes per unit of time). The objective functional to be minimized was defined by

$$J(u) = \int_0^{t_f} \left(E(t) + Au^2(t)\right) dt$$

Then the authors proved the (1) existence and positivity of solutions, (2) existence of an optimal control and (3) characterization of the optimal control. Then they concluded that the model $PEDCB$ illustrated how efficient strategies can be applied to minimize the number of people susceptible to become pre-diabetic and automatically how to minimize the number of people developing diabetes without and with complications.

El Mansouri et al. [60]

In 2023, El Mansouri et al. [60] proposed the SWRIEDC model, which describes the dynamics of the obesity epidemic in a specific population by highlighting the negative role of overweight in type 2 diabetes's progression. The population is divided into seven compartments, namely: Obesity-prone individuals (S), overweight or moderately obese (W), obese (I), recovered people (R), pre-diabetics (E), diabetics without complications (D) and diabetics with complications (C). Figure 21 shows the flow chart of the model [60]:

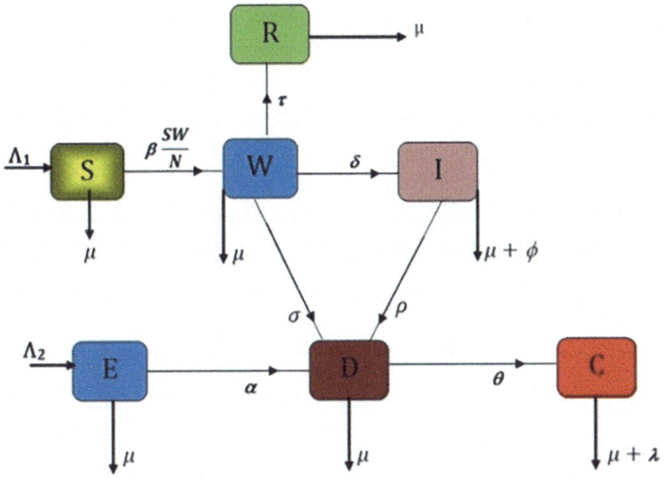

Fig. 21 Schematic illustration of obesity-diabetes model *Source* From Asmaidi and Suryanto [60], under the Creative Commons Attribution International License (CC BY 4.0). http://creativecommons.org/licenses/by/4.0/

The proposed mathematical model is governed by the following system of differential equations:

Let N(t) = S(t) + W(t) + I(t) + R(t) + E(t) + D(t) + C(t): The total number at date t of individuals in the model population. The model is described by the following ordinary differential equations:

$$
\begin{aligned}
\frac{dS(t)}{dt} &= \Lambda_1 - \mu S(t) - \beta \frac{S(t)}{N} W(t) \\
\frac{dW(t)}{dt} &= \beta \frac{S(t)}{N} W(t) - (\tau + \sigma + \mu + \delta) W(t) \\
\frac{dR(t)}{dt} &= \tau W(t) - \mu R(t) \\
\frac{dI(t)}{dt} &= \delta W(t) - (\phi + \rho + \mu) I(t) \\
\frac{dE(t)}{dt} &= \Lambda_2 - (\mu + \alpha) E(t) \\
\frac{dD(t)}{dt} &= \sigma W(t) + \rho I(t) + \alpha E(t) - (\mu + \theta) D(t) \\
\frac{dC(t)}{dt} &= \theta D(t) - (\mu + \lambda) C(t)
\end{aligned}
\tag{142}
$$

With initial conditions, $S(0) \geq 0, W(0) \geq 0, R(0) \geq 0, I(0) \geq 0, E(0) \geq 0, D(0) \geq 0$ and $C(0) \geq 0$

Remark
It is interesting to note that the authors assume being overweight has a negative effect on the susceptible population and represent this "interaction between susceptible and overweight" with the term $(\beta S(t)W(t)/N)$. However, they do not indicate any interaction between susceptible and obese individuals. The use of the interaction term S(t)W(t) appears to be adapted from infectious disease models, and the authors do not provide a detailed explanation for this choice.

Nasir and Mat Daud [93]

Nasir and Mat Daud extended the DC model of Boutayeb et al. [46] by adding a compartment $D_p(t)$, representing the number of post-treatment diabetics at time t. Their model is the following [93]:

$$\begin{aligned}
\frac{dD(t)}{dt} &= I - \alpha D(t) - (\mu + \mu_1)\, D(t), \quad D(0) > 0 \\
\frac{dD_c(t)}{dt} &= \alpha D(t) + \eta D_p(t) - (\gamma + \mu + \mu_2)\, D_c(t), \quad D_c(0) > 0 \\
\frac{dD_p(t)}{dt} &= \gamma D_c(t) - \eta D_p(t) - (\mu + \mu_3)\, D_p(t), \quad D_p(0) > 0
\end{aligned} \tag{143}$$

In 2022, Nasir introduced optimal control in the presence of intervention effects on a previous model published by Nasir and Mat Daud [93]. The new model with optimal control is given by the following system of differential equations [91]:

$$\begin{aligned}
\frac{dD(t)}{dt} &= (1 - \xi_1 u(t))\, I - \alpha\, (1 - \xi_2 v(t))\, D(t) - (\mu + \mu_1)\, D(t), \; D(0) > 0 \\
\frac{dD_c(t)}{dt} &= \alpha\, (1 - \xi_2 v(t))\, D(t) + \eta\, (1 - \xi_3 v(t))\, D_p(t) - (\gamma + \mu + \mu_2)\, D_c(t), \; D_c(0) > 0, \\
\frac{dD_p(t)}{dt} &= \gamma D_c(t) - \eta\, (1 - \xi_3 v(t))\, D_p(t) - (\mu + \mu_3)\, D_p(t), \; D_p(0) > 0
\end{aligned} \tag{144}$$

The objective function is

$$J(u, v) = \min_{u,v \in W} \int_0^{t_f} \left[A_1 D(t) + A_2 D_c(t) + A_3 D_p(t) + B_1 u^2(t) + B_2 v^2(t) \right] dt,$$

In their conclusion, they indicated that the aim of their work was to use the optimal control method in order to reduce the burden of diabetes and its deadly complications. For this purpose, they introduced a controlled diabetics model and found an objective function that minimizes the total number of diabetics with the least relative cost. Then, they carried out numerical simulations to illustrate how complications of diabetes can be reduced through optimal control strategies. Finally, they stated that preventing the incidence of diabetes should be a priority.

Nasir [92]

In 2022, Nasir [92] stressed that Boutayeb et al. introduced the first DC diabetic model using ordinary differential equations in 2004 [46] and this model was used, modified and extended in many research frameworks dealing with other factors such as healthy and pre-diabetic individuals, awareness programs, genetic, as well as lifestyle and behavioral influence. In the same direction, the author proposed a five-state diabetic population model, assuming that the total population at time t(N(t)) can be divided into the following five subpopulations: non-diabetics (P(t)), pre-diabetic individuals (E(t)), undiagnosed diabetics (U(t)), diagnosed diabetics (D(t)) and diabetics with complications (C(t)). The rate of diabetes is given by $\Lambda(t) = \mathrm{U(t)} + \mathrm{D(t)} + \mathrm{C(t)}/\mathrm{N(t)}$ where $\mathrm{N(t)} = \mathrm{P(t)} + \mathrm{E(t)} + \mathrm{U(t)} + \mathrm{D(t)} + \mathrm{C(t)}$.

The model proposed is given by the following system of ordinary differential equations.

$$
\begin{aligned}
\frac{dP(t)}{dt} &= \chi - (\beta + \mu)P(t) + \rho E(t) \\
\frac{dE(t)}{dt} &= \beta P(t) - (\rho + \delta + \sigma + \mu)E(t) \\
\frac{dU(t)}{dt} &= \alpha\sigma E(t) - (v + \gamma + \mu + \mu_1)\, U(t) \\
\frac{dD(t)}{dt} &= (1 - \alpha)\sigma E(t) + \gamma U(t) - (\eta + \mu + \mu_2)\, D(t) \\
\frac{dC(t)}{dt} &= \delta E(t) + vU(t) + \eta D(t) - (\mu + \mu_3)\, C(t)
\end{aligned}
\tag{145}
$$

The author also modified the previous model by introducing an optimal control with prevention strategy to get the following model:

$$
\begin{aligned}
\frac{dP(t)}{dt} &= \chi - \beta\,(1 - \xi_1 u(t))\, P(t) + \rho E(t) - \mu P(t), \\
\frac{dE(t)}{dt} &= \beta\,(1 - \xi_1 u(t))\, P(t) - (\rho + \delta + \sigma\,(1 - \xi_2 u(t)) + \mu)\, E(t), \\
\frac{dU(t)}{dt} &= \alpha\sigma\,(1 - \xi_2 u(t))\, E(t) - (v + \gamma + \mu + \mu_1)\, U(t), \\
\frac{dD(t)}{dt} &= (1 - \alpha)\sigma\,(1 - \xi_2 u(t))\, E(t) + \gamma U(t) - (\eta + \mu + \mu_2)\, D(t), \\
\frac{dC(t)}{dt} &= \delta E(t) + vU(t) + \eta D(t) - (\mu + \mu_3)\, C(t),
\end{aligned}
\tag{146}
$$

with the functional to be minimized given by

$$J(u(t)) = \int_0^T \left[E(t) - P(t) + Au^2(t)\right] dt,$$

The author used Malaysian diabetes data to estimate different parameters, carried out stability and sensitivity analysis and implemented numerical simulations. They concluded that diabetes prevalence can be reduced by controlling the transition to pre-diabetes (β) and diabetes (σ), stressing that optimal control strategies, including action on lifestyle and the awareness of a healthy diet, should effectively reduce the prevalence of diabetes and its complications.

El Moutaouakil et al. [61]

El Moutaouakil et al. [61] considered the three-stage compartmental model of Derouich et al. [57], including pre-diabetic individuals (E(t)), diabetic individuals without complications (D(t)) and diabetic individuals with complications (C(t)).

$$\begin{aligned} \frac{dE(t)}{dt} &= I - (\mu + (\beta_3 + \beta_1)(1 - u(t)))\,E(t) \\ \frac{dD(t)}{dt} &= \beta_1(1 - u(t))E(t) - (\mu + \beta_2(1 - u(t)))\,D(t) + \gamma C(t) \\ \frac{dC(t)}{dt} &= \beta_3(1 - u(t))E(t) + \beta_2(1 - u(t))D(t) - (\mu + \gamma + v + \delta)C(t) \end{aligned} \tag{147}$$

The function minimized by Derouich et al. was of the following form:

$$\Gamma(u) = \int_0^T \left(D(t) + C(t) + Ku^2(t)\right) dt$$

Formulating some critics on the previous model, El Mouaquakil et al. proposed the following Multi-Objective Control Model [61]:

$$\begin{aligned} \frac{dE(t)}{dt} &= I - (\mu + (\beta_3 + \beta_1)(1 - u_1(t)))\,E(t) \\ \frac{dD(t)}{dt} &= \beta_1(1 - u_1(t))\,E(t) - (\mu + \beta_2(1 - u_2(t)))\,D(t) + \gamma C(t) \\ \frac{dC(t)}{dt} &= \beta_3(1 - u_1(t))\,E(t) + \beta_2(1 - u_2(t))\,D(t) - (\mu + \gamma + v + \delta)C(t) \end{aligned} \tag{148}$$

Discretizing the system (148) using a time step h lead the authors to the following system (149)

$$
\begin{aligned}
E_{i+1} &= Ih - \left(\mu + (\beta_3 + \beta_1)\left(1 - u_{1,i}\right)\right) hE_i + E_i \\
D_{i+1} &= \beta_1\left(1 - u_{1,i}\right) hE_i - \left(\mu + \beta_2\left(1 - u_{2,i}\right)\right) hD_i + \gamma hC_i + D_i \qquad i = 0, \ldots, d-1 \\
C_{i+1} &= \beta_3\left(1 - u_{1,i}\right) hE_i + \beta_2\left(1 - u_{2,i}\right) hD_i - (\mu + \gamma + v + \delta)hC_i + C_i
\end{aligned}
\tag{149}
$$

where $\tilde{u}_p = \left(u_{p,0}, \ldots, u_{p,d}\right)$, $p = 1, 2$. Then, the Trapezoidal rule was used to estimate the computed values of the two objective functions

$$
\Gamma_1\left(u_1, u_2\right) \approx \widetilde{\Gamma}_1\left(\left[\widetilde{u}_1, \widetilde{u}_2\right]\right) = \frac{h}{2}\sum_{i=0}^{n-1}\left(C_{i+1} + C_i\right) + \frac{Kh^{u-1}}{2}\sum_{i=0}^{n}\left(u_{1,i+1}^2 + u_{1,i}^2\right)
$$

$$
\Gamma_2\left(u_1, u_2\right) \approx \widetilde{\Gamma}_2\left(\left[\widetilde{u}_1, \widetilde{u}_2\right]\right) = \frac{h}{2}\sum_{j=n}^{d-1}\left(D_{i+1} + D_i\right) + \frac{Kh}{2}\sum_{i=n}^{d-1}\left(u_{2,i+1}^2 + u_{2,i}^2\right)
$$

Finally, the multi-objective problem proposed was given by

$$
\begin{cases}
\text{Min } \widetilde{\Gamma}_1\left(\left[\widetilde{u}_1, \widetilde{u}_2\right]\right) \quad \text{Min } \widetilde{\Gamma}_2\left(\left[\widetilde{u}_1, \widetilde{u}_2\right]\right) \\
\left[\widetilde{u}_1, \widetilde{u}_2\right] \in [0, 1]^{2d} \\
(\widetilde{E}, \widetilde{D}, \widetilde{C}) \text{ solution of } (1.149)
\end{cases}
$$

Logaprakash and Monica [79]

In 2023, Logaprakash and Monica published a paper dealing with "Optimal control of diabetes model with the impact of endocrine-disrupting chemical: an emerging increased diabetes risk factor". Figure 22 below illustrates the flow between the seven compartments: healthy population (P), pre-diabetes population (S), diabetes population (D), diabetes population with complication (C), Exposed population (F) and Food exposed with EDC (F) [79].

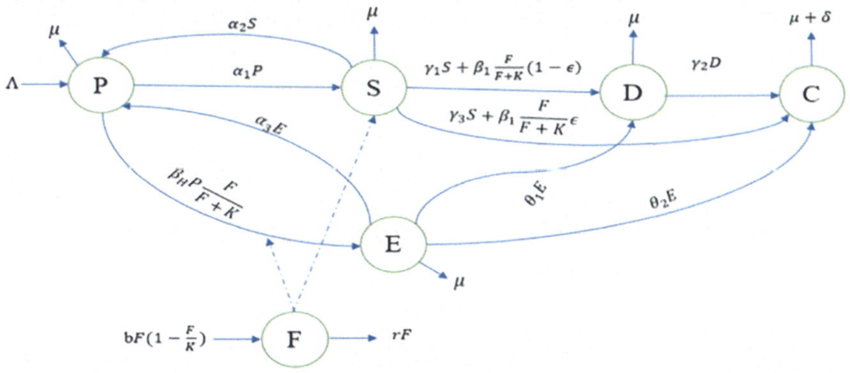

Fig. 22 Flow diagram of the model *Source* From Logaprakash and Monica [79], under the Creative Commons Attribution International License (CC BY 4.0). http://creativecommons.org/licenses/by/4.0/

The seven differential equations governing the optimal control model are the following:

$$\begin{aligned}
\frac{dP}{dt} &= \Lambda - \left(\alpha_1(1-u_1) + \beta_P(1-u_2)\frac{F}{F+K} + \mu\right)H + \alpha_2 S + (\alpha_3 + pu_3)E\\
\frac{dS}{dt} &= \alpha_1(1-u_1)H - \left(\gamma_1 + \gamma_3 + \alpha_2 + \beta_1(1-u_2)\frac{F}{F+K} + \mu\right)S,\\
\frac{dD}{dt} &= \left(\gamma_1 + (1-\epsilon)\beta_1(1-u_2)\frac{F}{F+K}\right)S - (\gamma_2 + \mu)D + \theta_1 E,\\
\frac{dC}{dt} &= \left(\gamma_3 + \epsilon\beta_1(1-u_2)\frac{F}{F+K}\right)S + \theta_2 E + \gamma_2 D - (\mu + \delta)C,\\
\frac{dE}{dt} &= \beta_P(1-u_2)\frac{F}{F+K}P - (\theta_1 + \theta_2 + \alpha_3 + \mu + pu_3)E,\\
\frac{dF}{dt} &= bF\left(1 - \frac{F}{K}\right) - u_4 F
\end{aligned} \tag{150}$$

with the objective functional to be minimized given by

$$J(u_1(t), u_2(t), u_3(t), u_4(t)) = \int_0^T \left(A_1 S + A_2 C + A_3 D + A_4 E + A_5 F + \frac{B_1 u_1^2}{2} + \frac{B_2 u_2^2}{2} + \frac{B_3 u_3^2}{2} + \frac{B_4 u_4^2}{2}\right) dt$$

The authors proved the existence of the optimal control, characterized it and carried out numerical simulations with different values of parameters in order to find the optimal control strategies that are more effective in controlling the prevalence of diabetes. Their conclusion stressed that: "This proposed model offers a different approach to understanding the prevalence of diabetes, particularly when the daily consumption of food is exposed to some harmful chemicals that lead to health problems. A suitable control strategy discussed includes intervention for exposed people, diabetes prevention, control of EDC concentration on daily consumption, and prevention of consuming EDC".

9 Compartmental Models Illustrating the Impact of Media Coverage on the Controlling of Diabetes

—— Agarwal and Pathak [3] ——

In 2014, Agarwal and Pathak [3] proposed a compartmental model dealing with the impact of media coverage on the controlling of diabetes. Their model was based on five compartments, namely: pre-diabetics (P(t)), diabetics without complications

(D(t)), diabetics with complications (C(t)), aware pre-diabetic individuals at time t (P_M(t)) and the cumulative density of awareness programs driven by the media in that region at time t ($M(t)$). The model is governed by five ordinary differential equations as follows:

$$\begin{aligned}
\frac{dP}{dt} &= A - (\lambda_1 + \lambda_3) P - \lambda P M + \lambda_0 P_M - \mu P,\\
\frac{dD}{dt} &= \lambda_1 P - \lambda_2 D + \gamma C - dD,\\
\frac{dC}{dt} &= \lambda_2 D + \lambda_3 P - (\gamma + d + \nu + \delta) C,\\
\frac{dP_M}{dt} &= \lambda P M - d P_M - \lambda_0 P_M,\\
\frac{dM}{dt} &= \mu(D + C) - \mu_0 M,
\end{aligned} \tag{151}$$

where

A is the incidence of pre-diabetes, the rates λ_1, λ_2 and λ_3 are the probabilities of developing diabetes, developing complications for a diabetic person and developing complications for a pre-diabetic person. The constant γ is the rate at which complications are cured and v is the rate at which diseased patients with complications become severely disabled. Other parameters are: d the mortality rate due to complications (d), the dissemination rate of awareness among susceptible class (λ), the rate with which awareness programs are being implemented (μ) while μ_0 is the depletion rate of these programs due to ineffectiveness, social problems in the population and λ_0 is the rate of transfer of aware class to susceptible class.

The authors considered (1) boundedness of the System, (2) existence of equilibrium points, (3) stability analysis (local stability and global stability) and (4) numerical simulation. Then they concluded stressing that "*the analysis demonstrates that an endemic equilibrium is locally as well as globally stable under certain conditions. The analysis showed that awareness programs through the media campaigning are helpful in decreasing the diabetic class in a population. By awareness programs, at least we could have prevented diabetes whenever possible and, where not possible, to minimize complications and maximize quality of life*".

10 Compartmental Models for Gestational Diabetes and Diabetes During Pregnancy

——— Daud et al. [55] ———

The model proposed by Daud et al. [55] is based on the following four compartments:
V(t): Diabetic non-pregnant women of childbearing age
N(t): Non diabetic non-pregnant women of childbearing age

V(t): Diabetic women during pregnancy
G(t): Pregnant women with complications of diabetes
The system of differential equations is given by

$$\begin{aligned}
\frac{dV}{dt} &= \Psi - \phi_1 V(t) + \eta K(t) + \tau G(t) \\
\frac{dN}{dt} &= \sigma - \phi_2 N(t) + \sigma V(t) + \chi K(t) + \gamma G(t), \\
\frac{dK}{dt} &= \alpha V(t) + \Omega N(t) - \phi_3 K(t) \\
\frac{dG}{dt} &= \pi K(t) - \phi_4 G(t),
\end{aligned} \tag{152}$$

where

$$\begin{aligned}
\phi_1 &= \mu + \xi + \alpha + \sigma \\
\phi_2 &= \varepsilon + \Omega + \mu + \kappa \\
\phi_3 &= \eta + \chi + \pi + \upsilon + \delta \\
\phi_4 &= \tau + \gamma + \theta + \beta
\end{aligned}$$

The different parameters and rates used in this model are the following:

Death due to causes other than diabetes (μ), diabetic pregnant women who die due to causes other than diabetes (α), diabetic pregnant women who die due to causes other than the complication of diabetes (Θ), diabetic non-pregnant women who die due to diabetes (κ), diabetic pregnant women who die due to diabetes (Δ), pregnant women who die due to diabetes complication (β), women entering the childbearing age without diabetes (ψ), women leaving the normal childbearing age population without diabetes (ξ), women entering the childbearing age population with diabetes (ϖ), women leaving the childbearing age population with diabetes (ε), non-diabetic women in childbearing age becoming pregnant with diabetes (α), non-diabetic women in childbearing age becoming diabetic (σ), diabetic women getting pregnant (Λ), women with gestational diabetes becoming non-diabetic after puerperium (η), diabetic pregnant women who continue being diabetic in non-pregnant state (χ), diabetic women developing various complications during pregnancy or postpartum period (π), Women with gestational diabetes with complications recover to become non-diabetic (τ), women with diabetic complication during pregnancy recover from complication (but continue to be diabetic in non-pregnant state (γ).

Lamlili et al. [78]

This model focusses on sensitivity analysis to provide policymakers with pragmatic and efficient strategies which can reduce the rate of developing diabetes by controlling the number of people becoming overweight.

$$\begin{cases} \frac{dP(t)}{dt} &= n - \alpha \mathrm{P} + w_2 \mathrm{W} \\ \frac{dW(t)}{dt} &= w_1 \mathrm{P} - \beta \mathrm{W} \\ \frac{dD(t)}{dt} &= d_1 \mathrm{P} + w_3 \mathrm{W} - \delta \mathrm{D} \\ \frac{dC(t)}{dt} &= c_1 \mathrm{P} + w_4 \mathrm{W} + d_2 \mathrm{D} - \varepsilon \mathrm{C} \end{cases} \tag{153}$$

with:

$$\alpha = (c_1 + d_1 + w_1 + \mu)$$
$$\beta = (w_2 + w_3 + w_4 + \mu)$$
$$\delta = (d_2 + \mu)$$
$$\varepsilon = (\mu + \nu)$$

11 Fractional-Order Model

Srivastava HM et al. [105]

In 2019, Srivastava HM et al. [105] et al. applied a fractional-order approach to the model (49)–(50) as follows:

$$D_t^{\alpha}[B(t)] = A(t) - (\delta + \varepsilon)B(t) + \lambda C(t) \tag{154}$$
$$D_t^{\beta}[C(t)] = A(t) - (\lambda + \mu + v + \varepsilon)C(t) + \delta B(t) \tag{155}$$
$$E(t) = B(t) + C(t)$$
$$D_t^{\alpha}[C(t)] = -(\delta + \psi)C(t) + \delta E(t) \tag{156}$$
$$D_t^{\beta}[E(t)] = A(t) - (\mu + v)C(t)) - \varepsilon E(t) \tag{157}$$

with $\psi = \lambda + \mu + v + \varepsilon$

The parameters of the model can be found in [45].

Dubey et al. [59]

In this paper, Dubey et al. start from the DC model published by Boutayeb et al. in 2004 [46]. They consider the incidence (A(t)) of Diabetes Mellitus as a constant and they define E(t) as the size of the whole population of people having diabetics with complications (C(t)) and without complications (B(t)). The proposed fractional mathematical model is given by [59]:

$$\begin{aligned} {}_0^{CF}D_t^{\alpha}(C(t)) &= -(\delta + \varphi)C(t) + \delta E(t), \quad t > 0, \qquad 0 < \alpha \le 1, \\ {}_0^{CF}D_t^{\beta}(E(t)) &= A(t) - (v + \mu)C(t) - \varepsilon E(t), \quad t > 0, \ 0 < \beta \le 1, \end{aligned} \tag{158}$$

with $C(0) = C_0$ and $E(0) = E_0$ and $E(t) = B(t) + C(t)$.

The authors used the perturbation technique to derive the solution for the problem and the fixed point theorem to show the existence and uniqueness of the coupled solution. Stability analysis and numerical simulations were also used to characterize the solution of the fractional-order model.

——— Albalawi et al. [11] ———

In 2022, Lamlili et al. published a 4-compartmental model dealing with the effect of overweight and obesity on the dynamics of a diabetic population [78]. This model was expanded by Albalawi and colleagues, using fractional-order derivatives instead of ordinary differential equations. The new model proposed is governed by the following equations [11].

$$\begin{aligned} {}^{C}D_0^{\sigma} A(t) &= \eta - (\alpha_1 + \beta_1 + r_1 + \mu)\, A(t) + \alpha_3 I(t) \\ {}^{C}D_0^{\sigma} I(t) &= \alpha_1 A(t) - (\alpha_2 + \alpha_3 + \alpha_4 + \mu)\, I(t) \\ {}^{C}D_0^{\sigma} D(t) &= \beta_1 A(t) + \alpha_2 I(t) - (\beta_2 + r_2 + \mu)\, D(t) \\ {}^{C}D_0^{\sigma} C(t) &= \alpha_4 I(t) + \beta_2 D(t) - (\delta + \mu) C(t) \end{aligned} \tag{159}$$

where
A(t) is the number of people following a healthy lifestyle (not overweight/obese),
I(t) is the number of people not following a healthy lifestyle (overweight/obese),
D(t) is the number of people with diabetes without any complication,
C(t) is the number of people with diabetes suffering from complications,

$$\mathrm{N(t) = A(t) + I(t) + D(t) + C(t)}$$

The authors showed the existence and uniqueness of the solution followed by stability, numerical simulation and discussion.

——— Rana [101] ———

Rana modified the DC model of Boutayeb et al. [46] by using the fractional variational iteration method and the fractional homotopy perturbation method. The proposed model was the following [101]:

$$\begin{aligned} \frac{\mathbf{d}^{\alpha}\mathbf{C}(\mathrm{t})}{\mathrm{dt}^{\alpha}} &= -(\lambda + \theta)C + \lambda \mathbf{N}, \quad \mathbf{t} > 0 \\ \frac{\mathrm{d}^{\alpha} N(t)}{\mathrm{dt}^{\alpha}} &= \mathbf{I} - (\upsilon + \delta)C - \mu N, \end{aligned} \tag{160}$$

The author solved the problem using fractional variational iteration method (FVIM) as follows:

$$C_{n+1}(t) = C_n(t) + \int_0^t \lambda \left(\frac{d^\alpha C(t)}{d\xi^\alpha} + (\lambda + \theta)C - \lambda N \right) (d\xi)^\alpha,$$
$$N_{n+1}(t) = N_n(t) + \int_0^t \lambda \left(\frac{d^\alpha N(t)}{d\xi^\alpha} - I + (v + \delta)C + \mu N \right) (d\xi)^\alpha$$

where λ is Lagrange's multiplier. By variational theory, λ must satisfy

$$\left. \frac{d^\alpha \lambda}{d\xi^\alpha} \right|_{\xi = x} = 0 \quad \text{and,} \quad 1 + \lambda|_{\xi = x} = 0.$$

Then, taking Laplace transform on both sides of (160), the fractional homotopy perturbation method (FHPTM) was used to find a solution of the following equations:

$$L[u(x,t)] = p^{-1}h_1(x) - p^{-\alpha}L\left[g_1(x,t)\right] + p^{-\alpha}L\left[S_1(u,v) + Q_1(u,v)\right]$$
$$L[v(x,t)] = p^{-1}h_2(x) - p^{-\alpha}L\left[g_2(x,t)\right] + p^{-\alpha}L\left[S_2(u,v) + Q_2(u,v)\right]$$

The author also carried out the numerical implementation of FVIM and concluded that "both methods are strong numerical computational methods for solving non-integro-differential equations".

Karaagac et al. [71]

Karaagac and colleagues proposed "a computational technique for the Caputo fractal-fractional diabetes mellitus model without genetic factors". The authors explained that their model dealt with the dynamics within and between four subpopulations representing respectively the class of sensitive people (q(t)) (the authors stress that it is a class of population members who risk being infected by a disease), the exposed individuals (y(t)), the class of individuals who have the disease but cannot access the treatment (z(t)), and finally the class of individuals who have the disease and have access to treatment (p (t)) (see flow chart below) [71].

The mathematical model proposed was the following:

$$\begin{aligned} \frac{dq(t)}{dt} &= \kappa - \mu q(t) - \beta q(t)y(t) \\ \frac{dy(t)}{dt} &= \beta q(t)y(t) - (\mu + 1)y(t) \\ \frac{dz(t)}{dt} &= \alpha\gamma y(t) - (\mu + \delta)z(t) \\ \frac{dp(t)}{dt} &= (1 - \alpha\gamma)y(t) - (\mu + \sigma)p(t) \\ x(t) &= q(t) + y(t) + z(t) + p(t) \end{aligned} \tag{161}$$

The different parameters are defined as follows:

κ: the number of individuals born in the population at a specific time range

μ: the number of deaths occurs in the population in a specific time range
δ: the rate of individuals who have been infected with a particular disease and eventually died of it (case fatality rate)
σ: the rate of death from disease with the treatment
$\alpha\gamma$: rate of transfer of disease from latent individuals to sick individuals without treatment
β: the rate of transfer of infectious disease from hidden individuals to sensitive individuals

Under fractal-fractional-order derivatives, the model is governed by the following equations:

$$\begin{aligned} {}^{FFP}D^{\zeta,\epsilon}x(t) &= \kappa - \mu x(t) - \delta z(t) - \sigma p(t) \\ {}^{FFP}D^{\zeta,\epsilon}y(t) &= \beta(x(t) - y(t) - z(t) - p(t))y \\ &\quad - (\mu + 1)y(t) \\ {}^{FFP}D^{\zeta,\epsilon}z(t) &= \alpha\gamma y(t) - (\mu + \delta)z(t) \\ {}^{FFP}D^{\zeta,\epsilon}p(t) &= (1 - \alpha\gamma)y(t) - (\mu + \sigma)p(t) \end{aligned} \tag{162}$$

Important remark
As already stressed for similar models, this model is very confusing. While the authors indicate that the model is dedicated to diabetes, their introduction of the model and its parameters uses the terminology of infective disease (δ: the rate of individuals who have been infected, β: the rate of transfer of infectious disease) and more importantly, the interaction between subpopulations, usually used in term of infection by contact!

Saima et al. [102]

Saima et al. converted Boutayeb's DC model [46] into the susceptible diabetes complexity model attributable to the susceptible person category (SDC). The SDC framework is summarized as follows [102]:

$$\begin{cases} \frac{d\mathbb{S}}{d\tau} = \gamma\mathbb{S} + \gamma(1 - \rho)(\mathbb{D} + \mathbb{C}) - \beta\mathbb{S}\mathbb{D} - \mu\mathbb{S} \\ \frac{d\mathbb{D}}{d\tau} = \gamma\mathbb{S}(\mathbb{D} + \mathbb{C}) - (\lambda + \mu)\mathbb{D} + \gamma\mathbb{C} \\ \frac{d\mathbb{C}}{d\tau} = \lambda\mathbb{D} - (\mu + \delta + \eta)\mathbb{D} + \gamma\mathbb{C} \end{cases} \tag{163}$$

Then the previous model (163) is expanded to fractional-order systems by inserting the ABC fractional operator : ${}^{ABC}_{\tau}\mathbf{D}^{\phi}_{a_1}$ for the classical time derivative $\mathrm{d}/\mathrm{d}\tau$, leading to the following ABC-FD model:

$$\begin{aligned}
&{}^{ABC}_{\ \tau}{}^{\pi}\mathbf{D}^{\phi}_{a_1}\mathbb{S}(\tau) = \gamma\mathbb{S} + \gamma(1-\rho)(\mathbb{D}+\mathbb{C}) - \beta\mathbb{D}\mathbb{D} - \mu\mathbb{S},\\
&{}^{ABC}_{\ \tau}\mathbf{D}^{\phi}_{a_1}\mathbb{D}(\tau) = \gamma\mathbb{S}(\mathbb{D}+\mathbb{C}) - (\lambda+\mu)\mathbb{D} + \gamma\mathbb{C},\\
&{}^{AB}(\tau) = \lambda\mathbb{D} - (\mu+\delta+\eta)\mathbb{D} + \gamma\mathbb{C}.
\end{aligned} \tag{164}$$

Important remark

As indicated in Eqs. 163 and 164, the authors used a term SD showing interaction between individuals in the susceptible class (S) and those who have diabetes (D). No convincing explanation is given by the authors concerning this formulation rather used to express the rate of infection in the case of communicable diseases.

Singh et al. modified the DC model of Boutayeb et al. [46] by changing the first order time derivative to the novel propounded CF arbitrary order derivative to obtain the following model [104]:

$$\begin{aligned}
&{}^{CF}_{0}D^{\sigma}_{\eta}C = -(\lambda+\theta)C + \lambda N, \quad \eta > 0,\\
&{}^{CF}_{0}D^{\sigma}_{\eta}N = I - (\nu+\delta)C - \mu N.
\end{aligned} \tag{165}$$

In this paper, the authors considered the existence of the solution by the Picard-Lindelof approach, followed by stability analysis of the iterative scheme and numerical results. They concluded that: "The results of this study are very helpful for medical practitioners dealing with diabetes and related issues. Thus, we have concluded that the CF derivative is useful in the description of physical, chemical, biological, medical, social and engineering processes".

——— Ahmad et al. [5] ———

Starting from the DC model of Boutayeb et al. [46], Ahmad et al. used the AtanganaBaleanu fractional-order derivative to solve the following time-fractional model [5]:

$$\begin{aligned}
&{}^{ABC}_{0}D^{\kappa}C(t) = -\lambda C - \theta C + \lambda N\\
&{}^{ABC}_{0}D^{\kappa}N(t) = I - \nu C - \delta C - \mu N
\end{aligned} \tag{166}$$

with initial conditions C(0) = C0 = 0 and N(0) = N0 = 0

——— Moya et al. [86] ———

Moya et al. [86] proposed a mathematical model with fractional order for overweight/obesity and its impact on the diagnosis of diabetes. As indicated by the flow chart (Fig. 23), the model is based on four compartments: susceptible, overweight, obesity and diabetes.

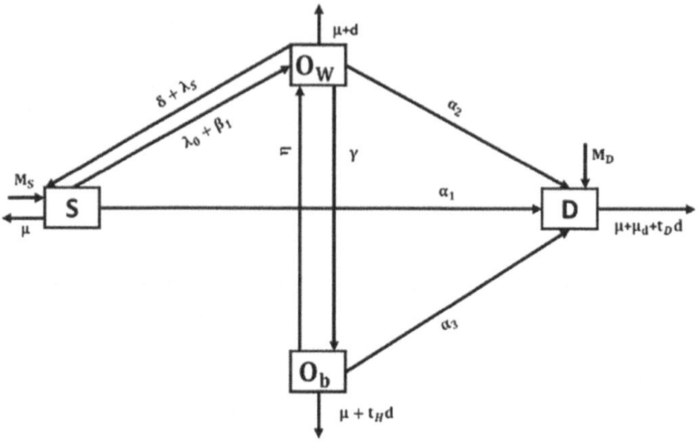

Fig. 23 Chart flow of the model. *Source* From Moya et al. [86], under the Creative Commons Attribution International License (CC BY 4.0). http://creativecommons.org/licenses/by/4.0/

The proposed mathematical model is governed by the following equations:

$$\begin{aligned} {}^{c}\mathbb{D}_t^{\alpha} S &= M_S^{\alpha} + (\delta^{\alpha} + \lambda_S)\, O_w - \left(\mu^{\alpha} + \alpha_1^{\alpha} + \beta_1^{\alpha} + \lambda_O\right) S \\ {}^{c}\mathbb{D}_t^{\alpha} O_w &= \left(\lambda_O + \beta_1^{\alpha}\right) S + \eta^{\alpha} O_b - \left(\lambda_S + \gamma^{\alpha} + \mu^{\alpha} + d^{\alpha} + \delta^{\alpha} + \alpha_2^{\alpha}\right) O_w \\ {}^{c}\mathbb{D}_t^{\alpha} O_b &= \gamma^{\alpha} O_w - \left(\eta^{\alpha} + \mu^{\alpha} + t_H d^{\alpha} + \alpha_3^{\alpha}\right) O_b \\ {}^{c}\mathbb{D}_t^{\alpha} D &= M_D^{\alpha} + \alpha_1^{\alpha} S + \alpha_2^{\alpha} O_w + \alpha_3^{\alpha} O_b - \left(\mu_d^{\alpha} + \mu^{\alpha} + t_D d^{\alpha}\right) D \end{aligned} \tag{167}$$

——— Balakrishnan et al. [37] ———

Balakrishnan et al. extend a model published by Anusha in 2021 [18]. They introduced time delay to describe short-term memory of prior time and fractional order to explain long-term temporal memory. Using the fractional derivative of Caputo order D^{α}, the modified fractional-order diabetes model with delay takes the following form [37]:

$$\begin{aligned} D^{\alpha} S(t) &= \Lambda - \sigma S - \upsilon S + lR \\ D^{\alpha} I(t) &= \sigma S - \beta_1 I - \rho I T - \upsilon I \\ D^{\alpha} T(t) &= \beta_2 I(t-\tau) + \rho I T - \frac{b u_r T}{1 + \gamma_1 u_r T} - \upsilon T \\ D^{\alpha} R(t) &= \frac{b u_r T}{1 + \gamma_1 u_r T} - \upsilon R - lR \end{aligned} \tag{168}$$

With Initial values $S(0), T(0), R(0) > 0, I(s) = \chi(s) > 0, s \in [-\tau, 0]$ and treatment function given by

$$\frac{b u_r T}{1 + \gamma_1 u_r T}$$

The authors explain that "u_r is the treatment effect/treatment control parameter, γ_1 represents the saturation constant that is associated with the control of treatment, and b/γ_1 represents the supply of medical resources per unit time. The rate of treatment increases as imbalanced glucose level leads to diabetes populations growing, reaching their maximum and then remaining constant despite new cases".

The authors showed that the system (S) has a unique (positive bounded) solution in $[0, +\infty)$ and has a Diabetes-free equilibrium $E_0\left(S_0^* = \Lambda/\sigma + v, 0, 0, 0\right)$ and interior equilibrium $E_1(S^*, I^*, T^*, R^*)$. They also carried out stability analysis and found that E_0 is unconditionally locally asymptotically stable while E_1 is locally asymptotically stable under suitable conditions.

Introducing the awareness rate ζ_1 and a control u_s, the authors also considered the following optimal control model:

$$
\begin{aligned}
D^\alpha S(t) &= \Lambda - \sigma S - vS + lR + \zeta_1 u_s S \\
D^\alpha I(t) &= \sigma S - \beta_1 I - \rho IT - vI \\
D^\alpha T(t) &= \beta_2 I(t-\tau) + \rho IT - vT - \frac{bu_r T}{1 + \gamma_1 u_r T} \\
D^\alpha R(t) &= \frac{bu_r T}{1 + \gamma_1 u_r T} - vR - lR
\end{aligned}
\tag{169}
$$

with an objective function defined by

$$
J(u_s, u_r) = \int_0^{t_f} \left[Au_s^2(t) + BI(t) + Cu_r^2(t)\right] dt,
$$

Finally, the authors examined the dynamics of their model using numerical simulations with different parameter values and then analyzed the system under various control scenarios.

El Quissari and El Moutaouakil [62]

El Quissari and El Moutaouakil considered the EDC model proposed by Boutayeb and Chetouani in 2007 [42], including the compartment (E) of pre-diabetes beside the usual DC model. Stressing that they use some properties on the fractional derivatives of Riemann-Liouville and Caputo, the authors proposed the following fractional-order model [62]:

$$
\begin{aligned}
&{}_0^C\mathrm{D}_t^\alpha E(t) = I - (\mu + \beta_3 + \beta_1) E(t) \\
&{}_0^C\mathrm{D}_t^\alpha D(t) = \beta_1 E(t) - (\mu + \beta_2) D(t) + \gamma C(t) \\
&{}_0^\alpha C(t) = \beta_3 E(t) + \beta_2 D(t) - (\mu + \gamma + \nu + \delta) C(t)
\end{aligned}
\tag{170}
$$

The authors proved the existence and uniqueness of a bounded solution of the fractional-order differential system (170). Then they introduced optimal control to obtain the following system:

$$
\begin{aligned}
{}_0^{C}D_t^{\alpha} E(t) &= I - (\mu + (\beta_3 + \beta_1)(1 - \omega(t)))\, E(t) \\
{}_0^{C}D_t^{\alpha} D(t) &= \beta_1(1 - \omega(t))E(t) - (\mu + \beta_2(1 - \omega(t)))\, D(t) + \gamma C(t) \\
{}_0^{C}D_t^{\alpha} C(t) &= \beta_3(1 - \omega(t))E(t) + \beta_2(1 - \omega(t))D(t) - (\mu + \gamma + \nu + \delta)C(t),
\end{aligned}
\tag{171}
$$

where ω is a control. The objective function is defined as

$$
J(\omega) = \int_0^T \left(D(t) + C(t) + \frac{A}{2}\omega^2(t) \right) dt \quad A > 0.
$$

We define the control set U as

$$
U = \{\omega : \omega \text{ is measurable}, \omega(t) \in [0, 1] \text{ for all } t \in [0, T]\}.
$$

Finally, the authors used a genetic algorithm to find a numerical solution of the fractional optimal control.

——— Dadhich et al. [53] ———

In 2024, Dadhich et al. [53] proposed a fractional-order model given by the following equations:

$$
\begin{aligned}
{}^{CF}D_t^{\eta} S(t) &= \lambda - (\pi + \mu + \beta E)S \\
{}^{CF}D_t^{\eta} E(t) &= (\pi + \beta E)S - (\mu + 1)E, \\
{}^{CF}D_t^{\eta} I(t) &= \alpha\gamma E - (\mu + \delta_1)\, I \\
{}^{CF}D_t^{\eta} I_T(t) &= (1 - \alpha\gamma)E - (\mu + \delta_2)\, I_T.
\end{aligned}
\tag{172}
$$

12 Stochastic Compartmental Models

——— Zhang et al. [113] ———

In order to describe the dynamic behavior of diabetes mellitus, Zhang et al. [113] extended, in 2019, the model {(49)–(50)} to a stochastic model, proposing a class of Ito SDEs in the form of:

$$
dC(t) = (-(\lambda(t) + \theta(t))C(t) + \lambda(t)N(t))dt + g_1(t)dW_t, \tag{173}
$$

$$
dN(t) = (I(t) - (v(t) + \delta(t))C(t) - \mu(t)N(t))dt + g_2(t)dW_t \tag{174}
$$

where $t \in \mathbb{R}^+$;

With $g_1(t)$ and $g_2(t)$ the uncertain influences and the other parameters remain as defined in the model {(49)–(50)}.

Proving theoretical results and carrying out numerical analysis, the authors concluded that "the main result of this article is the numerical simulation of stochastic diabetes mellitus model. The results show that the methods are effective and the numerical results can match the results of theoretical analysis and reality".

——— Anusha and Arthithan [19] ———

In 2024, Anusha and Arthithan [19] extended their deterministic model published in 2021 to a stochastic model with white noise. They also introduced optimal control in their previous model.

13 Modeling Co-existence of Diabetes and Communicable Diseases

13.1 Co-existence of Diabetes and Tuberculosis

——— Pan et al. [96] ———

In 2015, the effect of diabetes on tuberculosis was analyzed by Pan et al. [96] on the basis of tuberculosis control data in 13 countries with high tuberculosis. The authors used the following TB status: Susceptible, Fast latent infection, Fast latent reinfection, Slow latent infection, Infectious pulmonary status and Recovered. The following set of equations:

1. **Susceptible** (S and S_{dm})

$$d(S)/dt = g - \beta * I * S - S * u - S * x_{dm} \tag{175}$$
$$d(S_{dm})/dt = x_{dm} * S - \beta * I * S_{dm} - S_{dm} * u * RR_u \tag{176}$$

2. **Fast latent infection** (F and F_{dm})

$$d(F)/dt = \beta * I * S - F * j1 - F * k - F * xdm - F * u \tag{177}$$
$$\begin{aligned} d(F_{dm})/dt = {} & F * x_{dm} + \beta * I * S_{dm} - F_{dm} * j1 * RR_j - F_{dm} * k \\ & - F_{dm} * u * RR_u \end{aligned} \tag{178}$$

3. **Fast latent reinfection** (Fi and Fi_{dm})

$$\begin{aligned} d(Fi)/dt = {} & \beta * I * (L + R) - Fi * j1 * (1 - im) - Fi * k \\ & - Fi * x_{dm} - Fi * u \end{aligned} \tag{179}$$
$$\begin{aligned} d(Fi_{dm})/dt = {} & Fi * x_{dm} + \beta * I * (L_{dm} + R_{dm}) - Fi_{dm} * j1 * (1 - im) * RR_j \\ & - Fi_{dm} * k - F_{dm} * u * RR_u \end{aligned} \tag{180}$$

4. **Slow latent infection** (L and L_{dm})

$$d(L)/dt = F * k + Fi * k + c * Ip - \beta * I * L - L * j2 - L * x_{dm} - L * u \quad (181)$$

$$d(L_{dm})/dt = L * x_{dm} + F_{dm} * k + Fi_{dm} * k + c * Ip_{dm} - \beta * I * L_{dm} - L_{dm} * RR_j - L_{dm} * u * RR_u \quad (182)$$

5. **Infectious pulmonary status** (Ip and Ip_{dm})

$$d(Ip)/dt = F * j1 + Fi * j1 * (1 - im) + L * j2 + R * q - Ip * c - CDR * INC * treat - Ip * utb - Ip * x_{dm} \quad (183)$$

$$d(Ip_{dm})/dt = x_{dm} * Ip + F_{dm} * j1 * RR_j + Fi_{dm} * j1 * (1 - im) * RR_j + L_{dm} * j2 * RR_j + R_{dm} * q * RR_q - c * Ip_{dm} - CDR * INC_{dm} * treat - Ip_{dm} * utb * RR_u tb \quad (184)$$

6. **Recovered** (R and R_{dm})

$$d(R)/dt = CDR * INC * treat - R * q - \beta * I * R - R * x_{dm} - R * u \quad (185)$$

$$d(R_{dm})/dt = R * x_{dm} + CDR * INCdm * treat - R_{dm} * q * RR_q - \beta * I * R_{dm} - R_{dm} * u * RR_u \quad (186)$$

Implementing different scenarios, the authors found that "lowering the prevalence of diabetes by an absolute level of 6.6–13.8% could accelerate the decline of tuberculosis incidence by an absolute level of 11.5–25.2% and tuberculosis mortality by 8.7–19.4%".

More importantly, they stated that reducing diabetes incidence by 35% by 2025 could avert 7.8 million tuberculosis cases and 1.5 million tuberculosis deaths by 2035.

Sweilam et al. [106]

Assuming that the total population N is a constant in time, Sweilam et al. used optimal control for a fractional tuberculosis infection model including the impact of diabetes and resistant strains. The proposed model with four controls is given by the following 14 equations [106]

$$
\begin{aligned}
{}_{a}^{ABC}D_t^{\alpha} S_1 = {} & A^{\alpha} - \left(\mu^{\alpha} + \alpha_D^{\alpha} + \lambda_T\right) S_1 && (187)\\
{}_{a}^{ABC}D_t^{\alpha} S_2 = {} & \alpha_D^{\alpha} S_1 - \left(\mu^{\alpha} + \theta\lambda_T\right) S_2, && (188)\\
{}_{a}^{ABC}D_t^{\alpha} E_{1s} = {} & (1-\xi)(1-P_1)\lambda_T S_1 + \sigma_{31}\left(1-\delta_1^{\alpha} R_{1s}\right) - \left(1-r_1^{\alpha}\right)\left(k_1^{\alpha}\right.\\
& \left.+\sigma_1\lambda_T\right) E_{1s} - \left(\xi + \alpha_D^{\alpha} + \mu^{\alpha}\right) E_{1s}, && (189)\\
{}_{a}^{ABC}D_t^{\alpha} E_{1R} = {} & \xi(1-P_1)\lambda_T S_1 + \xi E_{1s} + \sigma_{32}\left(1-\delta_1^{\alpha}\lambda_T R_{1R}\right) - \left(1-r_1^{\alpha}\right)\\
& \times\left(k_1^{\alpha} + \sigma_1\lambda_T\right) E_{1R} - \left(\alpha_D^{\alpha} + \mu^{\alpha}\right) E_{1R}, && (190)\\
{}_{a}^{ABC}D_t^{\alpha} E_{2s} = {} & (1-\xi)(1-P_2)\theta\lambda_T S_2 + \sigma_{41}\left(1-\delta_2^{\alpha}\right)\theta\lambda_T R_{2s} + \alpha_D^{\alpha} E_{1s}\\
& - \left(1-r_2^{\alpha}\right)\left(k_2^{\alpha} + \sigma_2\theta\lambda_T\right) E_{2s} - \left(\xi + \mu^{\alpha}\right) E_{2s}, && (191)
\end{aligned}
$$

$$
\begin{aligned}
{}_{a}^{ABC}D_t^{\alpha} E_{2R} = {} & \xi(1-P_2)\theta\lambda_T S_2 + \xi E_{2s} + \sigma_{42}\left(1-\delta_2^{\alpha}\right)\theta\lambda_T R_{2R} + \alpha_D^{\alpha} E_{1R}\\
& - \left(1-r_2^{\alpha}\right)\left(k_2^{\alpha} + \sigma_2\theta\lambda_T\right) E_{2R} - \mu^{\alpha} E_{2R}, && (192)\\
{}_{a}^{ABC}D_t^{\alpha} I_{1s} = {} & (1-\xi)P_1\lambda_T S_1 + \left(1-r_1^{\alpha}\right)\left(k_1^{\alpha} + \sigma_1\lambda_T\right) E_{1s} + \delta_{11}^{\alpha} R_{1s}\\
& - \left(\tau_1\alpha_D^{\alpha} + \eta_1\xi + \gamma_{11}^{\alpha} + \mu^{\alpha} + d_1^{\alpha} + \omega_1 u_1\right) I_{1s}, && (193)\\
{}_{a}^{ABC}D_t^{\alpha} I_{1R} = {} & \xi P_1\lambda_T S_1 + \left(1-r_1^{\alpha}\right)\left(k_1^{\alpha} + \sigma_1\lambda_T\right) E_{1R} + \eta_1\xi I_{1s} + \delta_{12}^{\alpha} R_{1R}\\
& - \left(\tau_1\alpha_D^{\alpha} + \gamma_{12}^{\alpha} + \mu^{\alpha} + d_1^{\alpha} + \omega_2 u_2\right) I_{1R} && (194)\\
{}_{a}^{ABC}D_t^{\alpha} I_{2s} = {} & (1-\xi)P_2\theta\lambda_T S_2 + \left(1-r_2^{\alpha}\right)\left(k_{21}^{\alpha} + \sigma_2\theta\lambda_T\right) E_{2s} + \tau_1\alpha_D^{\alpha} I_{1s}\\
& +\delta_{21}^{\alpha} R_{2s} - \left(\eta_2\xi + \gamma_{21}^{\alpha} + \mu^{\alpha} + d_2^{\alpha} + \omega_3 u_3\right) I_{2s}, && (195)
\end{aligned}
$$

$$
\begin{aligned}
{}_{a}^{ABC}D_t^{\alpha} I_{2R} = {} & \xi P_2\theta\lambda_T S_2 + \left(1-r_2^{\alpha}\right)\left(k_2^{\alpha} + \sigma_2\theta\lambda_T\right) E_{2R} + \eta_2\xi I_{2s} + \tau_1\alpha_D^{\alpha} I_{1R}\\
& +\delta_{22}^{\alpha} R_{2R} - \left(\gamma_{22}^{\alpha} + \mu^{\alpha} + d_2^{\alpha} + \omega_4 u_4\right) I_{2R}, && (196)
\end{aligned}
$$

$$
\begin{aligned}
{}_{a}^{ABC}D_t^{\alpha} R_{1s} = {} & \gamma_{11}^{\alpha} I_{1s} + \omega_1 u_1 I_{1s} - \sigma_{31}\left(1-\delta_1^{\alpha}\right)\lambda_T R_{1s} && (197)\\
& - \left(\delta_{11}^{\alpha} + \xi + \alpha_D^{\alpha} + \mu^{\alpha}\right) R_{1s}
\end{aligned}
$$

$$
\begin{aligned}
{}_{a}^{ABC}D_t^{\alpha} R_{1R} = {} & \gamma_{21}^{\alpha} I_{1R} + \omega_2 u_2 I_{1R} + \xi R_{1s} - \sigma_{32}\left(1-\delta_1^{\alpha}\lambda_T R_{1R}\right) && (198)\\
& - \left(\delta_{12}^{\alpha} + \alpha_D^{\alpha} + \mu^{\alpha}\right) R_{1R},
\end{aligned}
$$

$$
\begin{aligned}
{}_{a}^{ABC}D_t^{\alpha} R_{2s} = {} & \gamma_{21}^{\alpha} I_{2s} + \omega_3 u_3 I_{2s} + \alpha_D^{\alpha} R_{1s} - \sigma_{41}\theta\left(1-\delta_2^{\alpha}\right)\lambda_T R_{2s} && (199)\\
& - \left(\delta_{21} + \xi + \mu^{\alpha}\right) R_{2s},
\end{aligned}
$$

$$
\begin{aligned}
{}^{ABC}_{a}D_t^{\alpha} R_{2R} = \quad & \gamma_{22}^{\alpha} I_{2R} + \omega_4 u_4 I_{2R} + \xi R_{2s} + \alpha_D R_{1R} \\
& -\sigma_{42}\theta \left(1 - \delta_2^{\alpha}\right) \lambda_T R_{2R} - \left(\delta_{22}^{\alpha} + \mu^{\alpha}\right) R_{2R}
\end{aligned}
\tag{200}
$$

where

$$
\lambda_T = \beta \frac{I_{1s} + \varepsilon I_{1R} + \varepsilon_1 I_{2s} + \varepsilon_2 I_{2R}}{N}
$$

The authors concluded that "Four optimal control variables, u_1, u_2, u_3 and u_4, are introduced to reduce the number of individuals infected. It is concluded that the proposed fraction-order model can potentially describe more complex dynamics than the integer model and can easily include the memory effects present in many real-world phenomena".

Agwu et al. [4]

Agwu et al. [4] proposed a mathematical model to describe the interactions between the diabetic population and the diabetic population co-infected with tuberculosis. The model is governed by the following equations:

$$
\begin{aligned}
\frac{dS}{dt} &= \Lambda - (\lambda + \alpha_1 + \mu) S \\
\frac{dS_D}{dt} &= \alpha_1 S - \theta \lambda S_D - \mu S_D \\
\frac{dL}{dt} &= \lambda (S + \xi_1 R_T) - (k_1 + \sigma_1 \lambda) L - \mu L \\
\frac{dI}{dt} &= (k_1 + \sigma_1 \lambda) L - \tau_1 \alpha_1 I - (r_1 + d_1 + \mu) I \\
\frac{dR_T}{dt} &= r_1 I - \lambda \xi_1 R_T - \psi_1 \alpha_1 R_T - \mu R_T \\
\frac{dL_D}{dt} &= \lambda (\theta S_D + \xi_2 R_D) - (k_2 + \sigma_2 \lambda) L_D - \alpha_2 L_D - \mu L_D \\
\frac{dI_D}{dt} &= (k_2 + \sigma_2 \lambda) L_D + \tau_1 \alpha_1 I - \tau_2 \alpha_2 I_D - (r_2 + d_2 + \mu) I_D \\
\frac{dR_D}{dt} &= r_2 I_D + \psi_1 \alpha_1 R_T + r_3 I_C - \lambda \xi_2 R_D - \mu R_D \\
\frac{dL_C}{dt} &= \alpha_2 L_D - (k_3 + \sigma_3 \lambda) L_C - (\delta_1 + \mu) L_C \\
\frac{dI_C}{dt} &= (k_3 + \sigma_3 \lambda) L_C + \tau_2 \alpha_2 I_D - (r_3 + d_3 + \delta_2 + \mu) I_C
\end{aligned}
\tag{201}
$$

with the corresponding initial conditions

$$S(0) \geq 0, \quad S_D(0) \geq 0, \quad L(0) \geq 0, \quad I(0) \geq 0, R_T(0) \geq 0, L_D(0) \geq 0, I_D(0) \geq 0,$$
$$R_D(0) \geq 0, \quad L_C(0) \geq 0, \quad I_C(0) \geq 0.$$

The authors analyzed the feasibility of the region, the positivity of the solution, basic reproductive number R0, equilibrium points and their stability. They found that the previous system has a disease-free equilibrium (DFE) given by

$$D_0 = \left(S^*, S_D^*, L^*, I^*, R_T^*, L_D^*, I_D^*, R_D^*, L_D^*, I_C^*\right) = \left(S^*, S_D^*, 0, 0, 0, 0, 0, 0, 0, 0\right)$$

where
$$S^* = \frac{\Lambda}{\mu + \alpha_1}, \quad S_{D^*} = \frac{\alpha_1 S^*}{\mu} = \frac{\alpha_1 \Lambda}{\mu (\mu + \alpha_1)}.$$

The authors also carried out a backward bifurcation analysis of the model, analysis of the optimal control model, numerical simulations and cost-effectiveness analysis and discussed nine control strategies.

Awad et al. [28–30]

In order to assess the direct and indirect impacts of the effects of diabetes mellitus (DM) on tuberculosis (TB), Awad and colleagues [29] constructed a population-based dynamical TB-DM mathematical model. This model was based on a literature review which allowed the authors to develop a conceptual framework as indicated by Fig. 24.

The proposed model which was calibrated using data on tuberculosis from India is governed by the following system of ordinary differential equations.

TB transmission dynamics for the population without DM

$$\frac{dS}{dt} = (1 - f_{DM})(\mu N + M_{\text{TBMorIncid}} + M_{DMMorIncid}) - (\lambda + \mu)S \tag{202}$$

$$\frac{dL^F}{dt} = p\lambda S + p(1-q)\lambda L^S + p(1-q)\lambda(R + T^{SP} + T^{SN} + T^{EP}) - (\omega_{LF} + \mu)L^F \tag{203}$$

$$\frac{dL^S}{dt} = (1-p)\lambda S + (1-p)(1-q)\lambda(R + T^{SP} + T^{SN} + T^{EP}) - (p(1-q)\lambda + \omega_{LS} + \mu)L^S \tag{204}$$

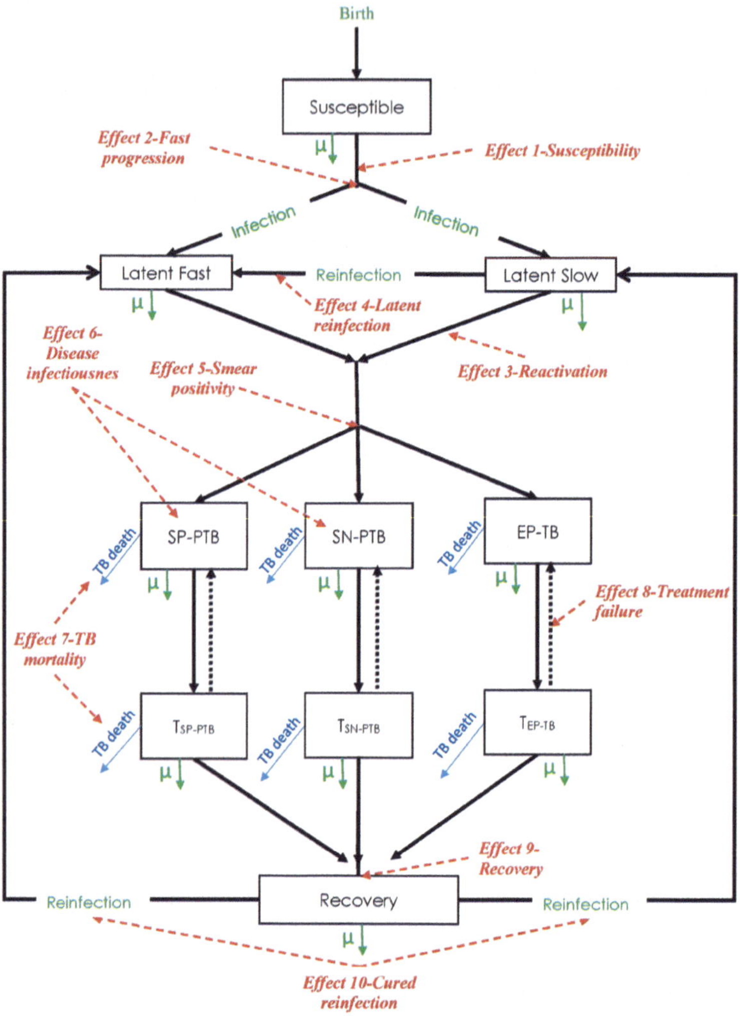

SP-PTB: smear-positive pulmonary TB; **SN-PTB**: smear-negative pulmonary TB; **EP-TB**: extra-pulmonary TB; **T_{SP-PTB}**: treated smear-positive pulmonary TB; **T_{SN-PTB}**: treated smear-negative pulmonary TB; **T_{EP-TB}**: treated extra-pulmonary TB; **μ**: natural death

Fig. 24 A conceptual framework for the TB-DM epidemiologic synergy. The black lines indicate the transitions within TB's natural history and treatment states. The red lines indicate the 10 plausible effects of DM on TB natural history and treatment outcomes. *Source* From Awad et al. [29], under the Creative Commons Attribution International License (CC BY 4.0). http://creativecommons.org/licenses/by/4.0/

$$\frac{dI^{SP}}{dt} = \alpha^{LFtoSP}\omega_{LF}L^{F} + \alpha^{LStoSP}\omega_{LS}L^{S} - (\xi_{SP} + z\xi_{SP} + \upsilon_{SP} + \mu)I^{SP} \tag{205}$$

$$\frac{dI^{SN}}{dt} = \alpha^{LFtoSN}\omega_{LF}L^{F} + \alpha^{LStoSN}\omega_{LS}L^{S} - (\xi_{SN} + z\xi_{SN} + \upsilon_{SN} + \mu)I^{SN} \tag{206}$$

$$\frac{dI^{EP}}{dt} = \alpha^{LFtoNP}\omega_{LF}L^{F} + \alpha^{LStoNP}\omega_{LS}L^{S} - (\xi_{NP} + z\xi_{NP} + \upsilon_{NP} + \mu)I^{EP} \tag{207}$$

$$\frac{dT^{SP}}{dt} = z\xi_{SP}I^{SP} - (\upsilon_{TSP} + \psi_{TSP} + (1-q)\lambda + \xi_{TSP} + \mu)T^{SP} \tag{208}$$

$$\frac{dT^{SN}}{dt} = z\xi_{SN}I^{SN} - (\upsilon_{TSN} + \psi_{TSN} + (1-q)\lambda + \xi_{TSN} + \mu)T^{SN} \tag{209}$$

$$\frac{dT^{EP}}{dt} = z\xi_{NP}I^{EP} - (\upsilon_{TNP} + \psi_{TNP} + (1-q)\lambda + \xi_{TNP} + \mu)T^{EP} \tag{210}$$

$$\begin{aligned}\frac{dR}{dt} &= \upsilon_{SP}S^{SP} + \upsilon_{SN}I^{SN} + \upsilon_{NP}I^{EP} + \upsilon_{TSP}S^{SP} + \upsilon_{TSN}T^{SN} + \upsilon_{TNP}T^{EP}\\ &\quad + \psi_{TSP}T^{SP} + \psi_{TSN}T^{SN} + \psi_{TNP}T^{EP} - ((1-q)\lambda + \mu)R\end{aligned} \tag{211}$$

TB transmission dynamics for the population with DM

$$\begin{aligned}\frac{dS^{DM}}{dt} &= f_{DM}(\mu N + M_{\text{TBMorIncid}} + M_{DMMorIncid})\\ &\quad - (E_1\lambda + \mu + \gamma)S^{DM}\end{aligned} \tag{212}$$

$$\begin{aligned}\frac{dL^{F_{DM}}}{dt} &= E_2 pE_1\lambda S^{DM} + pE_4(1-q)\lambda L^{S_{DM}} + pE_{10}(1-q)\lambda(R^{DM} + T^{SP_{DM}} + T^{SN_{DM}} + T^{EP_{DM}})\\ &\quad - (E_3\omega_{LF} + \mu + \gamma)L^{F_{DM}}\end{aligned} \tag{213}$$

$$\begin{aligned}\frac{dL^{S_{DM}}}{dt} &= (1 - E_2 p)E_1\lambda S^{DM} + (1-p)E_{10}(1-q)\lambda(R^{DM} + T^{SP_{DM}} + T^{SN_{DM}} + T^{EP_{DM}})\\ &\quad - (pE_4(1-q)\lambda + E_3\omega_{LS} + \mu + \gamma)L^{S_{DM}}\end{aligned} \tag{214}$$

$$\begin{aligned}\frac{dI^{SP_{DM}}}{dt} &= \alpha^{(LFtoSP)_{DM}}E_3\omega_{LF}L^{F_{DM}} + \alpha^{(LStoSP)_{DM}}E_3\omega_{LS}L^{S_{DM}}\\ &\quad - (E_8 z\zeta_{SP} + E_9\upsilon_{SP} + E_7\xi_{SP} + \mu + \gamma)I^{SP_{DM}}\end{aligned} \tag{215}$$

$$\begin{aligned}\frac{dI^{SN_{DM}}}{dt} &= \alpha^{(LFtoSN)_{DM}}E_3\omega_{LF}L^{F_{DM}} + \alpha^{(LStoSN)_{DM}}E_3\omega_{LS}L^{S_{DM}}\\ &\quad - (E_8 z\zeta_{SN} + E_9\upsilon_{SN} + E_7\xi_{SN} + \mu + \gamma)I^{SN_{DM}}\end{aligned} \tag{216}$$

$$\frac{dI^{EP_{DM}}}{dt} = \alpha^{(LFtoNP)_{DM}} E_3\omega_{LF} L^{F_{DM}} + \alpha^{(LStoNP)_{DM}} E_3\omega_{LS} L^{S_{DM}} - (E_8 z\zeta_{NP} + E_9 \upsilon_{NP} + E_7\xi_{NP} + \mu + \gamma) I^{EP_{DM}} \tag{217}$$

$$\frac{dT^{SP_{DM}}}{dt} = E_8 z\zeta_{SP} I^{SP_{DM}} - (E_9 \upsilon_{TSP} + E_9\psi_{TSP} + E_{10}(1-q)\lambda + E_7\xi_{TSP} + \mu + \gamma) T^{SP_{DM}} \tag{218}$$

$$\frac{dT^{SN_{DM}}}{dt} = E_8 z\zeta_{SN} I^{SN_{DM}} - (E_9 \upsilon_{TSN} + E_9\psi_{TSN} + E_{10}(1-q)\lambda + E_7\xi_{TSN} + \mu + \gamma) T^{SN_{DM}} \tag{219}$$

$$\frac{dT^{EP_{DM}}}{dt} = E_8 z\zeta_{NP} I^{NP_{DM}} - (E_9 \upsilon_{TNP} + E_9\psi_{TNP} + E_{10}(1-q)\lambda + E_7\xi_{TNP} + \mu + \gamma) T^{EP_{DM}} \tag{220}$$

$$\frac{dR^{DM}}{dt} = E_9(\upsilon_{SP} I^{SP_{DM}} + \upsilon_{SN} I^{SN_{DM}} + \upsilon_{NP} I^{EP_{DM}} + \upsilon_{TSP} T^{SP_{DM}} + \upsilon_{TSN} T^{SN_{DM}} + \upsilon_{TNP} T^{EP_{DM}}) \tag{221}$$

$$+ E_9(\psi_{TSP} T^{SP_{DM}} + \psi_{TSN} T^{SN_{DM}} + \psi_{TNP} T^{EP_{DM}}) - (E_{10}(1-q)\lambda + \mu + \gamma) R^{DM}$$

With:

$$M_{TBMorIncid} = \xi_{SP} I^{SP} + \xi_{SN} I^{SN} + \xi_{NP} I^{EP} + \xi_{TSP} T^{SP} + \xi_{TSN} T^{SN} + \xi_{TNP} T^{EP},$$

$$M_{DMMorIncid} = \gamma(I^{SP_{DM}} + I^{SN_{DM}} + I^{EP_{DM}} + T^{SP_{DM}} + T^{SN_{DM}} + T^{EP_{DM}}),$$

$$\begin{aligned} \alpha^{(LFtoSP)_{DM}} &= \alpha^{(LStoSP)_{DM}} = F_1 \alpha^{LFtoSP}, \\ \alpha^{(LFtoSN)_{DM}} &= \alpha^{(LStoSN)_{DM}} = F_2 \alpha^{LFtoSN}, \\ \alpha^{(LFtoEP)_{DM}} &= \alpha^{(LStoEP)_{DM}} = \alpha^{LFtoEP}, \end{aligned}$$

$$F_1 = \left(1 + \frac{E_5-1}{E_5+1}\right) \frac{\alpha_a^{(LFtoSP)} + \alpha_a^{(LFtoSN)}}{\left[\alpha_a^{(LFtoSP)}\left(1+\frac{E_5-1}{E_5+1}\right) + \alpha_a^{(LFtoSN)}\left(1-\frac{E_5-1}{E_5+1}\right)\right]}, \text{ and}$$

$$F_2 = \left(1 - \frac{E_5-1}{E_5+1}\right) \frac{\alpha_a^{(LFtoSP)} + \alpha_a^{(LFtoSN)}}{\left[\alpha_a^{(LFtoSP)}\left(1+\frac{E_5-1}{E_5+1}\right) + \alpha_a^{(LFtoSN)}\left(1-\frac{E_5-1}{E_5+1}\right)\right]}.$$

Extending and adapting the previous model, Awad and colleagues published two other deterministic compartmental mathematical models to forecast the impact of diabetes mellitus on tuberculosis disease incidence and mortality in India [30] and to investigate the impact of diabetes mellitus on tuberculosis epidemiology in Indonesia [28].

Pinto and Carvalho [99]

Pinto and Carvalho published a fractional-order modeling dealing with diabetes mellitus and TB co-existence. In order to analyze the clinical implications of diabetes mellitus in the dynamics of tuberculosis transmission, the authors proposed a model consisting of 14 classes representing a population of non-diabetics and diabetics affected by the dynamics of TB with the following seven status: susceptible individuals (S_1 and S_2), fast latent TB infected individuals (E_1 F and E_2 F), fast latent TB re-infected individuals (E_1Fr and E_2Fr), slow latent TB infected individuals (E_1 S and E_2 S), active TB infected individuals (I1 and I2), TB recovered individuals (R_1 and R_2), and MDR-TB strains infected individuals (TR_1 and TR_2) [99]. The proposed fractional-order model is given by the system:

$$\begin{aligned}
\frac{d^\alpha S_1}{dt^\alpha} &= \Lambda^\alpha - \left(\mu^\alpha + \alpha_D^\alpha + \lambda_T\right) S_1 \\
\frac{d^\alpha E_{1F}}{dt^\alpha} &= \lambda_T S_1 - \left(k^\alpha + \phi_1^\alpha + \alpha_D^\alpha + \mu^\alpha\right) E_{1F} \\
\frac{d^\alpha E_{1kr}}{dt^\alpha} &= \lambda_T \left(E_{1S} + R_1\right) - \left(\phi_1^\alpha (1-\sigma) + k^\alpha + \alpha_D^\alpha + \mu^\alpha\right) E_{1Fr} \\
\frac{d^\alpha E_{1S}}{dt^\alpha} &= k^\alpha \left(E_{1F} + E_{1Fr}\right) + c^\alpha I_1 - \left(\lambda_T + \phi_2^\alpha + \alpha_D^\alpha + \mu^\alpha\right) E_{1S}
\end{aligned} \tag{222}$$

$$\begin{aligned}
\frac{d^\alpha I_1}{dt^\alpha} &= \phi_1^\alpha E_{1F} + \phi_1^\alpha (1-\sigma) E_{1Fr} + \phi_2^\alpha E_{1S} + \delta_1^\alpha R_1 + a^\alpha T_{R1} \\
&\quad - \left(\mu^\alpha + d_1^\alpha + \tau \alpha_D^\alpha + \gamma^\alpha + \sigma_T^\alpha + c^\alpha\right) I_1 \\
\frac{d^\alpha R_1}{dt^\alpha} &= \gamma^\alpha I_1 - \left(\lambda_T + \delta_1^\alpha + \alpha_D^\alpha + \mu^\alpha\right) R_1 \\
\frac{d^\alpha T_{R1}}{dt^\alpha} &= \sigma_T^\alpha I_1 - \left(\tau_1 \alpha_D^\alpha + a^\alpha + \mu^\alpha + d_1^\alpha\right) T_{R1}
\end{aligned} \tag{223}$$

$$\begin{aligned}
\frac{d^\alpha S_2}{dt^\alpha} &= \alpha_D^\alpha S_1 - \left(\lambda_T + \mu^\alpha + d^\alpha\right) S_2 \\
\frac{d^\alpha E_{2F}}{dt^\alpha} &= \lambda_T S_2 + \alpha_D^\alpha E_{1F} - \left(k^\alpha + \theta_1 \phi_1^\alpha + \mu^\alpha + d^\alpha\right) E_{2F} \\
\frac{d^\alpha E_{2Fr}}{dt^\alpha} &= \lambda_T \left(E_{2S} + R_2\right) + \alpha_D^\alpha E_{1Fr} - \left(\theta_1 \phi_1^\alpha (1-\sigma) + k^\alpha + \mu^\alpha + d^\alpha\right) E_{2Fr}
\end{aligned} \tag{224}$$

$$\begin{aligned}
\frac{d^\alpha E_{2S}}{dt^\alpha} &= k^\alpha \left(E_{2F} + E_{2Fr}\right) + \alpha_D^\alpha E_{1S} + c^\alpha I_2 - \left(\lambda_T + \theta_1 \phi_2^\alpha + \mu^\alpha + d^\alpha\right) E_{2S} \\
\frac{d^\alpha I_2}{dt^\alpha} &= \theta_1 \phi_1^\alpha E_{2F} + \theta_1 \phi_1^\alpha (1-\sigma) E_{2Fr} + \theta_1 \phi_2^\alpha E_{2S} + \tau \alpha_D^\alpha I_1 + \delta_2^\alpha R_2 + a^\alpha T_{R2} \\
&\quad - \left(\mu^\alpha + d_2^\alpha + \gamma^\alpha + \sigma_T^\alpha + c^\alpha\right) I_2
\end{aligned} \tag{225}$$

$$\begin{aligned}\frac{d^{\alpha}R_2}{dt^{\alpha}} &= \gamma^{\alpha}I_2 + \alpha_D^{\alpha}R_1 - \left(\lambda_T + \delta_2^{\alpha} + \mu^{\alpha} + d^{\alpha}\right)R_2\\ \frac{d^{\alpha}T_{R_2}}{dt^{\alpha}} &= \sigma_T^{\alpha}I_2 + \tau_1\alpha_D^{\alpha}T_{R1} - \left(a^{\alpha} + \mu^{\alpha} + d_2^{\alpha}\right)T_{R2}\end{aligned} \tag{226}$$

The authors calculated the reproduction number of the model, found the disease-free and the endemic equilibria and analyzed their stability. They have also obtained numerical results and used them for discussion. Finally, they have drawn conclusions from a biological point of view.

Jajarmi et al. [69]

Jajarmi et al. [69] studied the population dynamics resulting from co-existence of tuberculosis and diabetes. They proposed a mathematical model using numerical methods, fractional modeling and optimal control. The proposed fractional model (in the ABC sense) was based on a human population subdivided according to non-diabetic (index 1) and diabetic (index 2), co-existing with one of the seven classes due to tuberculosis status: susceptible (S_1 and S_2), fast latent infected (E_1F and E_2F), fast latent re-infected (E_1F_r and E_2F_r), slow latent infected (E_1S and E_2 S), (active tuberculosis infected (I_1 and I_2), recovered (R_1 and R_2) and multidrug-resistant tuberculosis strains infected populations (TR_1 and TR_2).

The mathematical model is governed by the following system of 14 equations:

$$\begin{aligned}
{}^{ABC}_{t_0}\mathscr{D}_t^{\rho}S_1(t) &= \Lambda^{\rho} - \left(\mu^{\rho} + \alpha_D^{\rho} + \lambda_T(t)\right)S_1(t),\\
{}^{ABC}_{t_0}\mathscr{D}_t^{\rho}S_2(t) &= \alpha_D^{\rho}S_1(t) - \left(\lambda_T(t) + \mu^{\rho} + d^{\rho}\right)S_2(t),\\
{}^{ABC}_{t_0}\mathscr{D}_t^{\rho}E_{1Fr}(t) &= \lambda_T(t)\left(E_{1S}(t) + R_1(t)\right)\\
&\quad - \left(\phi_1^{\rho}(1-\sigma)k^{\rho} + \alpha_D^{\rho} + \mu^{\rho}\right)E_{1Fr}(t),\\
{}^{ABC}_{t_0}\mathscr{D}_t^{\rho}E_{2Fr}(t) &= \lambda_T(t)\left(E_{2S}(t) + R_2(t)\right)\\
&\quad + \alpha_D^{\rho}E_{1Fr}(t) - \left(\theta_1\phi_1^{\rho}(1-\sigma) + k^{\rho} + \mu^{\rho} + d^{\rho}\right)E_{2Fr}(t),\\
{}^{ABC}_{t_0}\mathscr{D}_t^{\rho}E_{1S}(t) &= k^{\rho}\left(E_{1F}(t) + E_{1Fr}(t)\right) + c^{\rho}I_1(t)\\
&\quad - \left(\lambda_T(t) + \phi_2^{\rho} + \alpha_D^{\rho} + \mu^{\rho}\right)E_{1S}(t),\\
{}^{ABC}_{t_0}\mathscr{D}_t^{\rho}E_{2S}(t) &= k^{\rho}\left(E_{2F}(t) + E_{2Fr}(t)\right) + \alpha_D^{\rho}E_{1S}(t) + c^{\rho}I_2(t)\\
&\quad - \left(\lambda_T(t) + \theta_1\phi_2^{\rho} + \mu^{\rho} + d^{\rho}\right)E_{2S}(t),\\
{}^{ABC}_{t_0}\mathscr{D}_t^{\rho}R_1(t) &= \gamma^{\rho}I_1(t)\\
&\quad - \left(\lambda_T(t) + \delta_1^{\rho} + \alpha_D^{\rho} + \mu^{\rho}\right)R_1(t),\\
{}^{ABC}_{t_0}\mathscr{D}_t^{\rho}R_2(t) &= \gamma^{\rho}I_2(t) + \alpha_D^{\rho}R_1(t)\\
&\quad - \left(\lambda_T(t) + \delta_2^{\rho} + \mu^{\rho} + d^{\rho}\right)R_2(t),
\end{aligned}$$

$$
\begin{aligned}
{}^{ABC}_{t_0}\mathscr{D}_t^{\rho} T_{R1}(t) &= \sigma_T^{\rho} I_1(t) \\
&\quad - \left(\tau_1 \alpha_D^{\rho} + a^{\rho} + \mu^{\rho} + d_1^{\rho}\right) T_{R1}(t), \\
{}^{ABC}_{t_0}\mathscr{D}_t^{\rho} T_{R2}(t) &= \sigma_T^{\rho} I_2(t) + \tau_1 \alpha_D^{\rho} T_{R1}(t) \\
&\quad - \left(a^{\rho} + \mu^{\rho} + d_2^{\rho}\right) T_{R2}(t), \\
{}^{ABC}_{t_0}\mathscr{D}_t^{\rho} E_{1F}(t) &= \lambda_T(t) S_1(t) \\
&\quad - \left(k^{\rho} + \phi_1^{\rho} + \alpha_D^{\rho} + \mu^{\rho}\right) E_{1F}(t), \\
{}^{ABC}_{t_0}\mathscr{D}_t^{\rho} E_{2F}(t) &= \lambda_T(t) S_2(t) + \alpha_D^{\rho} E_{1F}(t) \\
&\quad - \left(k^{\rho} + \theta_1 \phi_1^{\rho} + \mu^{\rho} + d^{\rho}\right) E_{2F}(t), \\
{}^{ABC}_{t_0}\mathscr{D}_t^{\rho} I_1(t) &= \phi_1^{\rho} E_{1F}(t) + \phi_1^{\rho}(1-\sigma) E_{1Fr}(t) + \phi_2^{\rho} E_{1S}(t) \\
&\quad + \delta_1^{\rho} R_1(t) + a^{\rho} T_{R1}(t) \\
&\quad - \left(\mu^{\rho} + d_1^{\rho} + \tau \alpha_D^{\rho} + \gamma^{\rho} + \sigma_T^{\rho} + c^{\rho}\right) I_1(t), \\
{}^{ABC}_{t_0}\mathscr{D}_t^{\rho} I_2(t) &= \theta_1 \phi_1^{\rho} E_{2F}(t) + \theta_1 \phi_1^{\rho}(1-\sigma) E_{2FT}(t) + \theta_1 \phi_2^{\rho} E_{2S}(t) \\
&\quad + \tau \alpha_D^{\rho} I_1(t) + \delta_2^{\rho} R_2(t) + a^{\rho} T_{R2}(t) \\
&\quad - \left(\mu^{\rho} + d_2^{\rho} + \gamma^{\rho} + \sigma_T^{\rho} + c^{\rho}\right) I_2(t)
\end{aligned}
\tag{227}
$$

Through their model, the authors discussed the impact of diabetes mellitus on the epidemiology of tuberculosis. They showed existence of a non-negative and bounded solution, carried out stability analysis and numerical simulations with optimal control. Finally, they concluded that: "Indeed, our proposed model depicted that a trend of increasing diabetes mellitus causes higher tuberculosis prevalence and incidence and escalates tuberculosis multidrug resistance".

——— Dhara et al. [58] ———

In 2018, highlighting that studies strongly support the association between diabetes and tuberculosis Dhara et al. [58] extended the DC model to the following DCT model illustrated by Fig. 25.

The DCT model is governed by the following system of equations:

$$
\frac{dD}{dt} = I - (\lambda_1 + \mu_1) D(t) + \gamma_1 C(t) \tag{228}
$$

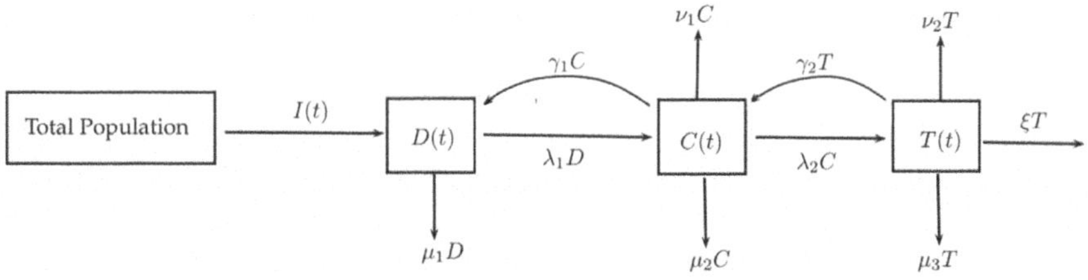

Fig. 25 Population dynamics associating diabetes with and without complications with tuberculosis

$$\frac{dC}{dt} = \lambda_1 D(t) - (\gamma_1 + \mu_2 + \upsilon_1 + \lambda_2)C(t) + \gamma_2 T(t) \tag{229}$$

$$\frac{dT}{dt} = \lambda_2 C(t) - (\mu_3 + \gamma_2 + \xi)T(t) \tag{230}$$

$$\lambda_1 = \lambda_1(t) = \frac{C(t)}{D(t) + C(t)} = \frac{C(t)}{N(t)} \tag{231}$$

$$\lambda_2 = \lambda_2(t) = \frac{T(t)}{D(t) + C(t) + T(t)} \tag{232}$$

where $D(t)$, $C(t)$ and $T(t)$ stand for diabetics without complications, diabetics with complications and diabetics with complications and tuberculosis infection at time t respectively and λ_1 and λ_2 are the rates of developing a complication in diabetic individuals and developing a tuberculosis infection with complicated diabetic individuals.

——— Danumjaya & Dhara [54] ———

Association between diabetes and tuberculosis was also considered by Danumjaya & Dhara [54] in 2019 who used non-standard finite difference schemes on the previous model {(228)–(232)}. Numerical and stability analysis were performed.

——— Moualeu et al. [84] ———

In 2012, Moualeu et al. [84] published an other model analyzing the impact of diabetes on the dynamical transmission of tuberculosis.

13.2 Co-existence of Diabetes and COVID-19

——— Anusha and Athithan [17] ———

In a preprint published by Anusha and Athithan [17], a mathematical model for the co-existence of diabetes and COVID-19 disease was proposed. The total population N(t) was subdivided into five compartments, namely: Diabetes susceptible class $S_d(t)$, Diabetes class D(t), COVID-19 Susceptible class $S_s(t)$, COVID-19 Infected class $I_c(t)$ and COVID-19 Recovered class $R_c(t)$.

Before developing their mathematical model, the authors made the following assumptions: (1) All compartments are mutually exclusive, (2) total population is given by $N(t) = S_d(t) + D(t) + S_c(t) + I_s(t) + R_c(t)$, (3) the population of our model is homogeneously mixed, (4) a simple mass-action type interaction for the disease transmission is considered, (5) diabetes is a non-communicable disease caused by an imbalanced level of glucose. So there is no transmission of diabetes between two people, (6) if either the production rate of insulin in the pancreas is lower or the body cannot use the produced insulin effectively, then diabetes susceptible individual (S) moves to diabetes class (D) at a rate of α, (7) the COVID-19 susceptible population (S) is composed of individuals who have not yet been infected by the COVID-19 but may be infected through contact with COVID-19 infected people, (8) since COVID-19 is not age-specific, transmission is considered from S_d to I_s at the rate of α_2, from S_s to I_s at the rate of β and from D to I_c at the rate of ρ, (9) γ is the rate at which infective individuals in I_c recovered from the disease enter to the recovered class R_c, (**10**) the individuals are losing their immunity against COVID-19 and hence, moving to COVID-19 susceptible class, (11) the general people may stay in the recovered class itself but the diabetes patients were automatically moved to their native diabetes class.

Based on these assumptions, the model of co-infection of diabetes and COVID-19 is governed by the following differential equations:

$$\begin{aligned}
\frac{dS_d}{dt} &= \Phi - \alpha_1 S_d - \alpha_2 S_d I_c - \mu S_d \\
\frac{dD}{dt} &= \alpha_1 S_d + \delta R_c - \rho D I_c - \mu D \\
\frac{dS_c}{dt} &= \Lambda + \ell R_c - \beta S_c I_c - \mu S_c \\
\frac{dI_c}{dt} &= \beta S_c I_c + \rho D I_c + \alpha_2 S_d I_c - \gamma I_c - \mu I_c - \mu_1 I_c \\
\frac{dR_c}{dt} &= \gamma I_c - \ell R_c - \delta R_c - \mu R_c
\end{aligned} \tag{233}$$

with initial conditions

$$S_d(0) > 0, \quad D(0) \geq 0, \quad S_c(0) > 0, \quad I_c(0) \geq 0, \quad R_c(0) \geq 0.$$

Okyere and Ackora-Prah [94]

Okyere and Ackora-Prah used the DC model of Boutayeb et al. [46] in a mixed six compartmental model SEIRCD translating the dynamics of diabetes under COVID-19 [94]. Figure 26 indicates the flow between different compartments of the model.

The authors assumed that the population is homogeneously mixed, with no restriction of age, mobility or other social factors. As indicated in the flow chart of the model, the following parameters were used:

Individuals are recruited into the susceptible population at the rate Ω and they become diabetic at the rate λ. On the other side, susceptible individuals may be infected when they interact with those in class I or C. The infected people become exposed to the disease and hence join the class E at the rate β. Then exposed individuals will join the compartments I and C at rate $\alpha\phi$ and $(1-\alpha)\phi$ respectively. The other parameters stand for natural mortality rate (μ), the disease-induced death rate of COVID-19- infected people (δ), induced death rate of COVID-19 patients with diabetes ($\delta 1$) and susceptible with diabetes ($\delta 2$), the rate at which COVID-19 patients with diabetes recover from COVID-19 (γ_1), the rate at which COVID-19 infected people recover (γ).

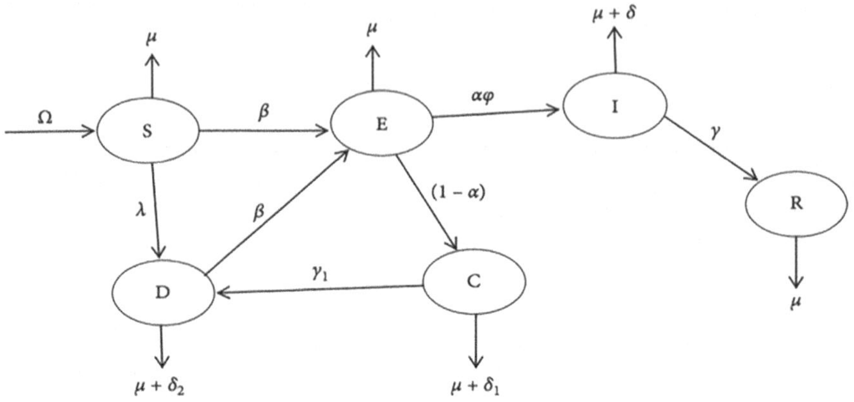

Fig. 26 Schematic diagram of diabetes-COVID-19 model *Source* Okyere and Ackora-Prah [94], under the Creative Commons Attribution International License (CC BY 4.0). http://creativecommons.org/licenses/by/4.0/

The proposed model is described by the following ordinary differential equations:

$$\begin{aligned}
\frac{dS}{dt} &= \Omega - \beta\left(\frac{I+C}{N}\right)S - (\mu+\lambda)S,\\
\frac{dD}{dt} &= \lambda S + \gamma_1 C - \beta\left(\frac{I+C}{N}\right)D - (\delta_2+\mu)\,D,\\
\frac{dE}{dt} &= \beta\left(\frac{I+C}{N}\right)S + \beta\left(\frac{I+C}{N}\right)D - (\phi+\mu)E,\\
\frac{dI}{dt} &= \alpha\phi E - (\delta+\mu+\gamma)I,\\
\frac{dC}{dt} &= (1-\alpha)\phi E - (\delta_1+\mu)\,C,\\
\frac{dR}{dt} &= \gamma I - \mu R.
\end{aligned} \tag{234}$$

Including the time-dependent controls into the previous system, they get the following system:

$$
\begin{aligned}
\frac{\mathrm{d}S}{\mathrm{d}t} &= \Omega - (1-u_1)\,\beta\left(\frac{I+C}{N}\right)S - (\mu+\lambda)S - u_2 S,\\
\frac{\mathrm{d}D}{\mathrm{d}t} &= \lambda S + \gamma_1 C - (1-u_1)\,\beta\left(\frac{I+C}{N}\right)D - (\delta_2+\mu)\,D - u_2 D,\\
\frac{\mathrm{d}E}{\mathrm{d}t} &= (1-u_1)\,\beta\left(\frac{I+C}{N}\right)S + (1-u_1)\,\beta\left(\frac{I+C}{N}\right)D - (\phi+\mu)E,\\
\frac{\mathrm{d}I}{\mathrm{d}t} &= \alpha\phi E - (\delta+\mu+\gamma)I,\\
\frac{\mathrm{d}C}{\mathrm{d}t} &= (1-\alpha)\phi E - (\delta_1+\mu)\,C,\\
\frac{\mathrm{d}R}{\mathrm{d}t} &= \gamma I - \mu R.
\end{aligned}
\tag{235}
$$

The authors showed positivity and boundedness of the solution, found the basic reproductive number and equilibrium points and examined stability of the equilibrium points and sensitivity analysis of the model.

They concluded that "numerical simulation of the optimal control model suggests the lockdown and vaccination when implemented reduces COVID-19 infection in the population".

Kouidere et al. [74]

Kouidere et al. [74] published a mathematical model dealing with the impact of COVID-19 on people with diabetes. These authors also used the DC model of Boutayeb et al. [46] in a mixed eight compartmental model SEIRQ$_T$PCD, including susceptible people that may be infected with COVID-19 disease (S), the infected individuals without symptoms (E), the infected individuals with symptoms (I), the recovered people after being infected with the COVID-19 disease (R), the individuals who are not infected with the virus in the quarantine period (Q_T), the pre-diabetics people (P), the diabetics without complications (D) and the diabetics with complications (C) (Fig. 27).

The mixed model proposed is governed by the following system of ordinary differential equations:

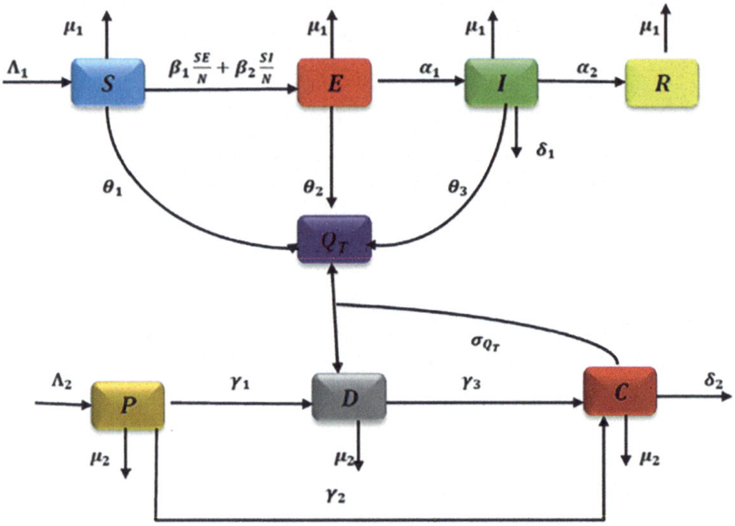

Fig. 27 Compartments model *Source* From Kouidere et al. [74] [60], under the Creative Commons Attribution International License (CC BY 4.0). http://creativecommons.org/licenses/by/4.0/

$$\begin{aligned}
\frac{dS(t)}{dt} &= \Lambda_1 - (\mu + \theta_1)\, S(t) - \beta_1 \frac{S(t)E(t)}{N} - \beta_2 \frac{S(t)I(t)}{N} \\
\frac{dE(t)}{dt} &= \beta_1 \frac{S(t)E(t)}{N} + \beta_2 \frac{S(t)I(t)}{N} - (\mu + \alpha_1 + \theta_2)\, E(t) \\
\frac{dI(t)}{dt} &= \alpha_1 E(t) - (\alpha_2 + \theta_3 + \mu + \delta_1)\, I(t) \\
\frac{dR(t)}{dt} &= \alpha_2 I(t) - \mu R(t) \\
\frac{dQ_T(t)}{dt} &= \theta_1 S(t) + \theta_2 E(t) + \theta_3 I(t) - \mu Q_T(t) \\
\frac{dP(t)}{dt} &= \Lambda_2 - (\gamma_1 + \gamma_2 + \mu_2)\, P(t) \\
\frac{dD(t)}{dt} &= \gamma_1 P(t) - (\gamma_3 + \mu_2)\, D(t) - \sigma_Q D(t) \\
\frac{dC(t)}{dt} &= \gamma_1 P(t) + \gamma_2 D(t) - \mu_2 C(t) + \sigma_Q D(t) - \delta_2 C(t)
\end{aligned} \tag{236}$$

The authors showed positivity, existence and uniqueness of the solution of the previous system. Then strategies with three controls $u(t)$, $v(t)$ and $w(t)$ for $t \in [0; T]$ were considered in the following way: "The first control can be interpreted as the proportion to be subjected to sensitization and prevention. So, we note that $u(t)$ is to diagnose and awareness program to susceptible people at time t. The second control $v(t)$ is the proportion of quarantine and health monitoring of people with COVID-19 disease and its complications. The third control w(t) is the proportion awareness program for diabetics without complications for diet tracking and regular diet in the quarantine period at time". Consequently, the optimal control model is

$$\begin{aligned}
\frac{dS(t)}{dt} &= \Lambda_1 - (\mu + \theta_1) S(t) - \beta_1(1 - u(t))\frac{S(t)E(t)}{N} - \beta_2(1 - u(t))\frac{S(t)I(t)}{N} \\
\frac{dE(t)}{dt} &= \beta_1(1 - u(t))\frac{S(t)E(t)}{N} + \beta_2(1 - u(t))\frac{S(t)I(t)}{N} - (\mu + \alpha_1 + \theta_2) E(t) \\
\frac{dI(t)}{dt} &= \alpha_1 E(t) - (\alpha_2 + \theta_3 + \mu + \delta_1) I(t) - v(t)I(t) \\
\frac{dR(t)}{dt} &= \alpha_2 I(t) - \mu R(t) + v(t)I(t) \\
\frac{dQ_T(t)}{dt} &= \theta_1 S(t) + \theta_2 E(t) + \theta_3 I(t) - \mu Q_T(t) \\
\frac{dP(t)}{dt} &= \Lambda_2 - (\gamma_1 + \gamma_2 + \mu_2) P(t) \\
\frac{dD(t)}{dt} &= \gamma_1 P(t) - (\gamma_3 + \mu_2) D(t) - \chi_{Q_T}(t)\sigma_Q(1 - w(t))D(t) \\
\frac{dC(t)}{dt} &= \gamma_2 P(t) + \gamma_3 D(t) - \mu_2 C(t) + \chi_{Q_T}(t)\sigma_Q(1 - w(t))D(t) - \delta_2 C(t)
\end{aligned} \tag{237}$$

The problem is to minimize the objective functional

$$J(u, v, w) = E(T) + I(T) - D(T) + \int_0^T \left[E(t) + I(t) - D(t) + \frac{A}{2}u^2(t) + \frac{B}{2}v^2(t) + \frac{G}{2}w^2(t)\right] dt$$

Anand et al. [15]

Anand et al. [15] published a mathematical model describing the interactions between the susceptible population, Covid-19 infected, diabetic population and Covid-19 infected, non-diabetic population. At any time t, the authors considered the total susceptible population ($x(t)$), the infected diabetic population ($y_1(t)$) and the infected non-diabetic population ($y_2(t)$)

Parameter	Significance
a	Contact (exposure) rate of y_1 with x
b	Contact (exposure) rate of y_2 with x
a_1	Rate of infection of exposed y_1
b_1	Rate of infection of exposed y_2
α	Fraction of population infected by y_1 and identified as diabetic
β	Fraction of diabetic population infected by y_2 and identified as diabetic
α_1	Quarantine + Treatment rate of y_1
α_2	Quarantine + Treatment rate of y_2
p	Fraction of recovered population of y_1 becoming susceptible again
q	Fraction of recovered population of y_2 becoming susceptible again
d	Mortality rate of x
d_1	Mortality rate of y_1
d_2	Mortality rate of y_2

The model proposed is governed by the following ordinary differential equations:

$$\begin{aligned} x' &= f(x) - dx - af_1(x)y_1 - bf_2(x)y_2 + p\alpha_1 y_1 + q\alpha_2 y_2, \\ y_1' &= a_1\alpha f_1(x)y_1 + b_1\beta f_2(x)y_2 - d_1 y_1 - \alpha_1 y_1, \\ y_2' &= a_1(1-\alpha)f_1(x)y_1 + b_1(1-\beta)f_2(x)y_2 - d_2 y_2 - \alpha_2 y_2 \end{aligned} \tag{238}$$

with initial conditions

$$x(t=0) = x_0, \quad y_1(t=0) = y_{10}, \quad y_2(t=0) = y_{20}.$$

The authors showed non-negativity and boundedness of solutions, derived the Basic reproduction number and then dealt with disease equilibria and stability analysis. They have also carried out numerical experiments and discussed different situations through examples involving various values of parameters (Fig. 28).

14 Other Compartmental Models

Awad et al. [31]

In 2022, Awad et al. [31] used an age-structured mathematical model to forecast the type 2 diabetes burden and the impact of key risk factors (obesity, smoking and physical inactivity). The model was parametrized using data on type 2 diabetes from the following natural history studies: Qatar's 2012 STEPwise survey, the Global Health Observatory and the International Diabetes Federation Diabetes Atlas, among other data sources.

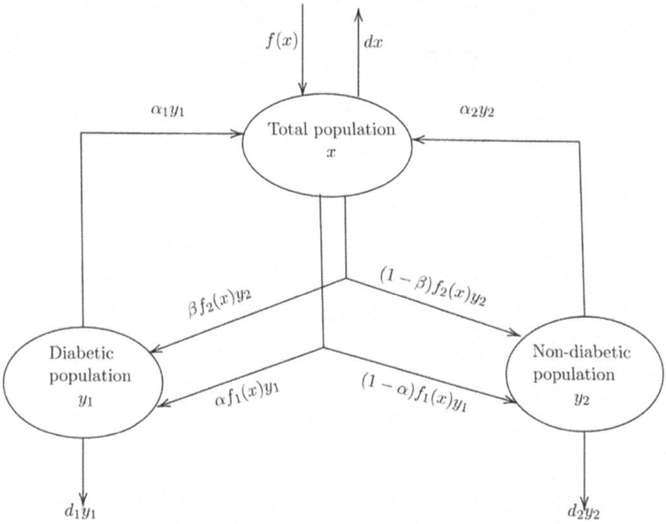

Fig. 28 Flow of populations and pictorial representation *Source* From Anand et al. [15], under the Creative Commons Attribution International License (CC BY 4.0). http://creativecommons.org/licenses/by/4.0/

Considering the population growth due births $b(t)$, the rate at which people leave the population through natural mortality and/or emigration (μ(t, a)) and the immigration rate (m(t, a)), the stochastic model proposed was parameterized by the following functions:

$$b(t) = a_0 e^{-\left(\frac{t-t_0}{b_0}\right)^2}$$

and

$$\mu(t,a) = f(a)\frac{a_1 e^{-\left(\frac{t-t_1}{b_1}\right)^2}}{\left[1 + e^{-b_2(a-a_2)}\right]}$$

and

$$b(t) = a_3 e^{-\left(\frac{a-a_4}{b_3}\right)^2} a_5 e^{-\left(\frac{t-t_2}{b_4}\right)^2}$$

Here, the parameters $a_0, a_1, a_2, a_3, a_4, a_5, t_0, t_1, t_2, b_0, b_1, b_2, b_3, b_4$ and $f(a)$ were obtained by fitting the model to the demographic data for each nationality group obtained from Qatar's Planning and Statistics Authority; the Gulf Labour Markets, Migration and Population Programme; and the Population Division of the United Nations Department of Economic and Social Affairs.

Using their mathematical model, the authors forecasted that type 2 diabetes prevalence will increase from 7 to 14% between 2021 and 2050. During the same period, obesity prevalence will also increase from 8.2 to 12.5%, smoking will decline from 28.3 to 26.9%, and physical inactivity will increase from 23.1 to 26.8%. The pro-

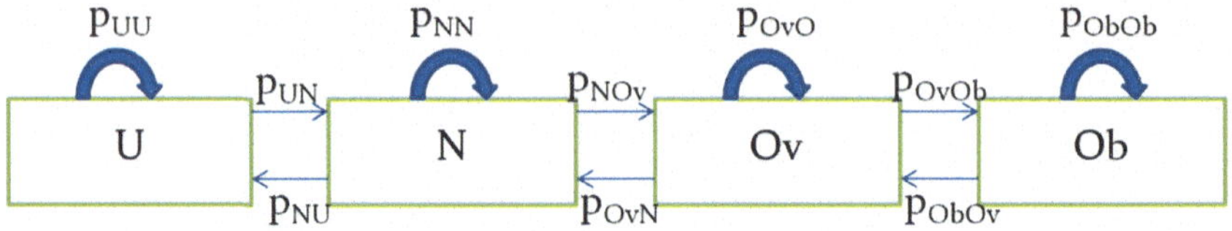

Fig. 29 Biomass in compartments: Underweight (U), Normal weight (N), Overweight (Ov) and Obesity (Ob)

portion of incident T2DM cases attributed to obesity increased from 21.9 to 29.9%, while the contribution of smoking and physical inactivity decreased from 7.1 to 6.0% and from 7.3 to 7.2%, respectively.

The authors carried out Sensitivity analysis and Uncertainty analysis, stressing that, substantial variability across various nationality groups residing in Qatar. For example, in Qataris and Egyptians, the burden of type 2 diabetes was mainly due to obesity, while in other nationality groups, it appeared to be multifactorial.

——— Boutayeb W et al. [49] ———

In 2014, following a paper by Walpole et al. who estimated the adult human biomass worldwide [108], Boutayeb W et al. [49] proposed a mathematical model illustrating simple and efficient ways to reduce overweight/obesity in Morocco (Fig. 29).

——— Jones et al. [70] ———

Jones et al. [70] proposed a model of the growth of diabetes and its trend between 1980 and 2050, stressing characteristics dynamics of diabetics population such as inability of diabetes control, delays between primary prevention and improvement in diabetes income.

——— Awad et al. [23] ———

Awad and colleagues have extended, modified or adapted the previous model to deal with "Preventing type 2 diabetes mellitus in Qatar by reducing obesity, smoking and physical inactivity" [27], "Characterizing the type 2 diabetes mellitus epidemic in Jordan up to 2050" [25], "diabetes risk score for Qatar: to identify individuals living with diabetes" [23], "diabetes risk score for Qatar: to identify individuals at high risk for diabetes" [24], "Forecasting the type 2 diabetes mellitus epidemic and the role of key risk factors in Oman up to 2050" [22], "Impact of trends and gender disparity in obesity on future type 2 diabetes in turkey" [14] and "Epidemiological impact of public health interventions against diabetes in Qatar" [10].

——— Sisman and Merdan [67] ———

Sisman and Merdan considered the same mathematical model (36) in a paper published in 2022 and using the fuzzy set and fuzzy difference equations in discrete fuzzy and crisp environment [67].

Bassey [38]

In 2019, Bassey [38] published a paper on "Optimal Control Model for Dual Treatment of Delayed Type-II Diabetes Infection in Human Population". However, we consider that, based on the title of this paper, the author assumes that diabetes is an infectious disease which is an aberration and nonsense. Consequently, we did not dive into this tide.

Ph.D theses [12, 20, 48, 56, 112]

More detailed information on mathematical models used in diabetes population dynamics can be found in the Ph.D theses defended by Derouich in 2001 [56], Zadeng in 2014 [112], Boutayeb in 2015 [48], Ali Merina in 2017 [12] and Asiri in 2018 [20].

15 Conclusion

In this chapter, we systematically reviewed a large number of publications dealing with the challenging global increasing trend of diabetes in all countries, including developed and developing of them. We analyzed the large number of papers selected and classified them according to different mathematical tools (ordinary differential equation, partial differential equation, delay differential equation, discrete models, numerical approximation, optimal control, fractional order and stochastic modeling). It is important to stress that the quasi totality of the papers included in this review, used, extended or improved the simple idea of using compartmental models for non-communicable diseases through the pioneer D-C model launched in 2004 and aiming to reduce the incidence and prevalence of diabetes and more particularly its deadly complications.

This chapter will surely be a reference for senior and junior researchers using mathematical modeling to understand the diabetes populations dynamics and propose pragmatic rules to health decision-makers seeking efficient and cost-effective strategies to halt the dreadful burden caused by the increasing trend of diabetes and its complications.

References

1. Adamu, I., Haruna, Y., Garba, E.: A mathematical model for attenuating the spread of diabetes and its management in a population. Global J. Math. Sci. **11**(1–2), 1–14 (2012)
2. Adamu, I., Momoh, A., Tahir, A.: Stability analysis of the mathematical model for the dynamics of diabetic population under the combine effect of birth rate and treatment. Int. J. Sci. Technol. **5**(1), 26–35 (2016)
3. Agarwal, M., Pathak, R.: Impact of media coverage on the controlling of diabetes. J. Multidiscip. Eng. Sci. Technol. **1**(5), 259–267 (2014)

4. Agwu, C.O., Omame, A., Inyama, S.C.: Analysis of mathematical model of diabetes and tuberculosis co-infection. Int. J. Appl. Comput. Math. **9**, 36 (2023)
5. Ahmad, A., Farman, M., Akgül, A.: Analysis and simulation of fractional-order diabetes model. Adv. Theory Nonlinear Anal. Appl. **4**, 483–497 (2020). https://doi.org/10.31197/atnaa.778506
6. Akinsola, V., Oluyo, T.: Mathematical model of the complications and control of diabetes mellitus. Int. J. Math. Comput. Appl. Res. **4**(5), 1–14 (2014)
7. Akinsola, V., Oluyo, T.: A note on the divergence of the numerical solution of the mathematical model for the burden of diabetes and its complications using Euler method. Int. J. Math. Comput. Appl. Res. **5**(3), 93–100 (2015)
8. Akinsola, V., Oluyo, T.: Mathematical analysis with numerical solutions of the mathematical model for the complications and control of diabetes mellitus. J. Stat. Manag. Syst. 1–25 (2019)
9. Akinsola, V., Oluyo, T.: Analytic solution of mathematical model of the complications and control of diabetes mellitus using fundamental matrix method. J. Interdisc. Math. **23**(4), 877–884 (2020)
10. Alareeki, A., Awad, S., Critchley, J., El-Nahas, K., Al-Hamaq, A., Alyafei, S., Al-Thani, M., Abu-Raddad, L.: Epidemiological impact of public health interventions against diabetes in Qatar: mathematical modelling analyses. Front. Public Health **11**, 1167807 (2023). https://doi.org/10.3389/fpubh.2023.1167807
11. Albalawi, K., Malik, K., Alkahtani, B., Goswami, P.: An analysis of the effects of lifestyle changes by using a fractional-order population model of the overweight/obese diabetic population. Fractal Fract. **7**(12), 839 (2023). https://doi.org/10.3390/fractalfract7120839
12. Ali Merina, H.: Modélisation mathématique et simulation numérique des complications du diabète. Ph.D. thesis, Université de Mostaganem (2017)
13. AlShurbaji, M., Kader, L., Hannan, H., Mortula, M., Husseini, G.: Comprehensive study of a diabetes mellitus mathematical model using numerical methods with stability and parametric analysis. Int. J. Environ. Res. Public Health **20**(2), 939 (2023)
14. Anakök, G., Awad, S., Çaglayan, ., Huangfu, P., Abu-Raddad, L., Unal, B., et al.: Impact of trends and gender disparity in obesity on future type 2 diabetes in turkey: a mathematical modelling analysis. BMJ Open **12**, e053541 (2022). https://doi.org/10.1136/bmjopen-2021-053541
15. Anand, M., Danumjaya, P., Rao, S.: A nonlinear mathematical model on the covid-19 transmission pattern among diabetic and non-diabetic population. Math. Comput. Simul. **210**, 346–369 (2023). https://doi.org/10.1016/j.matcom.2023.03.016
16. Andima, R., Mutuku, W., Farai, N., Awuor, K., Oke, A.: Mathematical modelling of diabetes under a constrained hospitalisation resources. Open Access Libr. J. **9**, e9371 (2022)
17. Anusha, S., Athithan, S.: Mathematical modelling co-existence of diabetes and covid-19: deterministic and stochastic approach (2021)
18. Anusha, S., Athithan, S.: Mathematical modelling of diabetes and its restrain. Int. J. Modern Phys. C **32**, 2150114 (2021). https://doi.org/10.1142/S0129183121501146
19. Anusha, S.M., Athithan, S.: Dynamical behaviour of a stochastic mathematical model and optimal control for type 2 diabetes. Math. Model. Eng. Probl. **11**(3), 792–796 (2024). https://doi.org/10.18280/mmep.110324
20. Asiri, K.: Mathematical models of diabetes in the united states population. Ph.D. thesis, Morgan State University (2018)
21. Asmaidi, A., Suryanto, E.: Mathematics modelling of diabetes mellitus type seiit by considering treatment and genetic factors. Jurnal Inovasi Teknologi dan Rekayasa **3**(1), 29–39 (2018)
22. Awad, S., Al-Mawali, A., Al-Lawati, J., Morsi, M., Critchley, J., Abu-Raddad, L.: Forecasting the type 2 diabetes mellitus epidemic and the role of key risk factors in Oman up to 2050: mathematical modelling analyses. J. Diab. Invest. **12**, 1162–1174 (2021). https://doi.org/10.1111/jdi.13452
23. Awad, S., Dargham, S., Toumi, A.: Diabetes risk score for qatar: a mathematical modelling approach to identify individuals living with diabetes (2020). https://doi.org/10.1016/j.diabres.2017.11.015

24. Awad, S., Dargham, S., Toumi, A., Dumit, E., El-Nahas, K., Al-Hamaq, A., et al.: A diabetes risk score for Qatar utilizing a novel mathematical modelling approach to identify individuals at high risk for diabetes. Sci. Rep. **11**, 1–10 (2021). https://doi.org/10.1038/s41598-021-81385-3
25. Awad, S., Huangfu, P., Dargham, S., Ajlouni, K., Batieha, A., Khader, Y., et al.: Characterizing the type 2 diabetes mellitus epidemic in Jordan up to 2050. Sci. Rep. **10**, 1–10 (2020)
26. Awad, S., O'Flaherty, M., Critchley, J., Abu-Raddad, L.: Forecasting the burden of type 2 diabetes mellitus in Qatar to 2050: a novel modelling approach. Diab. Res. Clin. Pract. **137**, 100–108 (2018). https://doi.org/10.1016/j.diabres.2017.11.015
27. Awad, S., O'Flaherty, M., El-Nahas, K., Al-Hamaq, A., Critchley, J., Abu-Raddad, L.: Preventing type 2 diabetes mellitus in Qatar by reducing obesity, smoking, and physical inactivity: mathematical modelling analyses. Popul. Health Metrics **17**, 1–13 (2019). https://doi.org/10.1186/s12963-019-0200-1
28. Awad, S.F., Critchley, J.A., Abu-Raddad, L.J.: Impact of diabetes mellitus on tuberculosis epidemiology in Indonesia: a mathematical modelling analysis. Tuberculosis **134**, 102164 (2022). https://doi.org/10.1016/j.tube.2022.102164
29. Awad, S.F., Dargham, S.R., Omori, R., Pearson, F., Critchley, J.A., Abu-Raddad, L.J.: Analytical exploration of potential pathways by which diabetes mellitus impacts tuberculosis epidemiology. Sci. Rep. **9**, 8494 (2019). https://doi.org/10.1038/s41598-019-44916-7
30. Awad, S.F., Huangfu, P., Ayoub, H.H., Pearson, F., Dargham, S.R., Critchley, J.A., Abu-Raddad, L.J.: Forecasting the impact of diabetes mellitus on tuberculosis disease incidence and mortality in India. J. Global Health **9**, 020415 (2019). https://doi.org/10.7189/jogh.09.020415
31. Awad, S.F., Toumi, A.A., Al-Mutawaa, K.A., Alyafei, S.A., Ijaz, M.A., Khalifa, S.A., Kokku, S.B., Mishra, A.C., Poovelil, B.V., Soussi, M.B., et al.: Type 2 diabetes epidemic and key risk factors in Qatar: a mathematical modeling analysis. BMJ Open Diab. Res. Care **10**(2), e002704 (2022)
32. Aye, P.: Mathematical analysis of the effect of control on the dynamics of diabetes mellitus and its complications. Earthline J. Math. Sci. **6**(2), 375–395 (2021). https://doi.org/10.34198/ejms.6221.375395
33. Aye, P., Adeyemo, K., Oke, A., Omotoye, A.: Analysis of mathematical model for the dynamics of diabetes mellitus and its complications. Appl. Math. Comput. Intell. **10**, 57–77 (2021)
34. Aye, P., Akinwande, N., Kayode, D., Adebileje, A.: Global stability analysis for the mathematical model of dynamics of diabetes mellitus and its complications. J. Niger. Assoc. Math. Phys. **48** (2018)
35. Aye, P.O.: Stability analysis of mathematical model for the dynamics of diabetes mellitus and its complications in a population. Data Anal. Appl. Math. (DAAM) **3**(1), 20–27 (2022)
36. Ayoade, A., Agboola, S.: A linear mathematical model for the transmission dynamics of diabetes mellitus. Math. Comput. Sci. **4**(2), 25–32 (2023)
37. Balakrishnan, G.P., Chinnathambi, R., Rihan, F.: A fractional-order control model for diabetes with restraining and time-delay. J. Appl. Math. Comput. **69**(4), 3403–3420 (2023). https://doi.org/10.1007/s12190-023-01885-5
38. Bassey, B.E.: Optimal control model for dual treatment of delayed type-ii diabetes infection in human population. Open Sci. J. Math. Appl. **7**(1), 34–49 (2019)
39. Bernard, S., César, T., Nuiro, S.P., Piétrus, A.: Unexistence of limit cycle in an optimal control problem of a population of diabetics. Revista de Matemática Teoría y Aplicaciones **25**(2), 239–259 (2018)
40. Bernard, S., Nuiro, S.P., Pietrus, A.: Diabetes, complications and limit cycles. Appl. Math. E-Notes **15**, 197–206 (2015)
41. Boutayeb, A.: Diabetes: From Prevention to Management; How to Deal with Its Different Aspects and Risk Factors?, chap. 4, pp. 33–47. BOHR Publishers (2024). https://doi.org/10.54646/DFPMDDARF.2024.02
42. Boutayeb, A., Chetouani, A.: A population model of diabetes and pre-diabetes. Int. J. Comput. Math. **84**(1), 57–66 (2007)

43. Boutayeb, A., Chetouani, A., Achouyab, A., Twizell, E.: A non-linear population model of diabetes mellitus. J. Appl. Math. Comput. **21**(1–2), 127–139 (2006)
44. Boutayeb, A., Derouich, M.: Age structured models for diabetes in east Morocco. Math. Comput. Simul. **58**(3), 215–229 (2002)
45. Boutayeb, A., Lamlili, M.E.N., Boutayeb, W.: A Review of Compartmental Mathematical Models Used in Diabetology, pp. 217–250. Springer, Cham (2020). https://doi.org/10.1007/978-3-030-34702-4_14
46. Boutayeb, A., Twizell, E., Achouayb, K., Chetouani, A.: A mathematical model for the burden of diabetes and its complications. Biomed. Eng. Online **3**(1), 20 (2004)
47. Boutayeb, A., Twizell, E.H.: An age structured model for complications of diabetes mellitus in Morocco. Simul. Model. Pract. Theory **12**(1), 77–87 (2004)
48. Boutayeb, W.: Modélisation bio-informatique et bio-mathématique appliquée au diabète. Ph.D. thesis, University Mohamed Premier, Oujda, Morocco (2015)
49. Boutayeb, W., Boutayeb, A., Lamlili, M.E., Derouich, M.: A mathematical model of overweight/obesity in Morocco using human biomass. Int. J. Latest Res. Sci. Technol. **3**(6), 65–67 (2014)
50. Boutayeb, W., Lamlili, M.E., Boutayeb, A., Derouich, M.: A simulation model for the dynamics of a population of diabetics with and without complications using optimal control. In: International Conference on Bioinformatics and Biomedical Engineering, pp. 589–598. Springer (2015)
51. Boutayeb, W., Lamlili, M.E., Boutayeb, A., Derouich, M.: The dynamics of a population of healthy people, pre-diabetics and diabetics with and without complications with optimal control. In: Proceedings of the Mediterranean Conference on Information & Communication Technologies 2015, pp. 463–471. Springer (2016)
52. Cetinkaya, I.: An application of nonstandard finite difference method to a model describing diabetes mellitus and its complications. J. New Theory **45**, 105–119 (2023)
53. Dadhich, S., Yadav, M.P.: Kritoka: a fractional-order differential equation model of diabetes mellitus seiit. Int. J. Math. Ind. (2024). https://doi.org/10.1142/S2661335224500011
54. Danumjaya, P., Dhara, M.: Stability preserving non-standard finite difference schemes for diabetes with tuberculosis infectious model. Lett. Biomath. 1–18 (2019)
55. Daud, A.A.M., Toh, C.Q., Saidun, S.: Development and analysis of a mathematical model for the population dynamics of diabetes mellitus during pregnancy. Math. Models Comput. Simul. **12**(4), 620–630 (2020)
56. Derouich, M.: Modeling and simulation of models with and without age structure: application to diabetes and dengue fever. Ph.D. thesis, University Mohamed Premier, Oujda, Morocco (2001)
57. Derouich, M., Boutayeb, A., Boutayeb, W., Lamlili, M.: Optimal control approach to the dynamics of a population of diabetics. Appl. Math. Sci. **8**(56), 2773–2782 (2014). https://doi.org/10.12988/ams.2014.43155
58. Dhara, M., Baths, V., Danumjaya, P.: Mathematical modeling and dynamics of tuberculosis infection among diabetic patients in India. J. Anal. 1–13 (2018)
59. Dubey, R., Goswami, P.: Mathematical model of diabetes and its complication involving fractional operator without singular kernel. Discrete Contin. Dyn. Syst. - S **14**(7), 2151–2161 (2021). https://doi.org/10.3934/dcdss.2020144
60. El Mansouri, A., Souni, I., Khajji, B., Labzai, A., Belami, M., Tidli, Y.: Optimal control in a mathematical model of spread of the obesity epidemic and its impact on diabetes. Commun. Math. Biol. Neurosci. 74 (2023). https://doi.org/10.28919/cmbn/8041
61. El Moutaouakil, K., El Ouissari, A., Palade, V., Charroud, A., Olaru, A., Baïzri, H., Chellak, S., Cheggour, M.: Multi-objective optimization for controlling the dynamics of the diabetic population. Mathematics **11**(13), 2957 (2023). https://doi.org/10.3390/math11132957
62. El Ouissari, A., El Moutaouakil, K.: Genetic algorithm to fractional optimal control of a diabetic patient. Ufa Math. J. **15**(3), 129–147 (2023)
63. Enagi, A.I., Bawa, M., Sani, A.M.: Mathematical study of diabetes and its complication using the homotopy perturbation method. Int. J. Math. Comput. Sci. **12**(1), 43–63 (2017)

64. Ferdous, A.: An ordinary differential equation model for assessing the impact of lifestyle intervention on type 2 diabetes epidemic. Healthc. Anal. **4**, 100271 (2023)
65. Hill, J., Nielsen, M., Fox, M.H.: Understanding the social factors that contribute to diabetes: a means to informing health care and social policies for the chronically ill. Perm. J. **17**(2), 67 (2013)
66. IDF: Diabetes Atlas 2021. https://diabetesatlas.org/atlas/tenth-edition/ (2021)
67. Şişman, S., Merdan, M.: A mathematical model of susceptible diabetes complication (sdc) model in discrete time fuzzy and crisp environment. Cumhur. Sci. J. **43**(4), 656–664 (2022). https://doi.org/10.17776/csj.1127855
68. Şisman, S., Merdan, M.: Global stability of susceptible diabetes complication (sdc) model in discrete time. Sigma J. Eng. Nat. Sci. **39**(3), 290–312 (2021). https://doi.org/10.14744/sigma.2021.00018
69. Jajarmi, A., Ghanbari, B., Baleanu, D.: A new and efficient numerical method for the fractional modelling and optimal control of diabetes and tuberculosis co-existence. Chaos **29**(9), 093111 (2019). https://doi.org/10.1063/1.5112177
70. Jones, A.P., Homer, J.B., Murphy, D.L., Essien, J.D., Milstein, B., Seville, D.A.: Understanding diabetes population dynamics through simulation modeling and experimentation. Am. J. Public Health **96**(3), 488–494 (2006)
71. Karaagac, B., Owolabi, K., Pindza, E.: A computational technique for the caputo fractal-fractional diabetes mellitus model without genetic factors. Int. J. Dyn. Control **11**, 2161–2178 (2023). https://doi.org/10.1007/s40435-023-01131-7
72. Kompas, S., Aldila, D., Handari, B.: Modelling the spread of diabetes transmission through social contact. In: International Conference on Science and Applied Science (ICSAS2020), AIP Conference Proceedings, vol. 2296, pp. 020098–1–020098–10 (2020). https://doi.org/10.1063/5.0030458
73. Kouidere, A., Balatif, O., Ferjouchia, H., Boutayeb, A., Rachik, M.: Optimal control strategy for a discrete time to the dynamics of a population of diabetics with highlighting the impact of living environment. Discrete Dyn. Nat. Soc. **2019** (2019)
74. Kouidere, A., Youssoufia, E.L., L., Ferjouchia, H., Balatif, O., Rachik, M.: Optimal control of mathematical modelling of the spread of the covid-19 pandemic with highlighting the negative impact of quarantine on diabetics people with cost-effectiveness. Chaos Solitons Fractals **145**, 110777 (2021)
75. Kouidere, A., Khajji, B., Balatif, O., Rachik, M.: A multi-age mathematical modeling of the dynamics of population diabetics with effect of lifestyle using optimal control. J. Appl. Math. Comput. **67**(1), 375–403 (2021)
76. Kouidere, A., Labzai, A., Ferjouchia, H., Balatif, O., Rachik, M.: A new mathematical modeling with optimal control strategy for the dynamics of population of diabetics and its complications with effect of behavioral factors. J. Appl. Math. **2**, 1–12
77. Kouidere, A., Labzai, A., Khajji, B., Ferjouchia, H., Balatif, O., Boutayeb, A., Rachik, M.: Optimal control strategy with multi-delay in state and control variables of a discrete mathematical modeling for the dynamics of diabetic population. Commun. Math. Biol. Neurosci. **2020**, Article–ID (2020)
78. Lamlili E.N., M., Boutayeb, W., Boutayeb, A.: The dynamics of a population of healthy adults, overweight/obese and diabetics with and without complications in Morocco (2022)
79. Logaprakash, P., Monica, C.: Optimal control of diabetes model with the impact of endocrine-disrupting chemical: an emerging increased diabetes risk factor. Math. Model. Numer. Simul. Appl. **3**(4), 318–334 (2023). https://doi.org/10.53391/mmnsa.1397575
80. Makanda, G.: A mathematical model on the effect of non-adherence to drugs on diabetes control. Int. J. Biol. Biomed. Eng. **13**, 70–77 (2019)
81. Mat Daud, A., Toh, C., Saidun, S.: Development and analysis of a mathematical model for the population dynamics of diabetes mellitus during pregnancy. Math. Model. Comput. Simul. **12**(4), 620–630 (2020)
82. Modu, G.U., Hadejia, Y.A., Ahmed, I., Kumam, W., Thounthong, P.: Analysis of linear and nonlinear mathematical models for monitoring diabetic population with minor and major complications. Thai J. Math. **19**(3), 1004–1027 (2021)

83. Mollah, S., Biswas, S.: Optimal control for the complication of type 2 diabetes: the role of awareness programs by media and treatment. Int. J. Dyn. Control **11**, 877–891 (2023)
84. Moualeu, D., Bowong, S., Tewa, J., Emvudu, Y.: Analysis of the impact of diabetes on the dynamical transmission of tuberculosis. Math. Model. Nat. Phenom. **7**(3), 117–146 (2012)
85. Moya, E., Pietrus, A., Bernard, S.: Mathematical model for the study of obesity in a population and its impact on the growth of diabetes. Math. Model. Anal. **28**(4), 611–635 (2023). https://doi.org/10.3846/mma.2023.17510
86. Moya, E., Pietrus, A., Bernard, S., Nuiro, S.: A mathematical model with fractional order for obesity with positive and negative interactions and its impact on the diagnosis of diabetes. J. Math. Sci. Model. **6**(3), 133–149 (2023). https://doi.org/10.33187/jmsm.13339842
87. Nasif, H., Al-Nasir, S.: Discrete optimal control mathematical model of diabetes population. Iraqi J. Sci. **64**(4), 1925–1934 (2023). https://doi.org/10.24996/ijs.2023.64.4.30
88. Nasir, H.: Hopf bifurcation analysis for a diabetic population model with two delays and saturated treatment. Physica Scripta **96**(12), 125013 (2021)
89. Nasir, H.: Modeling the diabetic population in Malaysia using a functional rate of unhealthy lifestyle influence. J. Stat. Manag. Syst. **24**(4), 755–778 (2021)
90. Nasir, H.: A time-delay model of diabetic population: dynamics analysis, sensitivity, and optimal control. Physica Scripta **96**(11), 115002 (2021)
91. Nasir, H.: Optimal control of a three-state diabetic population model with interventions. J. Sci. Technol. **14**(1), 23–28 (2022). https://doi.org/10.30880/jst.2022.14.01.002. http://penerbit.uthm.edu.my/ojs/index.php/jst
92. Nasir, H.: Stability analysis and optimal control of a five-state diabetic population model. J. Stat. Manag. Syst. **25**(1), 245–267 (2022). https://doi.org/10.1080/09720510.2021.1881218
93. Nasir, H., Mat Daud, A.: Population models of diabetes mellitus by ordinary differential equations: a review. Math. Popul. Stud. **29**(3), 95–127 (2021)
94. Okyere, S., Ackora-Prah, J.: A mathematical model of transmission dynamics of sars-cov-2 (covid-19) with an underlying condition of diabetes. Int. J. Math. Math. Sci. (2022). https://doi.org/10.1155/2022/7984818
95. de Oliveira, S.R., Raha, S., Pal, D.: Global asymptotic stability of a non-linear population model of diabetes mellitus. In: International Conference on Differential & Difference Equations and Applications, pp. 351–357. Springer (2017)
96. Pan, S.C., Ku, C.C., Kao, D., Ezzati, M., Fang, C.T., Lin, H.H.: Effect of diabetes on tuberculosis control in 13 countries with high tuberculosis: a modelling study. Lancet Diab. Endocrinol. **3**(5), 323–330 (2015)
97. Permatasari, A., Tjahjana, R., Udjiani, T.: Existence and characterization of optimal control in mathematics model of diabetics population. In: Journal of Physics: Conference Series, vol. 983, p. 012069. IOP Publishing (2018)
98. Permatasari, A., Tjahjana, R., Udjiani, T.: Global stability for linear system and controllability for nonlinear system in the dynamics model of diabetics population. In: Journal of Physics: Conference Series, vol. 1025, p. 012086. IOP Publishing (2018)
99. Pinto, C.M.A., Carvalho, A.R.M.: Diabetes mellitus and tb co-existence: clinical implications from a fractional order modelling. Appl. Math. Model. **68**, 219–243 (2019). https://doi.org/10.1016/j.apm.2018.11.029
100. PRISMA: Prisma 2020 extensions translations endorsement. https://www.prisma-statement.org/prisma-2020-flow-diagram
101. Rana, P.: Mathematical model on diabetes mellitus using fractional approach; fractional homotopy perturbation method and fractional variational iteration method: a comparison. Int. Multidiscip. Res. J. **9**(2), 1–9 (2022). https://doi.org/10.53573/rhimrj.2022.v09i02.001
102. Saima, R., Jarad, F., Taghreed, M.: A study of behaviour for fractional order diabetes model via the nonsingular kernel. AIMS Math. **7**(4), 5072–5092 (2022). https://doi.org/10.3934/math.2022282
103. Side, S., Astari, G., Pratama, M., Sanusi, W.: Numerical solution of diabetes mellitus model without genetic factors with treatment using Runge Kutta method. J. Phys. Conf. Ser. **1244**(1), 1–9 (2019)

104. Singh, J., Kumar, D., Baleanu, D.: On the analysis of fractional diabetes model with exponential law. Adv. Differ. Equ. **2018**, 231 (2018). https://doi.org/10.1186/s13662-018-1680-1
105. Srivastava, H.M., Shanker Dubey, R., Jain, M.: A study of the fractional-order mathematical model of diabetes and its resulting complications. Math. Methods Appl. Sci. 1–14 (2019)
106. Sweilam, N.H., Al-Mekhlafi, S.M., Baleanu, D.: Optimal control for a fractional tuberculosis infection model including the impact of diabetes and resistant strains. J. Adv. Res. **17**, 125–137 (2019). https://doi.org/10.1016/j.jare.2019.01.007
107. Vanitha, R., Porchelvi, R.: A linear population model for diabetes mellitus. Bull. Pure Appl. Sci.-Math. Stat. **36**(2), 311–315 (2017)
108. Walpole, S.C., Prieto-Merino, D., Edwards, P., Cleland, J., Stevens, G., Roberts, I.: The weight of nations: an estimation of adult human biomass. BMC Public Health **12**(1), 439 (2012)
109. WHO: Diabetes. http://www.who.int/news-room/factsheets/detail/diabetes (2018)
110. Widyaningsih, P., Affan, R.C., Saputro, D.R.S.: A mathematical model for the epidemiology of diabetes mellitus with lifestyle and genetic factors. In: Journal of Physics: Conference Series, vol. 1028, p. 012110. IOP Publishing (2018)
111. Yusuf, T.: Optimal control of incidence of medical complications in a diabetic's patients' population. FUTA J. Res. Sci. **1**, 180–189 (2015)
112. Zadeng, D.: Mathematical modelling of diabetes mellitus. Ph.D. thesis, Department of Mathematics, School of Physical Sciences, Mizoram University (2014)
113. Zhang, Z., Qingyi Zhan, Q., Xie, X.: Numerical study on stochastic diabetes mellitus model with additive noise. Comput. Math. Methods Med. 1–8 (2019)

Mathematical Modelling for Glucose Homeostasis and Associated Mechanisms: A State-of-the-Art Review

Mohamed E. N. Lamlili, Wiam Boutayeb, and Abdesslam Boutayeb

Abstract The literature dealing with mathematical modelling for diabetes is abundant. During the last decades, a variety of models have been devoted to different aspects of diabetes, including glucose and insulin dynamics, management and complications prevention, cost and cost-effectiveness of strategies and epidemiology of diabetes in general. Several reviews are published regularly on mathematical models used for specific aspects of diabetes. In the present chapter we propose a selected review of mathematical models dealing with many aspects of diabetes and using various tools. The review includes, side by side, models which are simple and/or comprehensive; deterministic and/or stochastic; continuous and/or discrete; using ordinary differential equations, partial differential equations, optimal control theory, integral equations, matrix analysis and computer algorithms. This review concentrates specifically on mathematical models dealing with glucose homeostasis, using HbA1c and interaction between beta-cell mass, alpha cell mass, insulin, glucagon, glucose, insulin receptors, Free Fatty Acids (FFA), growth hormone and leptin.

1 Introduction

According to the World Health Organisation (WHO), the term diabetes mellitus describes a metabolic disorder of multiple aetiology characterised by chronic hyperglycemia with disturbances of carbohydrate, fat and protein metabolism, resulting from defect in insulin secretion, insulin action, or both. The three main types of diabetes are: (1) type1 diabetes (previously known as insulin-dependent, juvenile or childhood-onset), characterized by deficient insulin production and requiring daily administration of insulin; (2) type2 diabetes (formerly called non-insulin-dependent, or adult-onset); and (3) gestational diabetes which occurs during pregnancy [98]. Statistics released by the International Diabetes Federation (IDF) in 2021 indicate

M. E. N. Lamlili · W. Boutayeb
Mohammed First University, Oujda, Morocco

A. Boutayeb (✉)
Emirates Aviation University, Dubai, UAE
e-mail: x.boutayeb@gmail.com

A. Boutayeb and M. E. N. Lamlili (eds.), *Computational Mathematics and Modelling for Diabetes*, Industrial and Applied Mathematics,
https://doi.org/10.1007/978-981-96-1925-2_2

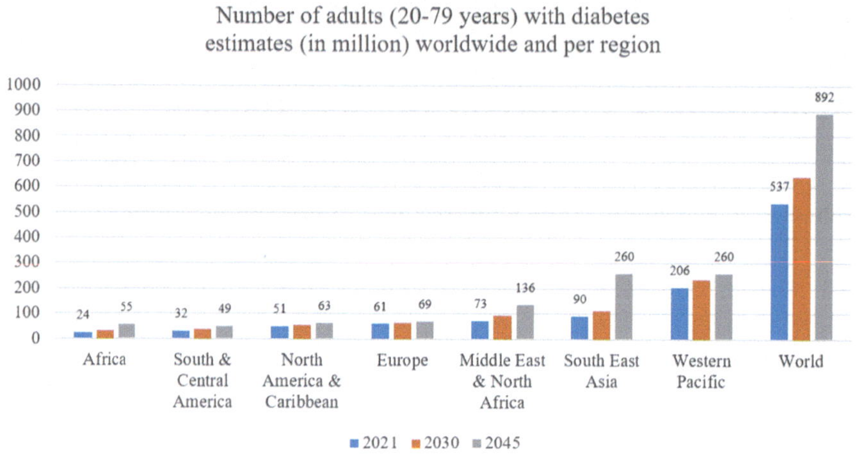

Fig. 1 Evolution of the number of adults with diabetes in the world and by regions *Source* IDF Diabetes Atlas 2021 10th edition

that diabetes is a global emergency. Indeed, one in ten adults (20–79 years) live with diabetes but one in two is undiagnosed. Three in four people with diabetes live in low- or middle-income countries and nearly 7 million deaths were globally attributed to diabetes in 2021. Moreover, Fig. 1 and Table 1 show an increasing trend in the burden of diabetes [52].

During the last decades the number of mathematical models proposed to deal with different aspects of diabetes has grown considerably as shown by a multitude of reviews [3, 10, 20, 25, 59, 66, 72, 86]. Mathematical modelling for diabetes generally concentrates on different aspects of diabetes such as the homeostasis of glucose; glycated haemoglobin (HbA1C); computer algorithms and devices; insulin and chemical reactions; dynamical systems of populations of diabetes with and without complications; and socioeconomic burden of diabetes. However, the majority of mathematical models are dedicated to the dynamics of glucose in relation with

Table 1 The high burden of diabetes

1 in 10 adults (20–79 years) has diabetes	537 million
1 in 2 adults (20–79 years) is undiagnosed	240 million
1 in 9 adults (20–79 years) has impaired glucose tolerance	541 million
1 in 18 adults (20–79 years) has impaired fasting glucose	319 million
3 in 4 people with diabetes live in low- and middle-income countries	403 million
1 in 6 live births are affected by hyper glycaemia in pregnancy	21 million
Children and adolescents below 20 years with type 1 diabetes	1.2 million
Deaths attributed to diabetes	6.7 million
11.5% of global health expenditure spent on diabetes	USD 966 billion

Source IDF Diabetes Atlas 2021 10th edition

insulin, beta-cells, alpha-cells, glucagon, free fatty acids, growth hormone and leptin. Consequently, this part of mathematical modelling will constitute a cornerstone of this review. Finally, other less prominent aspects of mathematical modelling are glycated haemoglobin (HbA1C); insulin and chemical reactions; computer software and devices and other.

Another chapter is devoted to a systematic review of mathematical models dealing with diabetes population dynamics (see Chap. 1 in this book).

2 Glucose Homeostasis and Insulin Dynamics

Himsworth and Ker [50]

In 1939, Himsworth and Ker [50] introduced the first approach to measure the insulin sensitivity in vivo. Since then, mathematical models have been used to estimate insulin-glucose dynamics and more generally, to illustrate glucose homeostasis based on interaction between different factors such as insulin, beta cells, alpha cells, glucagon, free fatty acids, growth hormone, leptin and others.

Bolie [18]

Bolie is among the pioneers in this field. In 1961, using ordinary differential equations, he proposed the following simple model [18]:

$$\frac{dG}{dt} = a_1 G - a_2 I + p \tag{1}$$

$$\frac{dI}{dt} = a_3 G - a_4 I \tag{2}$$

where p, a_1, a_2, a_3, a_4 are parameters

Bergman and Cobelli [12–17, 92]

Simple and comprehensive models were proposed by different authors [2, 36, 83, 87], especially to deal with insulin sensitivity. However, the real start of modelling the glucose-insulin dynamics is thought to begin with the so-called minimal model proposed by the team of Bergman and Cobelli in the early eighties [17, 92]. The model was formulated as follows:

$$\frac{dG}{dt} = -(p_1 + X)G + p_1 G_b, \; G(0) = p_0 \tag{3}$$

$$\frac{dX}{dt} = -p_2 X + p_3(I - I_b), \; X(0) = 0 \tag{4}$$

$$\frac{dI}{dt} = p_4(G - p_5)^+ t - p_6(I - I_b), \; I(0) = p_7 + I_b \tag{5}$$

where, $(G(t) - p_5)^+ = G(t) - p_5$ if $G(t) > p_5$ and 0 otherwise.

X(t) denotes an auxiliary function representing insulin-excitable tissue glucose uptake activity, G_b, and I_b are the subject's baseline glyceamia and insulinimia. $b_0 - b_7$ are parameters.

The same authors have further published papers, completing, testing or validating the results of the minimal model [12, 13, 15, 16]. According to a paper published by Bergman in 2002 [14], more than 500 studies related to the minimal model could be found in the literature.

A large number of papers have been published, using modified versions of the glucose minimal model (3)–(4) for describing Oral Glucose Tolerance Test (OGTT) and meal tests, while insulin minimal models derived from (5) are still limited to Intravenous Glucose Tolerance Test (IVGTT). It should also be stated that the major contribution of the glucose minimal model (3)–(4) has been to provide means of estimating insulin sensitivity $S_I = p_3/p_2$, avoiding the glucose clamp.

Variant versions based on the minimal model were considered by different authors [38, 40, 53, 67, 78, 89].

——— Derouich and Boutayeb [38] ———

An example of this category was proposed by Derouich and Boutayeb who used a modified version of the minimal model to introduce parameters related to physical exercise [38]:

$$\frac{dG}{dt} = -(1+q_2)XG + (p_1+q_1)(G_b - G) \tag{6}$$

$$\frac{dX}{dt} = -p_2X + (p_3+q_3)(I - I_b) \tag{7}$$

where q_1, q_2, q_3 are parameters related to physical activity and defined as follows:

q_1 : the effect of physical exercise in accelerating the utilization of glucose by muscles and the liver insulin

q_2 : the effect of physical exercise in increasing the muscular and liver sensibility to the action of insulin.

q_3 : the effect of physical exercise in increasing the utilization of insulin. In other words, q_1 increases insulin effectiveness in enhancing glucose disposal and consequently improving insulin sensitivity to become: $S_I = (p_3 + q_3)(1 + p_2)/p_2$.

——— Kartono et al. [53] ———

Then the previous model was modified by Kartono et al in 2017 to become a new mathematical model representing the insulin infusion process in insulin therapy in T1DM and T2DM case with exercise as follows [53]:

$$\frac{dG}{dt} = -(1+q_2)XG + (p_1+q_1)(G_b - G) \tag{6}$$

$$\frac{dX}{dt} = -p_2X + (p_3+q_3)(I - I_b) \tag{7}$$

$$\frac{dI}{dt} = (\gamma + q_1 + q_4)(G_b - G) - (k + q_3 + q_5)(I - I_b) + U(t) \tag{8}$$

If $G(t) > G_b$, $I(t_0) = I_0$

$$\frac{dI}{dt} = (k + q_3 + q_5)(I - I_b) + U(t) \tag{9}$$

If $G(t) < G_b$, $I(t_0) = I_0$

Where

q_4: the effect of the physical exercise in increasing the disappearance of glucose;

q_5: the effect of the physical exercise in increasing the secretion of insulin from the pancreas.

Roy and Parker [78]

Roy and Parker [78] also modified the minimal model by adding effect of exercise on plasma glucose and insulin levels.

They used the maximum rate of oxygen consumption for an individual during exercise $VO_2^{\max}(\text{ml/kg/ min})$

Assuming that Oxygen consumption is approximately linearly proportional to energy expenditure, they used exercise model: $\frac{dPVO_2^{\max}}{dt}(t) + 0.8u_3(t)$, $PVO_2^{\max}(0) = 0$

Where $PVO_2^{\max}$ is the exercise level as experienced by the individual and $u_3(t)$ is the ultimate exercise intensity above the basal level

The parameter value 0.8 (1/min) was selected to achieve a settling time of approximately 5 min.

Then for the glucose and insulin dynamics, coupled with the major exercise effects, the authors proposed the following model

$$\frac{dI}{dt} = -nI(t) + p_4u_1(t) - I_e(t);\ I(0) = I_b = \frac{p_4}{n}u_{1b} \tag{10}$$

$$\frac{dX}{dt} = -p_2X + p_3(I(t) - I_b);\ X(0) = 0 \tag{11}$$

$$\frac{dG}{dt} = -p_1(G(t) - G_b) - X(t)G(t) + \frac{W}{Vol_G}(G_{prod}(t) - G_{gly}(t)) - \frac{W}{Vol_G}G_{up}(t) + \frac{u_2(t)}{Vol_G};\ G(0) = G_b \tag{12}$$

$$\frac{dG_{prod}}{dt} = a_1PVO_2^{\max}(t) - a_2G_{prod}(t);\ G_{prod}(0) = 0 \tag{13}$$

$$\frac{dG_{up}}{dt} = a_3PVO_2^{\max}(t) - a_4G_{up}(t)G_{up}(0) = 0 \tag{14}$$

$$\frac{dI_e}{dt} = a_5PVO_2^{\max}(t) - a_6I_e(t);\ I_e(0) = 0 \tag{15}$$

$I_e(t)$ ($\mu U/ml/$ min) is the rate of insulin removal from circulatory system due to exercise-induced physiological changes.

Variables $G_{up}(t)$ and $G_{prod}(t)$ represent the rate of glucose uptake and hepatic glucose production induced by exercise, respectively.

W (kg) represent the weight of the subject

Variable stands for the decline of the glycogenolysis rate during prolonged exercise due to the depletion of liver glycogen stores

The authors indicate that the rate of glycogenolysis starts to decrease when the energy expenditure exceeds a critical threshold value (A_{TH}) which is a function of exercise intensity and duration.

$$\frac{dE_x(t)}{dt} = (k + q_3 + q_5)(I - I_b)$$

——— Hussain and Zadeng. [51] ———

In 2014, Hussain and Zadeng proposed a simpler minimal model as follows [51]:

$$x' = -a_1 x - a_2 xy + a_3 \tag{16}$$

$$y' = b_1 x - b_2 y \tag{17}$$

With $x \geq 0$, $y \geq 0$

x and y representing glucose concentration and insulin concentration respectively

a_1 is the rate constant which represents insulin-independent glucose disappearance

a_2 is the rate constant which represents insulin-dependent glucose disappearance

a_3 is the glucose infusion rate

b_2 is the rate constant which represents insulin production due to glucose stimulation

b_2 is the rate constant which represents insulin degradation

Section 2 concentrates on mathematical models dealing with glucose homeostasis in relation with insulin dynamics. The mathematical models proposed in this section use mainly ordinary differential equations with a particular interest given to the minimal model and its modified versions. However, some authors indicated drawbacks of these models, including simplified equations and the use of a "remote insulin" compartment (X) beside the plasma insulin (I). Consequently, many authors proposed the use of delay equations as a more realistic way to model the insulin dynamics through different phases of time. The following section will deal especially with delay equations for the combined glucose-insulin dynamics.

3 Delayed Differential Equation to Model Glucose-Insulin Dynamics

Among the multitude of authors who dealt with dealy models for insulin-glucose dynamics, we cite De Gaetano [34], Giang et al. [41], Li and Kuang [60], Li et al. [64], Palumbo et al. [74] and Prager et al. [75].

Some authors [34, 35, 70] indicated that while the minimal model has minimal number of constants ($p_0 - p_7$), and has been indisputably useful in physiological research, it has the following drawbacks: The model, as originally proposed, is to be regarded as composed of two separate parts. The first part uses Eqs. (3) and (4) and the second part uses Eq. (5). For the last part, plasma glucose concentration is to be regarded as a known forcing function. In other words, the model parameter fitting has to be conducted in two steps: first, using the recorded insulin concentration as input data in order to derive the parameters in the two first equations, then using the recorded glucose as input data to derive the parameters in the third equation.

Some of the mathematical results produced by the minimal model are not realistic (problems of positive equilibrium and solutions not bounded).

The artificial non-observable variable X(t) is introduced to take account of the delay in the action of insulin.

De Gaetano and Arino [34, 35]

Taking into account these remarks and stressing that the glucose-insulin system is an integrated physiologic dynamical system which should be dealt with as a whole, De Gaetano and Arino [34, 35] proposed an aggregated delay differential model called a dynamical model:

After renaming the parameters, the dynamical model takes the form [34]:

$$\frac{dG}{dt} = -b_1 G - b_4 I G + b_7, \; G(0) = G_b + b_0 \tag{18}$$

$$\frac{dI}{dt} = -b_2 I + \frac{b_6}{b_5} \int\limits_{t-b_5}^{t} G(s)ds, I(0) = I_b + b_3 b_0 \tag{19}$$

with $G(t) = G_b$ for $-b_5 \leq t < 0$

Mukhopadhyay et al. [70]

Mukhopadhyay et al. [70] recalled that this model has been shown to allow simultaneous estimation of both insulin secretion and glucose uptake parameters, to have positive, bounded solutions, and to be globally asymptotically stable around the pre-injection equilibrium blood glucose and insulin concentrations. They proposed an extension by introducing a generic weight function W in the delay integral kernel for the pancreatic response to glucose. The new model obtained is as follows:

$$\frac{dG}{dt} = -b_1 G - b_4 I G + b_7, G(0) = G_b + b_0 \tag{20}$$

$$\frac{dI}{dt} = -b_2 I + b_6 \int\limits_{0}^{\$} W(S)G(t-s)ds, I(0) = I_b + b_3 b_0 \tag{21}$$

with $G(t) = G_b$ for $t < 0$

Li et al. [62]

A more general model was proposed by Li et al. [62]. The authors noted that, while the dynamical model solves the problems of the minimal model, it implicitly or explicitly made a few assumptions that may not be necessary or realistic. Specifically, some of the interaction terms are too special and thus too restrictive. For example, the term $b_4 I(t)G(t)$ assumes that mass action law applies here while a more popular, general and realistic alternative is to replace this term by $b_4 I(t)G(t)/(\alpha G(t)+1)$. The way the delay is introduced is also restrictive. Consequently, the model proposed is the following:

$$\frac{dX}{dt} = -F(G) - g(G, I) + b_7, G(0) = G_b + b_0 \tag{22}$$

$$\frac{dI}{dt} = -p(I) + q(LG), I(0) = I_b + b_3 b_0 \tag{23}$$

with $G(t) = G_b$ for $-b_5 \le t < 0$.

[28, 39, 60, 63, 66, 88, 93, 97]

Many authors dealing with delay differential equations in modelling the dynamics of glucose and insulin have used the following function forms [28, 39, 60, 63, 66, 88, 93, 97]:

$$f_1(G) = R_m / \left(1 + \exp\left(\left(C_1 - G/V_g\right)/a_1\right)\right) \tag{24}$$

$$f_2(G) = U_b \left(1 - \exp\left(-G/\left(C_2 V_g\right)\right)\right) \tag{25}$$

$$f_3(G) = G/\left(C_3 V_g\right) \tag{26}$$

$$f_4(I) = U_0 + (U_m - U_0) / \left(1 + \exp\left(-\beta \ln \frac{I\,(1/V_i + 1/(Et_i))}{C_4}\right)\right) \tag{27}$$

$$f_5(I) = R_g / \left(1 + \exp\left(\hat{\alpha}\left(I/V_p - C_5\right)\right)\right) \tag{28}$$

$f_1(G)$: insulin production stimulated by glucose production,
$f_2(G)$: insulin-independent glucose utilization,
$f_3(G) f_4(I)$: insulin-dependent glucose uptake (mostly due to fat and muscle cells),
$f_5(I)$: glucose production controlled by insulin concentration.
The values of the parameters may be found for example in [16].
The functions $f_1 - f_5$ are assumed to satisfy certain general assumptions defined in [88].

Sturis et al. [88]

In 1991, Sturis et al. published a paper on "computer model for mechanisms underlying ultradian oscillations of insulin and glucose". They proposed the following mathematical model [88]:

$$\mathrm{d}x/\mathrm{d}t = f_1(z) - E\,(x/V_1 - y/V_2) - x/t_1 \quad (\mathrm{mU/min}) \tag{29}$$

$$\mathrm{d}y/\mathrm{d}t = E\,(x/V_1 - y/V_2) - y/t_2 \quad (\mathrm{mU/min}) \tag{30}$$

$$\mathrm{d}z/\mathrm{d}t = f_5\,(h_3) + I - f_2(z) - f_3(z)f_4(y) \quad (\mathrm{mg/min}) \tag{31}$$

$$\mathrm{d}h_1/\mathrm{d}t = 3\,(x - h_1)\,/t_3 \tag{32}$$

$$\mathrm{d}h_2/\mathrm{d}t = 3\,(h_1 - h_2)\,/t_3 \tag{33}$$

$$\mathrm{d}h_3/\mathrm{d}t = 3\,(h_2 - h_3)\,/t_3 \tag{34}$$

Where $x(t)$, $y(t)$ and $z(t)$ denoted the amount of plasma insulin, the amount of insulin in the interstitial fluid and the amount of glucose in the glucose space respectively. The variables h_1, h_2 and h_3, represented the delay process between insulin and glucose production. The authors used functions of the forms described by Eqs. (24–28) with exact equations and parameter values as follows:

$$f_1(z) = 209/\,\{1 + \exp\,[-z/\,(300\,V_3) + 6.6]\} \tag{35}$$

$$f_2(z) = 72\left[1 - \exp\,(-z/144\,V_3)\right] \tag{36}$$

$$f_3(z) = 0.01z/V_3 \tag{37}$$

$$f_4(y) = 90/\left[1 + \exp\,(-1.772\log\{y\,[1/V_2 + 1/\,(Et_2)]\}\, + 7.76)\right] + 4 \tag{38}$$

$$f_5\,(h_3) = 180/\left[1 + \exp\,(0.29h_3/V_1 - 7.5)\right] \tag{39}$$

They used the following run parameter values : $I = 216$ mg/min, $t_1 = 6$ min, $t_2 = 100$ min, $t_3 = 36$ min, $E = 0.2$l/min, $V_1 = 3$ liters, $V_2 = 11$ liters and $V_3 = 10$ liters.

Drozdov et Khanina [39]

In 1995, Drozdov et Khanina refined the model proposed by Sturis et al in 1991 to obtain the following model [39]:

$$\frac{dx}{dt}(t) = f_1\left(\frac{0.1z(t)}{V_3}\right) - \left(\frac{E}{V_1} + \frac{1}{T_1}\right)x(t) + \frac{E}{V_2}y(t) \tag{40}$$

$$\frac{dy}{dt}(t) = \frac{E}{V_1}x(t) - \left(\frac{E}{V_2} + \frac{1}{T_2}\right)y(t) \tag{41}$$

$$\frac{dz}{dt}(t) = f_3\left(\frac{x(t-T)}{V_1}\right) - \frac{0.1z(t)}{V_3}f_2\left(\frac{y(t)}{V_2}\right) + (L - p_0) \tag{42}$$

Where $x(\mu)$ and $y(\mu)$ denote insulin in plasma and insulin in interstitial fluid respectively while glucose space is treated as unique with glucose amount z(mg).

The authors denoted by $c_x = x/V_1(\mu\mathrm{U/ml})$, $c_y = y/V_2(\mu\mathrm{U/ml})$, and $c_z = 0.1z/V_3(\mathrm{mg/dl})$ concentrations of insulin and glucose.

V_1 (1), V_2 (1) and V_3 (1) were standing for volume of plasma, volume of interstitial liquid and volume of the glucose compartment.

Adopting the estimations suggested by Sturis et al., the authors used the values: $V_1 = 3$, $V_2 = 11$ and $V_3 = 10$.

——— Tolić et al. [93] ———

In 2000, Tolić et al. used a modified version of the model proposed by Sturis et al. in order to examine the effects of an oscillatory supply of insulin compared to a constant supply at the same average rate. The term fl(G), was replaced by a term representing an exogenous sinusoidal insulin infusion and consequently the first equation became [93]:

$$\frac{dI_p}{dt} = m(1 + A\sin(2\pi t/T)) - E\left(\frac{I_p}{V_p} - \frac{I_i}{V_i}\right) - \frac{I_p}{t_p} \tag{43}$$

Then, in order to gain insight into certain features of the previous model of Sturis, they considered the following simplified model:

$$\frac{dI_p}{dt} = aI_p + bI_i + cG + d \tag{44}$$

$$\frac{dI_i}{dt} = eI_p + fI_i \tag{45}$$

$$\frac{dG}{dt} = gI_iG + hG + kx_3 + lx_3^2 + nx_3^3 + p \tag{46}$$

$$\frac{dx_1}{dt} = rI_p - rx_1 \tag{47}$$

$$\frac{dx_2}{dt} = rx_1 - rx_2 \tag{48}$$

$$\frac{dx_3}{dt} = rx_2 - rx_3 \tag{49}$$

In summary, they showed that interactions between the oscillatory insulin supply and the receptor dynamics can be of minute significance only. They indicated, however, that it is possible to interpret seemingly conflicting results of clinical studies in terms of their different experimental conditions with respect to the hepatic glucose release.

——— Li et al. [63] ———

In 2006, Li et al. proposed a model with two time delays. The first delay τ_1 denoting the total time delay from the time that the glucose concentration level is elevated to the moment that the insulin has been transported to the interstitial space and becomes 'remote insulin' while the second delay τ_2 deals with the the effect of hepatic glucose production measured from the time that insulin has become 'remote insulin' to the moment that a significant change of hepatic glucose production occurs. The model proposed was as follows [63].

$$\frac{dG}{dt}(t) = G_{in} - f_2(G(t)) - f_3(G(t))f_4(I(t)) + f_5(I(t - \tau_2)) \tag{50}$$

$$\frac{dI}{dt}(t) = f_1(G(t - \tau_1)) - d_i I(t) \tag{51}$$

$$I(0) = I_0 > 0, G(0) = G_0 > 0, G(t) = G_0, t \in -[\tau_1, 0], \\ I(t) = I_0, t \in [-\tau_2, 0]\ \tau_1, \tau_2 > 0 \tag{52}$$

Where G_{in} and d_i represent the glucose infusion rate and the insulin clearance rate respectively. Using the functions $f_1 - f_5$ given in Eqs. (24–28) and stressing that the shapes of the functions, instead of their forms, are more important, they experimentally determined values of parameters used for these functions. The authors carried out numerical analysis and bifurcation analysis and compared their results with those given by previous models.

Li and Kuang [60]

In 2007, Li and Kuang considered the same model (50–52) to study analytically and perform carefully designed numerical simulations by allowing two parameters to vary. They concluded that their analytical and numerical results confirm most current existing physiological observations and reveal more insightful information [60].

Chalishajar et al. [28]

In 2016, Chalishajar et al. published a "Review Study of Detection of Diabetes Models through Delay Differential Equations". They proposed a multiple delay differential equation model, made from combinations of the previous models. The model is as follows [28]:

$$\mathrm{d}G(t)/\mathrm{d}t = G_{in} - f_2(G(t)) - f_3(G(t)) f_4(I(t - \tau_2)) + f_5(I(t - \tau_0)) \tag{53}$$

$$\mathrm{d}I_p(t)/\mathrm{d}t = f_1(G(t - \tau_1)) - \mathrm{d}/\mathrm{d}t\,(I_i(t)) \tag{54}$$

$$\mathrm{d}x(t)/\mathrm{d}t = f_1(0.1z(t)/V_3) - ((E/V_1) + (1/T_1)\,x(t) + (E/V_2)\,y(t)) \tag{55}$$

$$\mathrm{d}y(t)/\mathrm{d}t = (E/V_1)\,x(t) - ((E/V_2) + (1/T_2))\,y(t) \tag{56}$$

$$\mathrm{d}z(t)/\mathrm{d}t = f_3(x(t - \tau)/V_1) - (0.1z(t)/V_3)\,f_2(y(t)/V_2) + (L - p_0) \tag{57}$$

The authors explained that the last three equations (dx(t)/dt, dy(t)/dt and dz(t)/dt) with the time delay in the dz(t)/dt equation, were taken from the model proposed by Drozdov and Khanina [39]. The time delay in the dG(t)/dt function τ_0 is seen in the f_5(I), or the glucose production controlled by insulin concentration, and represents the hepatic glucose production delay while the delay τ_2 is seen in the f_4(I), or the insulin dependent glucose uptake dependent upon glucose because multiplied by f_3(G), is the time delay for the insulin-dependent glucose utilization by the cells [97].

The delay in the $\mathrm{dI_p}$(t) term is represented by τ_1, which is the total time delay form the time that the glucose concentration level is elevated to the moment that the insulin has been transported to the interstitial space and becomes "remote insulin". The last delay in this system is the delay in the dz(t)/dt term, given by Drozdov and Khanina [39]. This delay represents the delay in glucose production, which affects the amount of glucose produced by the liver.

Other delay mathematical models can be found in [28, 65, 66].

Saber et al. [81]

In 2018, Saber et al. extended the model (16–17) proposed by Hussain and Zadeng by introducing time delay for glucose and insulin. The delay model was as follows [81]:

$$\frac{dx}{dt} = -a_1x(t) - a_2x(t-\tau)y(t-\tau) + a_3, t \in [0, T] \tag{58}$$

$$\frac{dy}{dt} = b_1x(t) - b_2y(t), t \in [0, T] \tag{59}$$

with initial data:
$x(\theta) = \phi(\theta), \theta < 0,$
$y(\theta) = \psi(\theta), \theta < 0$

where $x \geq 0$, $y \geq 0$, x represents glucose concentration, y represents insulin con -centration

Saber and Alzahrani [79, 80]

In 2019, Saber and Alzahrani analysed the dynamical behaviour of two fractional order models with time delay. The first fractional order model was the following [80]:

$$D^{\alpha}x(t) = -a_1x(t) - a_2x(t-\tau)y(t-\tau) + a_3, t \in [0, T] \tag{60}$$

$$D^{\alpha}y(t) = b_1x(t) - b_2y(t), \quad t \in [0, T], \tag{61}$$

with initial data:
$x(\theta) = \Phi(\theta), \quad \theta < 0,$

$y(\theta) = \psi(\theta), \quad \theta < 0.$

As a continuation of the precedent models considered by the team of Sayed, the second fractional order model was devoted to the stability and Hopf bifurcation of the fractional-order model for the interaction of glucose and insulin model with time delays. The model is governed by the following system of ordinary and delay differential equations [79].

$$D^{a}x(t) = -a_1x(t) - a_2x(t)y(t) + a_3, t \in [0, T] \tag{62}$$

$$D^{\alpha}y(t) = b_1x(t-\tau) - b_2y(t), t \in [0, T] \tag{63}$$

with
$x(\theta) = \phi(\theta), \theta < 0$
$y(\theta) = \psi(\theta), \theta < 0$

Shabestari et al. [84]

In 2018, Shabestari et al. proposed a mathematical model for the dynamics of the insulin concentration (x), the glucose concentration (y) and the number of β-cells (z) with two timedelays in insulin secretion (τ_g) and glucose drop (τ_i). Their mathematical model which was the compound of previous models published by Chuedoung et al. [29], Bajaj et al. [9] and Sarika et al. [82], was governed by the following system of delay differential equations [84].

$$\frac{dx}{dt} = r_1 y\left(t - \tau_g\right) z\left(t - \tau_g\right) - r_2 x + c_1 z\left(t - \tau_g\right) \tag{64}$$

$$\frac{dy}{dt} = \frac{R_3 N}{z} - R_4 x\left(t - \tau_i\right) + C_2 \tag{65}$$

$$\frac{dz}{dt} = R_5(y - \widehat{y})(T - z) + R_6 z(T - z) - R_7 z \tag{66}$$

where x(t) is the insulin concentration, y(t) is the glucose concentration, z(t) is the number of β-cells, and y is the difference between glucose fasting level and its basal level. τ_g is the delay in insulin secretion in response to blood glucose level increase, and τ_i is the delay in glucose drop due to increased insulin level. $r_1 y\left(t - \tau_g\right) z\left(t - \tau_g\right)$ shows the increase in insulin concentration in response to blood glucose increase with the time delay τ_g.$r_2 x$ is the rate of insulin decrease independent of glucose, and $c_1 z\left(t - \tau_g\right)$ is the increase of insulin level secreted by β-cells and is independent from other components.

The authors stressed that previous models cannot display the behaviour of aforementioned biological system with respect to time delays while their model considers two time lags in insulin-glucose regulatory system and consequently, it is more realistic and is capable of showing the behaviour of insulin-glucose regulatory system in different time delays.

Abdul-Basset et al. [1]

The model of Shabestari was slightly modified in 2020 by Abdul-Basset et al. who added the term $R_8 y^2(t) z(t)$ in the third equation to include the death rate of β-cells [1].

Bridgewater et al. [27]

Bridgewater et al. investigated the limit cycle in a two-delay model with a constant glucose infusion using a scheme with perturbation. The mathematical model considered was the following [27].

$$\dot{G}(t) = G_{in} - a_1 G - a_2 G I + \frac{a_3}{I(t - \tau_2) + K_2} \tag{67}$$

$$\dot{I}(t) = \frac{b_1 G(t - \tau_1)^2}{G(t - \tau_1)^2 + K_1^2} - b_2 I \tag{68}$$

Vosoughi et al. [96]

In 2022, Vosoughi et al. extended the model of Sturis by using parameterized data collected in their study to capture dynamics of glucose and insulin. They used global sensitivity analysis method to assess model robustness with respect to parameter perturbations. They concluded that their results "have shown that utilizing the dynamics of changes in blood glucose levels throughout the day can be used to prevent hypoglycemia and hyperglycemic in the diabetic patients" [96].

Paidipati and Rao [73]

Paidipati and Rao [73] assumed that the processes of arrival and disposal of both glucose molecules and insulin granules follows Poisson distribution.

They defined:

b as the mass of β cells.

$\lambda_{g1}, \lambda_{g2}$ as arrival rates of glucose molecules to the blood plasma per unit time through food intake and liver respectively;

$\mu_{g1}, \mu_{g2}, \mu_{g3}$ as rates of glucose disposal per unit time due to insulin dependent cells, insulin independent cells and to transform glucose as glycogen respectively.

$\lambda_{i1}, \lambda_{i2}$ as rates of insulin granules per unit time from Pancreas and External sources respectively;

μ_{i1}, μ_{i2} as disposal rates of insulin per unit time for the process of glucose absorption to the adipose tissues/muscle cells and for glucose conversion as glycogen respectively.

m, n as concentrations of insulin and glucose at a point of time.

Then they proposed the following stochastic model

$$\begin{aligned}\frac{dE_x(t)}{dt} &= -(\mu_{g1} + \mu_{g3})E_{xy} + (\mu_{g1} + \mu_{g3} - \mu_{g2})E_x \\ &\quad + (\mu_{g1} + \mu_{g3})E_y\end{aligned} \tag{69}$$

$$\begin{aligned}\frac{dE_y(t)}{dt} &= -(\mu_{i1} + \mu_{i2})E_{xy} + (\mu_{i1} + \mu_{i2})E_x \\ &\quad + (\mu_{i1} + \mu_{i2})E_y\end{aligned} \tag{70}$$

$$\begin{aligned}\frac{dV_x(t)}{dt} &= [2(\lambda_{g1} + \lambda_{g2}) + (\mu_{g1} + 3\mu_{g2} + \mu_{g3})]E_x \\ &\quad + [2(\mu_{g1} + \mu_{g3} - \mu_{g2})V_x \\ &\quad + 3(\mu_{g1} + \mu_{g3})E_{xy} - (\mu_{g1} + \mu_{g3})E_y\end{aligned} \tag{71}$$

$$\begin{aligned}\frac{dV_y(t)}{dt} &= [2(b\lambda_{i1} + \lambda_{i2}) + (\mu_{i1} + \mu_{i2})]E_y + [2(\mu_{i1} + \mu_{i2})V_y \\ &\quad + (\mu_{i1} + \mu_{i2})E_{xy} - (\mu_{i1} + \mu_{i2})E_x\end{aligned} \tag{72}$$

$$\frac{dE_{xy}(t)}{dt} = [(\lambda_{g1} + \lambda_{g2}) + (\mu_{g1} + \mu_{g2} + \mu_{g3})]E_y$$
$$+[b\lambda_{i1} + \lambda_{i2}) + (\mu_{i1} + \mu_{i2})]E_x + (\mu_{g1} + \mu_{g3})V_y$$
$$+(\mu_{i1} + \mu_{i2})V_x + [(\mu_{g1} + \mu_{g3}) + (\mu_{i1} + \mu_{i2})$$
$$-\mu_{g2})]E_{xy} \quad (73)$$

Where

$E_x(t)$ is average level of glucose at time 't'

$E_y(t)$ is average level of insulin at time 't'

$V_x(t)$ is the variance of glucose level at time 't'

$V_y(t)$ is the variance of insulin level at time 't'

$E_{xy}(t)$ is is average covariance of glucose and insulin levels at time 't'

Then the author defined constants $a_1 - a_8$; $b_1 - b_4$; $d_1 - d_3$; $H_1 - H_2$ and α by:

$a_1 = (\mu_{g1} + \mu_{g3})$; $a_2 = (\mu_{g1} + \mu_{g3} - \mu_{g2})$; $a_3 = (\mu_{i1} + \mu_{i2})$; $a_4 = 2(\lambda_{g1} + \lambda_{g2}) + (\mu_{g1} + 3\mu_{g2} + \mu_{g3})$ $a_5 = 2(b\lambda_{i1} + \lambda_{i2}) + (\mu_{i1} + \mu_{i2})$; $a_6 = (\lambda_{g1} + \lambda_{g2}) + (\mu_{g1} + \mu_{g2} + \mu_{g3})$; $a_7 = (b\lambda_{i1} + \lambda_{i2}) + (\mu_{i1} + \mu_{i2})$

$a_8 = (\mu_{g1} + \mu_{g3}) + (\mu_{i1} + \mu_{i2}) - \mu_{g2})$; $b_1 = \dfrac{2a_2a_3V_x - 6a_1a_3V_y}{a_1(a_3 + 3a_5)}$; $b_2 = \dfrac{(a_3a_4 + 3a_1a_3)}{a_1(a_3 + 3a_5)}$;

$b_3 = \dfrac{(2a_2V_x - a_1b_1)}{3a_1}$; $b_4 = \dfrac{(a_4 - a_1b_2)}{3a_1}$; $d_1 = \dfrac{(a_1b_1 - a_1b_3)}{(a_2 + a_1b_4 + a_1b_2)}$

$d_2 = G_0 + \dfrac{(a_1b_1 - a_1b_3)}{(a_2 + a_1b_4 + a_1b_2)}$; $d_3 = -a_3b_3 + (a_3b_4 + a_3)d_1$; $k = I_0 + \dfrac{d_3}{a_3} + d_2(1 + b_4)$

$\alpha = a_2 + a_1b_4 + a_1b_2$; $H_1 = a_7d_2 - a_6d_2(1 + b_4)$; $H_2 = (a_7d_1 + a_1V_y + a_3V_x - a_6\dfrac{d_3}{a_3})$

Finally, solving Eqs. (69)–(73), the author provided the following measures by fixing variability of glucose and insulin levels:

Average number of glucose granules in the blood stream: $E_x(t) = d_1 + \exp(\alpha t)d_2$

Average number of insulin granules in the blood stream: $E_y(t) = -\dfrac{d_3}{a_3} - d_2(1 + b_4)\exp(\alpha t) + k\exp(a_3t)$

Covariance of glucose molecules and insulin granules:
$\dfrac{H_1\exp(\alpha t)}{(\alpha - a_8)} + \dfrac{a_6k\exp(a_3t)}{(\alpha - a_8)} - \dfrac{H_2}{a_8} + p\exp(a_3t) + (d_1 + d_2\exp(\alpha t))[\dfrac{d_3}{a_3} + d_2(1 + b_4)\exp(\alpha t) - k\exp(a_3t)]$

While Sects. 2 and 3 included mathematical models based only on glucose and insulin, the next section will analyse the effect of another principal component acting on diabetes homeostasis, namely: the β-cells mass

4 The Dynamics of Glucose, Insulin and β-Cells

Topp et al. [94]

In 2000, introducing replication and mortality of beta-cells, Topp et al. proposed a model combining the dynamics of Beta-cells with the dynamics of glucose and insulin [94]

$$\frac{dG}{dt} = a - bG(t) - cI(t)G(t) \tag{74}$$

$$\frac{dI}{dt} = \frac{d\beta(t)G(t)^2}{(e + G(t)^2)} - fI(t) \tag{75}$$

$$\frac{d\beta}{dt} = (-g + hG(t) - iG(t)^2)\beta(t) \tag{76}$$

Hernandez et al. [49]

Hernandez et al. [49] extended the model of Topp et al. by adding surface insulin receptor dynamics to get the following model:

$$\frac{d\beta}{dt} = (-g + hG(t) - iG(t)^2)\beta(t) \tag{77}$$

$$\frac{dI}{dt} = \frac{d\beta(t)G(t)^2}{(1 + R)(e + G(t)^2)} - fI(t) - fR(t)I(t) \tag{78}$$

$$\frac{dG}{dt} = a - (b + cR(t)I(t))G(t) \tag{79}$$

$$\frac{dR}{dt} = j(1 - R(t)) - kI(t)R(t) - lR(t) \tag{80}$$

Pranay Goel [42]

Pranay Goel [42] proposed the topp-IR model wich is a modified version of Topp model by studying dependence on the insulin resistance parameter, S_I:

$$\frac{dI}{dt} = \sigma \frac{G^2}{\alpha\ G^2}\beta - kI \tag{81}$$

$$\frac{dG}{dt} = R - (E + S_I I)G \tag{82}$$

$$\frac{d\beta}{dt} = (-d_0 + r_1 G - r_2 G^2 - d_I I)\beta \tag{83}$$

And extended it to Topp-HS model, considering that insulin sensitivity changes in response to secretory capacity, $S_I \equiv S_I(\sigma)$:

$$\frac{dI}{dt} = \sigma \frac{G^2}{\alpha\, G^2}\beta - kI \tag{84}$$

$$\frac{dG}{dt} = R - (E + S_I(\sigma)I)G \tag{85}$$

$$\frac{d\beta}{dt} = (-d_0 + r_1 G - r_2 G^2 - d_I I)\beta \tag{86}$$

——— Boutayeb et al. [25] ———

Boutayeb et al. [25] also used an extension of the Topp model by adding the effect of genetic predisposition to diabetes.

$$\frac{dG}{dt} = a - bG(t) - \frac{cI(t)G(t)}{\alpha G(t) + 1} \tag{87}$$

$$\frac{dI}{dt} = a - bG(t) - \frac{d\beta(t)G(t)^2}{(e + G(t)^2)} - fI(t) \tag{88}$$

$$\frac{d\beta}{dt} = (1 - \varepsilon)r\beta(t)(1 - \frac{\beta(t)}{K}) + \varepsilon(-g + hG(t) - iG(t)^2)\beta(t) \tag{89}$$

More information on the model and its parameters can be found in the Ph.D thesis defended in 2015 by Boutayeb W. [22].

Although glucose and insulin produced by β-cells are the essential actors determining the daily curve of blood sugar and hence the needed actions to the efficient way of diabetes monitoring, other components are also considered in Sect. 5.

5 Modelling Glucose—Insulin—Glucagon—β Cells—α Cells—FFA—GH—Leptin

5.1 Effect of FFA on the Homeostasis of Diabetes

——— Boutayeb et al. [26] ———

Boutayeb et al. [26] added the effect of free fatty acids (FFA) to the models of Topp and Henandez to get the following model:

$$\frac{d\beta}{dt} = (-g + hG(t) - iG(t)^2)\beta(t) \tag{90}$$

$$\frac{dI}{dt} = \frac{d\beta(t)G(t)^2}{(1 + R)(e + G(t)^2)} - fI(t) - fR(t)I(t) \tag{91}$$

$$\frac{dG}{dt} = a - (b + cR(t)I(t))G(t) + m_1(F(t) - F_b) \tag{92}$$

$$\frac{dR}{dt} = j(1 - R(t))a - kI(t)R(t) - lR(t) \tag{93}$$

$$\frac{dF}{dt} = -m_2(F(t) - F_b) + m_3(G(t) - G_b) \tag{94}$$

5.2 *Effect of Growth Hormone on the Homeostasis of Diabetes*

The effect of growth hormone on glucose homeostasis was considered by different authors [4–8, 45, 54].

——— Al Ali [4–8] ———

Based on previous research showing that growth hormone increases glucose production in blood through gluconeogenesis and glycogenosis the team of Al Ali carried out research on the effect of growth hormone on the regulation of glucose homeostasis [4–8].

In 2019, Al Ali et al. extended the model of Boutayeb et al. (90–94) by adding growth hormone as a sixth equation. The model proposed was the following [6].

$$\begin{aligned}
\frac{d\beta}{dt} &= (-g + hG(t) - iG(t)^2)\beta(t) \\
\frac{dI}{dt} &= \frac{d\beta(t)G(t)^2}{(1+R)(e+G(t)^2)} - fI(t) - fR(t)I(t) \\
\frac{dG}{dt} &= a - (b + cR(t)I(t))G(t) + m_1(F(t) - F_b) + cG(t)H(t) \\
\frac{dR}{dt} &= j(1 - R(t))a - kI(t)R(t) - lR(t) \\
\frac{dF}{dt} &= -m_2(F(t) - F_b) + m_3(G(t) - G_b) + x(GH(t) - GH_b) \\
\frac{dGH}{dt} &= p - wGH(t) - s(F(t) - F_b) - zR(t)
\end{aligned} \tag{95}$$

In 2019, Al Ali et al. considered another model dealing with the effect of growth hormone on the dynamics of glucose and insulin in patients with type 1 diabetes. The proposed mathematical model was governed by four ordinary differential equations describing the interaction between insulin, glucose, free fatty acids and growth hormone in the case of type 1 diabetes. The authors carried out stability analysis and numerical simulation using different parameter values in order to illustrate the dynamics of glucose for pragmatic monitoring. The mathematical model proposed was the following [5].

$$\frac{dI}{dt} = \rho\frac{I_0}{(1+I)} - \delta I \tag{96}$$

$$\frac{dG}{dt} = a_1 + a_2 - (b + c_1 I)\,G + m_1\,(F - F_b) + c_2 GH \tag{97}$$

$$\frac{dF}{dt} = -m_2\,(F - F_b) + m_3\,(G - G_b) + x\,(GH - GH_b) \tag{98}$$

$$\frac{dGH}{dt} = p - wGH - s\,(F - F_b) \tag{99}$$

The first equation indicates that, following a subcutaneous bolus injection of I_0, insulin concentration I(t) in the plasma increases by a rate $\rho I//(1 + I(t))$ and decreases by a rate $\delta I(t)$. The second equation translates how the concentration of glucose increases by the amount of glucose produced by liver and kidney (a_1) and glucose from carbohydrate intake (a_2) and the rate from FFA ($m_1(F(t) - F_b)$). The glucose decreases by a rate independent of insulin (bG(t)) and a rate $c_1 I(t)G(t)$ representing the glucose intake due to insulin sensitivity. The third equation indicates that the FFA decreases by a rate ($m_2(F(t) - F_b)$) and increases by a rate ($m_3(G(t) - G_b)$). Finally, the fourth equation assumes that the rate of growth hormone increases by the amount of GH produced by the somatotropic cells (p), occurring in the pituitary gland. It decreases by the rate of GH clearance/absorption by the liver (ωGH) and the rate of uptake by fat cells ($s(F(t) - F_b)$).

In 2022, Al Ali et al. explored dynamical properties of a Type 1 diabetes model using sensitivity approaches. They used a simplified model with only insulin, glucose and growth hormone as variables and proposed the following model [7]:

$$\frac{dI}{dt} = \frac{\psi I_0 I}{1+I} - \delta I \tag{100}$$

$$\frac{dG_L}{dt} = a - (b + cI)G_L + cH \tag{101}$$

$$\frac{dH}{dt} = \rho - wH \tag{102}$$

Where I, G_L and H stand for insulin, glucose and growth hormones respectively. It is assumed that, overtime, blood insulin level (I) drops as glucose is absorbed by muscle, fat and liver cells and clears at a constant rate δ while glucose (G_L) level is increased by the growth hormone through suppression of glucose uptake by insulin, at a constant rate c. The parameter a represents the average glucose obtained from carbohydrate intake and body production.

In this model, it is assumed that growth hormone (H) in mode is increased by the rate of production by the somatotropic cells in the pituitary gland at a constant rate ρ and is decreased by the rate of w due to the absorption by the liver.

In 2022, Al Ali et al. used also experimental data for type 1 diabetes to compare the results yielded by the previous model and another model including β-cells mass as a fourth variable with the variables insulin (I), glucose (G_L) and growth hormone (G_H) as follows [8]:

$$\frac{d\beta}{dt} = \left(hG_L - iG_L^2 - g\right)\beta, \tag{103}$$

$$\frac{dI}{dt} = \frac{\beta d G_L^2}{e + G_L^2} - fI, \tag{104}$$

$$\frac{dG_L}{dt} = a - (b + cI)G_L + cG_H, \tag{105}$$

$$\frac{dG_H}{dt} = \rho - wG_H. \tag{106}$$

Among others, the authors used a Markov Chain Monte Carlo method based on a Bayesian framework to fit the model to determine the best model structure. Akaike information criteria (AIC) and Bayesian information criteria (BIC) approaches were used to assess the best model structure for type 1 diabetes. In fitting the model with no β-cells to glucose level, insulin absorption rate and insulin clearance rate were varied.

According to the authors, the Model fitting showed that the model with no β-cells provides a better representation of a type 1 diabetic glucose homeostasis system.

More details on the mathematical models including growth hormone as a variable affecting glucose homeostasis can be found in the Ph. D thesis defended by Hannah Al Ali in February 2022 at the Coventry University [4].

5.3 *Effect of Leptin on the Homeostasis of Diabetes*

—— Hamou-Maamar et al. [46, 47] ——

Hamou-Maamar et al. [47] used the Topp model to include body fat, adiposity hormone(Leptin) and its interactions with the dynamics of glucose, insulin and βcells. The model is given by:

$$\frac{dG}{dt} = R_0 - \left(E_{G0} + S_L L(t) + S_1 \frac{1 + a_L L(t)}{1 + \tau F(t)} I(t)\right) G(t) \tag{107}$$

$$\frac{dI}{dt} = \frac{\sigma\beta(t)}{1 + \sigma_L L(t)} \frac{G^2(t)}{\alpha + G^2(t)} - K_1 I(t) \tag{108}$$

$$\frac{d\beta}{dt} = (-d_0 + r_1 G(t) - r_2 G^2(t))\beta(t) \tag{109}$$

$$\frac{dL}{dt} = p_L F(t) - k_L L(t) \tag{110}$$

$$\frac{dF}{dt} = (p_I I(t) - d_L L(t) - k_F)F(t) \tag{111}$$

More information on the model and its parameters can be found in the Ph.D thesis defended in 2017 by Hamou-Maamar M. [46].

5.4 *Effect of Epinephrine on the Homeostasis of Diabetes*

Beside the usual behavioural risk factors such as tobacco use, alcohol, unhealthy diet, physical inactivity; and the metabolic risk factors like obesity, high blood pressure and raised lipids, people often raise the question about stress as a risk factor for the onset of diabetes and its control. In the review of modelling studies dealing with the progression of secondary diabetes, Yang et al. stressed that the impact of epinephrine (adrenaline) on glucose metabolism arouses increasing attention from researchers [99]. They reported on two models dealing with epinephrine and diabetes.

——— Mohammed et al. [68] ———

The first mathematical model was proposed by Mohammed et al. to study the glucose regulation under the influence of trauma, excitement and/or stress [68]. This mathematical model extended the well known model of Topp et al. by suggesting that epinephrine increases the rate of glucose concentration by a constant (Ge) while decreasing the rate of insulin by a constant rate of insulin (ρI). The model proposed took the form [68]:

$$\frac{dG}{dt} = R_0 + G_e - (E_{GO} + S_I I)\, G \tag{112}$$

$$\frac{dI}{dt} = \frac{\sigma G^2}{G^2 + \alpha}\beta - (\rho + k)I \tag{113}$$

$$\frac{d\beta}{dt} = \left(-d + r_1 G - r_2 G^2\right)\beta \tag{114}$$

where Ge(mg/dl/day) represents the effect of epinephrine to increase the rate of glucose concentration and ρ(/ day) stands for the rate constant of insulin suppression by epinephrine.

——— Moscardó et al. [69] ———

The second model reported by Yang et al. on the effect of epinephrine on diabetes was proposed by Moscardó et al. to investigate the counter regulatory action of epinephrine during hypoglycaemia in type 1 diabetes. This model extended the Bergman Minimal Model by adding terms to the glucose equation to account for the influence of epinephrine as follows [69]:

$$\frac{dI}{dt} = \frac{u_1(t)}{V_{ol}} - nI, \tag{115}$$

$$\frac{dX}{dt} = p_3 I - p_2 X, \tag{116}$$

$$\frac{dG}{dt} = p_4 + \frac{u_2(t)}{V_{ol}} - p_1 G - XG + p_a A(t) - p_h \max\left(G_{b2} - G, 0\right), \tag{117}$$

$$A(t) = \begin{cases} 0 & \text{if } G(t) > G_{th} \\ A_m(t) - A_{\text{basal}} & \text{if } G(t) \le G_{th} \end{cases}, \tag{118}$$

Where the variables G, I, and X stand for the blood glucose concentration (mg/dl), plasma insulin (mU/ml), and insulin action from a remote compartment (/min), respectively.

In the case of type 1 diabetes, insulin supply for patients comes from exogenous infusion. Consequently, the authors used the term $u_1(t)/Vol_I$, where $u_1(t)(\mu U/min)$ denotes the insulin infusion rate and Vol_I denotes the insulin distribution volume. The parameter p_3 represents the rate of insulin input in the compartment while n and p_2 stand for the clearance rates of plasma insulin and insulin from a remote compartment respectively.

In the glucose equation, p_4 represents the hepatic glucose production rate while p_1G and XG denote the glucose uptake independent of insulin and the glucose uptake rate under the influence of insulin respectively.

The terms $A_m(t)$ and A_{basal} stand for the plasma epinephrine concentration (ng/l) and the basal epinephrine concentration while G_{th} represents the glucose threshold activating the epinephrine response. The model assumes that the epinephrine effect would be activated only after the glucose falls below the glucose threshold G_{th}. The term phmax $(G_{b2} - G, 0)$ is added mainly for a good fit to data.

5.5 *Effect of Thyroid Disease on the Homeostasis of Diabetes*

Yang et al. [99]

Yang et al. proposed hyperthyroid-diabetes models dealing with the progression to secondary diabetes induced by hyperthyroidism. Stressing that excessive thyroid hormones, rather than suppressed TSH level, have major effects on beta-cell apoptosis and hyperglycaemia for patients with hyperthyroidism, they considered only the influence of thyroid hormones on the glucose regulation and neglected the impact of TSH on the glucose regulatory system in the model to avoid the introduction of additional parameters [99].

They proposed the following generalized diabetes progression model [99]:

$$\frac{dG}{dt} = G_{in} + p_1(X) - f_2(G) - C(I)GI, \tag{119}$$

$$\frac{dI}{dt} = f_1(G)p_2(X)\beta - kI, \tag{120}$$

$$\frac{d\beta}{dt} = (f_3(I) + p_3(X))\,\beta, \tag{121}$$

With
$P_1X = h_1X^{\alpha}$,
$p_2X = 1 + h_2X$,
$p_3X = q_1X\,(q_2 - X)$

$f_2(G) = g_1 G, C(I) = r_0 + \frac{r_1}{r_1+e^3}, f_1(G) = \frac{s_1 G^2}{G^2+c_3},$
$f_3(I) = \frac{m_1 I}{I^2+(m_2)^2} - m_3$

where
G(mg/dl), I(mU/ml) and β(mg) represent plasma glucose concentration, insulin concentration, and the mass of functional beta-cells at time t (days), respectively.

Gin is the sum of the average glucose uptake rate and the hepatic glucose production per day. The term $f_2(G)$ represents the insulin-independent uptake of glucose, while C(I) GI stands for the insulin-dependent uptake of glucose.

The term $f_1(G)$ represents the beta-cell secretory function per cell, and k (/day) denotes the insulin clearance rate.

The function f_3(I) represents the net growth rate of functional beta-cell mass that depends on the insulin level.

The interference factor X included into the model accounts for the progressive impact of the environmentally induced or epigenetic-related diabetogenic factor on the glucose regulation.

The function p_1(X) stands for the increased hepatic glucose production caused by the pathological factor while p_2(X) represents the impact of the factor on the insulin secretion rate and finally, p_3(X) expresses the abnormal response of beta-cells to a hostile environment that develops in a slow time scale.

The authors assume that the elevated hepatic glucose production p_1(X) is a power function of X, which can be determined by the extent to which the hyperthyroid factor impacts the glucose generation rate

The model also assumed that the beta-cell secretion function is linearly increasing with X and the functional β-cell mass undergoes the influence of X in a pattern of a logistic function.

Then they extended this generalised model by adding two equations governing the dynamics of FT4 and TSH to obtain the two hyperthyroid-diabetes progression models (with and without hyperthyroid treatment) as follows:

The hyperthyroid-diabetes model without hyperthyroid treatment is given by [99]:

$$\frac{dG}{dt} = G_{in} + p_1\left(\frac{FT4 - U}{U}\right) - f_2(G) - C(I)GI \tag{122}$$

$$\frac{dI}{dt} = f_1(G)p_2\left(\frac{FT4 - U}{U}\right)\beta - kI \tag{123}$$

$$\frac{d\beta}{dt} = \left[f_3(I) + p_3\left(\frac{FT4 - U}{U}\right)\right]\beta \tag{124}$$

$$\frac{dFT4}{dt} = \frac{a_1(t)TSH}{b_1 + TSH} - d_1\mathrm{FT4} \tag{125}$$

$$\frac{dTSH}{dt} = a_2 - \frac{a_2(FT4 - U)}{b_2 + FT4} - d_2\mathrm{TSH} \tag{126}$$

The Hyperthyroid-diabetes model under hyperthyroid treatment is given by [99]:

$$\frac{dG}{dt} = G_{in} + p_1\left(\frac{FT4 - U}{U}\right) - f_2(G) - C(I)GI, \tag{127}$$

$$\frac{dI}{dt} = f_1(G)p_2\left(\frac{FT4 - U}{U}\right)\beta - kI, \tag{128}$$

$$\frac{d\beta}{dt} = \left[f_3(I) + p_3\left(\frac{FT4 - U}{U}\right)\right]\beta, \tag{129}$$

$$\frac{dFT4}{dt} = \frac{a_1(t)TSH}{b_1 + TSH} - d_1\text{FT4} - \frac{d_3 D}{D + IC_{50}}FT4, \tag{130}$$

$$\frac{dTSH}{dt} = a_2 - \frac{a_2(FT4 - U)}{b_2 + FT4} - d_2\text{TSH}, \tag{131}$$

5.6 *Effect of Hypercortisolism-Induced Glycaemia*

——— Zavala et al. [76] ———

Glucocorticoids (GCs) are a class of steroid hormones that have profound effects on energy mobilization, especially glucose metabolism and facilitate the process of gluconeogenesis in the liver, while they reduce glucose uptake and utilization by antagonizing insulin effects in white adipose tissue and skeletal muscle. As a result, over-exposure to GCs leads to hyperglycaemia and insulin resistance [55, 76, 100].

On this basis, Zavala et al. developed a mathematical model investigating the impact of disrupted cortisol rhythms on the response to oral glucose tolerance tests (OGTT) [76]:

$$\frac{dG(t)}{dt} = F(t) + vf_e(G) - a\left[c_L f_L(G) + c_M f_M(G)\right]T - r_G G \tag{132}$$

$$\frac{dI(t)}{dt} = \varepsilon + \sigma S_\beta(G)h_Q(Q) - r_I I \tag{133}$$

$$\frac{dT}{dt} = (u + v_I f_I(I))(1 - T) - \left(d + v_Q f_Q(Q)\right)T \tag{134}$$

Where $G(t)$ and $I(t)$ stand for the blood concentrations of glucose and insulin at time t(min), respectively; the variable T ∈ (0, 1) denotes the fraction of translocatable GLUTs in the cell membrane of peripheral cells, and 1-T represents the fraction of GLUTs that docked inside the cell.

5.7 *Association Between Benign Prostatic Hyperplasia and Type 2 Diabetes*

Boutayeb et al. [21]

In 2023, Boutayeb et al. proposed a mathematical dealing with the association between Benign prostatic hyperplasia and type 2 diabetes/metabolic syndrome. The model was given by the following system of ordinary differential equations [21]:

$$\frac{dG}{dt} = a - bG - cIG + m_1(F - Fb) \tag{135}$$

$$\frac{d\beta}{dt} = \left(-iG^2 + hG - g\right)\beta \tag{136}$$

$$\frac{dI}{dt} = \frac{d\beta G^2}{G^2 + e} - If \tag{137}$$

$$\frac{dF}{dt} = -m_2(F - Fb) + m_3(G - Gb) \tag{138}$$

$$\frac{dV}{dt} = v_0 + v_1 G + v_2 I - v_3 V + v_4 F \tag{139}$$

The reader can find explanation on the association between benign prostatic hyperplasia and metabolic syndrome in Chap. 7 of this book.

6 Models of Subcutaneous Insulin Kinetics

As indicated by Nucci and Cobelli, the process of insulin absorption is complex and can be affected by temperature, the concentration of the insulin, and the location of the injection [72]. However, the chemical reaction can be illustrated by the following diagram

$$H \rightleftarrows 3D$$

where H stands for hexamer and D for dimer, p is the transform rate from one hexameric molecule to three dimeric molecules, and pq is the transform rate from three dimeric molecules to one hexameric molecule. A dimer can be divided into two monomericm molecules of insulin. Both forms of dimer and monomer can penetrate the capillary boundary and reach the blood stream [59].

Shimoda et al. [85]

To study the dynamics of a short acting insulin analog injected subcutaneously, Shimoda et al. proposed the following compartment-based model [85]

$$\frac{dx_1}{dt} = k_{12}x_1 + u(t) \tag{140}$$

$$\frac{dx_2}{dt} = k_{21}x_1 - (k_d + k_a)x_2 \tag{141}$$

$$\frac{dI}{dt} = \frac{k_a}{V_d}x_2 - k_e I \tag{142}$$

where I is the plasma insulin concentration, $u(t)$ is the input function, x_1 stands for subcutaneous insulin mass in the injection depot and x_2 is the subcutaneous insulin mass proximal to plasma. The parameters V_d, k_d and k_a represent the the volume of plasma, and degradation rate in the depot and plasma respectively.

Li and Kuang [61]

Using the process given in the reaction diagram and hypothesizing that the insulin absorption in depot is inversely proportional to the concentration of insulin, Li and Kuang proposed the following model [61]

$$\frac{dH}{dt} = -p(H - qD^3) \tag{143}$$

$$\frac{dD}{dt} = p(H - qD^3) - \frac{bD}{(1 + I(t))} \tag{144}$$

$$\frac{dI}{dt} = \frac{rbD}{(1 + I(t))} - d_i I \tag{145}$$

where $I(t)(U/ml)$ is the plasma insulin concentration at time $t \geq 0$, $H(t)(U/ml)$ is the concentration of insulin analogue in its hexameric form and $D(t)(U/ml)$ is the concentration of the insulin analogue in the dimeric form.

The constant $di(1/\min)$ in the third equation is the insulin degradation rate that has been assumed as linear. In the term $rbD(t)/(1 + I(t))$ of the third equation, $b(U/min)$ is a constant parameter and $r \leq 1$, indicates that only fractional molecules can become plasma insulin.

Wilinska et al. [59]

Assuming that the continuous injection is at a constant rate u, and modifying a model of Wilinska et al., Li and Johnson proposed the following model [59]

$$\frac{dH}{dt} = u - p(H - qD^3) \tag{146}$$

$$\frac{dD}{dt} = p(H - qD^3) - \frac{bD}{(1 + I(t))} \tag{147}$$

$$\frac{dI}{dt} = \frac{rbD}{(1 + I(t))} - d_i I \tag{148}$$

This system has a unique equilibrium state $E^* = (H^*, D^*, I^*)$ where:

$$I^* = \frac{ru}{d_i}$$
$$D^* = \frac{d_i I^*(1 + I^*)}{rb} = b^{-1}u(1 + d_i^{-1}ru)$$
$$H^* = \frac{u}{p} + q(D^*)^3 = p^{-1}u + qb^{-3}u^3(1 + d_i^{-1}ru)^3$$

——— Boutayeb et al. [19] ———

In 2019, restricting their model to type 1 diabetes needing injected insulin, Boutayeb et al. [19] proposed the following model:

$$\frac{dI}{dt} = \frac{eI_0}{(1 + I(t))} - dI(t) \tag{149}$$

$$\frac{dG}{dt} = a_1 + a_2 - (b + cI(t))G(t) + m_1(F(t) - F_b) \tag{150}$$

$$\frac{dF}{dt} = -m_2(F(t) - F_b) + m_3(G(t) - G_b) \tag{151}$$

In 2019, Al Ali et al. [5] extended this model by adding growth hormone dynamics as indicated earlier in model (95).

The way an individual daily monitors his/her diabetes is a determinant parameter to achieve a healthy quality of life. However, beside the daily control of blood glucose, diabetic people also require a regular check of their HbA_{1C}. Consequently, the next section is dedicated to mathematical models dealing with HbA_{1C}.

7 HbA_{1C}

The relationship between the glycated haemoglobin (HbA_{1c}) levels and risks for diabetic complications in patients with type1 (T1DM) and type 2 (T2DM) diabetes was established by the Diabetes Control and Complications Trial Research Group (DCCT) [31, 32] and the UK Prospective Diabetes Study Group (UKPDS) [43, 44]. The relationship between HbA_{1c} and the average blood glucose was explored by different authors [11, 30, 48, 71]. The A1c-derived average glucose (ADAG) examined the link between HbA_{1c} and the average glucose [30, 71]. Using a linear regression analysis on data from 11 centres in the US, Europe, Africa and Asia, the ADAG study provided a linear relation between HbA_{1c} and the estimated average glucose (eAG) [30].

$$eAG(mg/dl) = 28.7 * eHbA1c(\%) - 46.7 \tag{152}$$

Other mathematical models were proposed to deal with the relationship between HbA_{1c} and the average glucose (AG) [23, 24, 33, 56–58, 77, 90, 91, 95]. Temsch et al. proposed a model respecting the decreasing contribution of older glucose levels to current HbA_{1c} values using truncated Fourier series and convolution [91].

——— Dayanand et al. [33] ———

Dayanand et al. dealt with a comparison of calculated HbA_{1c} with measured HbA_{1c} by high Pressure Liquid Chromatography Method [33].

——— Ladyzyński et al. [56, 58] ———

Ladyzyński et al. proposed and used the following mathematical model [56].

$$eHbA1c(\%) = 91.5 * (1 - 1 - e - k \times LS \times eAGk \times LS \times eAG) + 2.15 \tag{153}$$

where LS is the life span (=120 days), eAG is the average glucose supposed constant over the life span and k is the haemoglobin glycation rate constant.

The same authors carried out a linear regression on the data of the simulated ADAG population to obtain the following equation [58].

$$eHbA1c(\%) = 0.0296 \; * \; eAG(mg/dl) + 2.419 \tag{154}$$

——— Rohlfing et al. [77] ———

Rohlfing et al. used a linear regression on DCCT data, ending up with a relation similar to the one provided by ADAG, but referring to publications indicating that plasma glucose (PG) in the preceding 30 days contribute to about 50% to the HbA_{1c} value and PG from 90 to 120 days earlier contribute only to about 10% [77].

——— Tahara and Shima [90] ———

Tahara and Shima suggested that 50% of HbA_{1c} is determined by the PG level during the preceding one month period and that 25% of HbA_{1c} is determined by the PG during a prior 1 month period, the remaining 25% of HbA_{1c} level is therefore determined by the PG level during the 2 months period before these 2 months [90].

——— Boutayeb et al. [23] ———

Following this, Boutayeb et al. proposed a "weighted" average glucose (wAG) [23]:

$$wAG = 20\% \; * \; AGM - 3 + 30\% \; * \; AGM - 2 + 50\% \; * \; AGM - 1 \quad \text{and } wHbA1c = 0.0296 * wAG + 2.419 \tag{155}$$

where $AGM - 1$, $AGM - 2$ and $AGM - 3$ represent the average glucose computed chronologically over the preceding months to the $wHbA1c$ test.

8 Other

——— Ph.D theses [4, 22, 37, 46] ———

More detailed information on mathematical models used in diabetes homeostasis can be found in the Ph.D theses defended by Derouich in 2001 [37], Boutayeb in 2015 [22], Hamou-Maamar in 2017 [46] and Al Ali in 2022 [4].

9 A Summary of the Main Models with Their Strengths and Weaknesses

The reader can find in the following tables (Tables 2, 3, 4, 5 and 6), information concerning the main models reviewed in this chapter, including their strength, weakness, reference and section of appearance.

Table 2 Modelling glucose and insulin dynamics, using mainly ordinary differential equations

Model	Ref	Section	Strengths	Weaknesses
Bolie	[18]	Sect. 2	Uses simple ordinary differential equations to capture basic glucose and insulin dynamics	This is a simplified model not account for many biological and environmental factors influencing glucose metabolism
Hussain/Zadeng	[51]	Sect. 2		

Table 3 Modelling the effect pf physical exercise on glucose and insulin dynamics

Model	Ref	Section	Strengths	Weaknesses
Derouich/ Boutayeb	[38]	Sect. 2	Incorporate the effect of physical exercise on glucose and insulin dynamics, making models more realistic and applicable to active patients	Exercise modeling remains simplified and does not account for all individual and environmental variables
Kartono	[53]	Sect. 2		
Roy/Parker	[78]	Sect. 2		

Table 4 Modelling glucose and insulin dynamics, using delay equations

Model	Ref	Section	Strengths	Weaknesses
De Gaetano	[34, 35]	Sect. 3	Integrate delays to capture physiological response lags in insulin action and glucose production, offering a more precise representation of biological dynamics	Delay models are mathematically more complex and require precise data for delay calibration, potentially limiting practical use
Mukhopadhyay	[70]	Sect. 3		
Li/Kuang	[62]	Sect. 3		
	[60, 63]	Sect. 3		
Sturis	[88]	Sect. 3		
Drozdov/ Khanina	[39]	Sect. 3		
Saber et al.	[79–81]	Sect. 3		
Shabestari et al.	[84]	Sect. 3		
Bridgewater et al.	[27]	Sect. 3		

Table 5 Modelling the dynamics of glucose, insulin, β-cells and effect of FFA, growth hormone, leptin ...

Model	Ref	Section	Strengths	Weaknesses
Topp et al.	[94]	Sect. 4	Integrate multiple biological factors such as beta cells, insulin, glucagon, free fatty acids, growth hormone, and Leptin. These models are more comprehensive and better reflect biological complexity. They particularly stress the relation between diabetes and many other diseases and conditions	These models combine long (β-cell mass, FFA, prostate volume V) and short (I, G) time scales. Increased complexity requires detailed data and advanced calibration methods, limiting accessibility and clinical application
Hernandez et al.	[49]	Sect. 4		
Pranay Goel	[42]	Sect. 4		
Boutayeb et al.	[25]	Sect. 4		
	[26]	Sect. 5.1		
Al Ali	[4–8]	Sect. 5.2		
Hamou-Maamar et al.	[46, 47]	Sect. 5.3		
Mohammed et al.	[68]	Sect. 5.4		
Moscardó et al.	[69]	Sect. 5.4		
Yang et al.	[99]	Sect. 5.5		
Zavala et al.	[76]	Sect. 5.6		
Boutayeb et al.	[21]	Sect. 5.7		

Table 6 Monitoring diabetes and modelling HbA_{1C}

Model	Ref	Section	Strengths	Weaknesses
Committee I.E	[30]	Sect. 7	Establishing a relationship between HbA_{1C} and the average blood glucose and hence providing guidelines to people living with diabetes who usually seek the best way of diabetes monitoring daily and for a longue period	These models require data regularly collected through daily monitors.
Nathan et al	[71]	Sect. 7		
Ladyzyński et al.	[56, 58]	Sect. 7		
Rohlfing et al.	[77]	Sect. 7		
Tahara/Shima	[90]	Sect. 7		
Boutayeb et al.	[23]	Sect. 7		

References

1. Abdul-Basset, A., Rahma, A.H.F., Sajad, J.: Hopf bifurcation and chaos in time-delay model of glucose-insulin regulatory system [j]. Chaos, Solitons Fractals **137**, 109845 (2020)
2. Ackerman, E., Gatewood, L.C., Rosevear, J.W., Molnar, G.D.: Model studies of blood-glucose regulation. Bull. Math. Biophys. **27**(1), 21–37 (1965)
3. Ajmera, I., Swat, M., Laibe, C., Le Novere, N., Chelliah, V.: The impact of mathematical modeling on the understanding of diabetes and related complications. CPT: Pharmacomet. Syst. Pharmacol. **2**(7), 1–14 (2013)
4. Al Ali, H.: Mathematical and data-driven modelling of glucose homeostasis system to understand diabetes pathways. Ph.D. thesis, Coventry University (2022)
5. Al Ali, H., Boutayeb, W., Boutayeb, A., Merabet, N.: A mathematical model for type 1 diabetes, on the effect of growth hormone. In: 2019 8th International Conference on Modeling Simulation and Applied Optimization (ICMSAO), pp. 15–17. IEEE (2019)
6. Al Ali, H., Boutayeb, W., Boutayeb, A., Merabet, N.: A mathematical model on the effect of growth hormone on the glucose homeostasis. ARIMA **30**, 31–42 (2019)
7. Al Ali, H., Daneshkhah, A., Boutayeb, A., Malunguza, N.J., Mukandavire, Z.: Exploring dynamical properties of a type 1 diabetes model using sensitivity approaches. Math. Comput. Simul. **201**, 324–342 (2022)
8. Al Ali, H., Daneshkhah, A., Boutayeb, A., Mukandavire, Z.: Examining type 1 diabetes mathematical models using experimental data. IJERPH **19**(2), 1–20 (2022)
9. Bajaj, J., Rao, G.S., Rao, J.S., Khardori, R.: A mathematical model for insulin kinetics and its application to protein-deficient (malnutrition-related) diabetes mellitus (pddm). J. Theor. Biol. **126**(4), 491–503 (1987)
10. Balakrishnan, N.P., Rangaiah, G.P., Samavedham, L.: Review and analysis of blood glucose (bg) models for type 1 diabetic patients. Ind. Eng. Chem. Res. **50**(21), 12041–12066 (2011)
11. Barua, A., Acharya, J., Ghaskadbi, S., Goel, P.: The relationship between fasting plasma glucose and hba1c during intensive periods of glucose control in antidiabetic therapy. J. Theor. Biol. **363**, 158–163 (2014)
12. Bergman, R.: The minimal model approach to quantification of factors controlling glucose disposal in man. Carbohydr. Metab. **269** (1981)
13. Bergman, R.: The minimal model: yesterday, today and tomorrow. The minimal model approach and determinants of glucose tolerance **7**, 3–50 (1997)
14. Bergman, R.N.: Pathogenesis and prediction of diabetes mellitus: lessons from integrative physiology. Mount Sinai J. Med., New York **69**(5), 280–290 (2002)

15. Bergman, R.N.: The minimal model of glucose regulation: a biography. In: Mathematical Modeling in Nutrition and the Health Sciences, pp. 1–19. Springer (2003)
16. Bergman, R.N., Finegood, D.T., Ader, M.: Assessment of insulin sensitivity in vivo. Endocr. Rev. **6**(1), 45–86 (1985)
17. Bergman, R.N., Ider, Y.Z., Bowden, C.R., Cobelli, C.: Quantitative estimation of insulin sensitivity. Am. J. Physiol. Endocrinol. Metab. **236**(6), E667 (1979)
18. Bolie, V.W.: Coefficients of normal blood glucose regulation. J. Appl. Physiol. **16**(5), 783–788 (1961)
19. Boutayeb, A., Boutayeb, W., Al Ali, H., Merabet, N.: A mathematical model on glucose homeostasis in type 1 diabetes. BCPT **124**(S3), 17–18 (2019)
20. Boutayeb, A., Chetouani, A.: A critical review of mathematical models and data used in diabetology. Biomed. Eng. Online **5**(1), 43 (2006)
21. Boutayeb, A., Derouich, M., Boutayeb, W., Lamlili, M.E.: A mathematical model illustrating the association between benign prostatic hyperplasia and type 2 diabetes/metabolic syndrome. Preprint
22. Boutayeb, W.: Modélisation bio-informatique et bio-mathématique appliquée au diabéte. Ph.D. thesis, University Mohamed Premier, Oujda, Morocco (2015)
23. Boutayeb, W., Boutayeb, A., Lamlili, M., El Mostafa, S.B., Zitouni, N.: Simulation of a computed hba 1c using a weighted average glucose. SpringerPlus **5**(1), 226 (2016)
24. Boutayeb, W., Lamlili, M., Ben El Mostafa, S., Zitouni, N., Boutayeb, A.: Seeking a low value of hba 1c may be dangerous under recurrent episodes of hypoglycaemia: a short report. Mediterr. J. Nutr. Metab. **8**(3), 213–216 (2015)
25. Boutayeb, W., Lamlili, M.E., Boutayeb, A., Derouich, M.: Mathematical modelling and simulation of β-cell mass, insulin and glucose dynamics: effect of genetic predisposition to diabetes. J. Biomed. Sci. Eng. **7**(06), 330 (2014)
26. Boutayeb, W., Lamlili, M.E., Boutayeb, A., Derouich, M.: The impact of obesity on predisposed people to type 2 diabetes: mathematical model. In: International Conference on Bioinformatics and Biomedical Engineering, pp. 613–622. Springer (2015)
27. Bridgewater, A., Huard, B., Angelova, M.: Amplitude and frequency variation in nonlinear glucose dynamics with multiple delays via periodic perturbation. J. Nonlinear Sci. **30**(3), 737–766 (2020)
28. Chalishajar, D., Geary, D.H., Cox, G., et al.: Review study of detection of diabetes models through delay differential equations. Appl. Math. **7**(10), 1087–1102 (2016)
29. Chuedoung, M., Sarika, W., Lenbury, Y.: Dynamical analysis of a nonlinear model for glucose-insulin system incorporating delays and β-cells compartment. Nonlinear Anal. Theory Methods Appl. **71**(12), e1048–e1058 (2009)
30. Committee, I.E., et al.: International expert committee report on the role of the a1c assay in the diagnosis of diabetes. Diab. Care **32**(7), 1327–1334 (2009)
31. Control, D., Group, C.T.R.: The effect of intensive treatment of diabetes on the development and progression of long-term complications in insulin-dependent diabetes mellitus. New England J. Med. **329**(14), 977–986 (1993)
32. Control, D., Group, C.T.R., et al.: The association between glycaemic exposure and long-term diabetic complications in the diabetes control and complications trial. Diabetes **44**, 968–83 (1995)
33. Dayanand, C., Vegi, P.K., Cutty, A.: Comparison of calculated with measured by high pressure liquid chromatography method. Int. J. Sci. Innov. Discov. **2**(6), 573–575 (2012)
34. De Gaetano, A., Arino, O.: Mathematical modelling of the intravenous glucose tolerance test. J. Math. Biol. **40**(2), 136–168 (2000)
35. De Gaetano, A., Arino, O.: A statistical approach to the determination of stability for dynamical systems modelling physiological processes. Math. Comput. Modell. **31**(4–5), 41–51 (2000)
36. Della Corte, M., Romano, S., Voeghelin, M., Serio, M.: On a mathematical model for the analysis of the glucose tolerance curve. Diabetes **19**(6), 445–449 (1970)
37. Derouich, M.: Modeling and simulation of models with and without age structure: Application to diabetes and dengue fever. Ph.D. thesis, University Mohamed Premier, Oujda, Morocco (2001)

38. Derouich, M., Boutayeb, A.: The effect of physical exercise on the dynamics of glucose and insulin. J. Biomech. **35**(7), 911–917 (2002)
39. Drozdov, A., Khanina, H.: A model for ultradian oscillations of insulin and glucose. Math. Comput. Modell. **22**(2), 23–38 (1995)
40. Finegood, D.T.: Application of the minimal model of glucose kinetics. The Minimal Model Approach and Determinants of Glucose Tolerance **7**, 51–122 (1997)
41. Giang, D.V., Lenbury, Y., De Gaetano, A., Palumbo, P.: Delay model of glucose-insulin systems: global stability and oscillated solutions conditional on delays. J. Math. Anal. Appl. **343**(2), 996–1006 (2008)
42. Goel, P.: Insulin resistance or hypersecretion? the βig picture revisited. J. Theor. Biol. **384**, 131–139 (2015)
43. Group, U.P.D.S., et al.: Tight blood pressure control and risk of macrovascular and microvascular complications in type 2 diabetes: Ukpds 38. BMJ: Br. Med. J. **317**(7160), 703 (1998)
44. Group, U.P.D.S.U., et al.: Intensive blood-glucose control with sulphonylureas or insulin compared with conventional treatment and risk of complications in patients with type 2 diabetes (ukpds 33). The Lancet **352**(9131), 837–853 (1998)
45. Gunawardane, K., Hansen, T.K., Christiansen, J.S., Jorgensen, J.O.L.: Normal physiology of growth hormone in adults. In: Endotext [Internet]. MDText. com, Inc. (2015)
46. Hamou-Maamar, M.: Modélisation de la dynamique de l'insuline-glucose et simulation numérique par la méthode des ondelettes. Ph.D. thesis, University Abdelhamid Ibn Badis Mostaganem, Mostaganem, Algeria (2017)
47. Hamou-Maamar, M., Belhamiti, O.: Leptin effect's on glucose and insulin kinetics: a mathematical model. Commun. Nonlinear Sci. Numer. Simul. **113**, 106591 (2022)
48. Hempe, J.M., Gomez, R., McCarter, R.J., Jr., Chalew, S.A.: High and low hemoglobin glycation phenotypes in type 1 diabetes: a challenge for interpretation of glycemic control. J. Diab. Complicat. **16**(5), 313–320 (2002)
49. Hernandez, R.D., Lyles, D.J., Rubin, D.B., Voden, T.B., Wirkus, S.A., et al.: A model of β-cell mass, insulin, glucose, and receptor dynamics with applications to diabetes. CORNELL UNIV., DEPT. OF BIOMETRICS TECHNICAL REPORT BU-1579-M (2001)
50. Himsworth, H., Kerr, R., et al.: Insulin-sensitive and insulin-insensitive types of diabetes mellitus. Clin. Sci. **4**, 119–152 (1939)
51. Hussain, J., Zadeng, D.: A mathematical model of glucose-insulin interaction. Sci. Vis. **14**(12), 84–88 (2014)
52. IDF: Diabetes Atlas 2021 10th edition. https://diabetesatlas.org
53. Kartono, A., Irawati, F., Setiawan, A.A., Syafutra, H., Sumaryada, T.: The effects of physical exercise on the insulin-dependent diabetes mellitus subjects using the modified minimal model. Int. J. Pharm. Pharm. Sci. **9**, 179–86 (2017)
54. Kim, S.H., Park, M.J.: Effects of growth hormone on glucose metabolism and insulin resistance in human. Ann. Pediatr. Endocrinol. Metab. **22**(3), 145 (2017)
55. Kuo, T., McQueen, A., Chen, T.C., Wang, J.C.: Regulation of glucose homeostasis by glucocorticoids. Glucocorticoid Signaling: From Molecules to Mice to Man pp. 99–126 (2015)
56. Ladyzynski, P., Foltynski, P., Bak, M.I., Sabalinska, S., Krzymien, J., Kawiak, J.: Validation of a hemoglobin a 1c model in patients with type 1 and type 2 diabetes and its use to go beyond the averaged relationship of hemoglobin a 1c and mean glucose level. J. Transl. Med. **12**(1), 328 (2014)
57. Ladyzyński, P., Wójcicki, J.M., Bak, M., Sabalińska, S., Kawiak, J., Foltyński, P., Krzymień, J., Karnafel, W.: Validation of hemoglobin glycation models using glycemia monitoring in vivo and culturing of erythrocytes in vitro. Ann. Biomed. Eng. **36**(7), 1188–1202 (2008)
58. Ladyzyński, P., Wójcicki, J.M., Bak, M.I., Sabalińska, S., Kawiak, J., Foltyński, P., Krzymień, J., Karnafel, W.: Hemoglobin glycation rate constant in non-diabetic individuals. Ann. Biomed. Eng. **39**(11), 2721–2734 (2011)
59. Li, J., Johnson, J.D.: Mathematical models of subcutaneous injection of insulin analogues: a mini-review. Discrete Continuous Dyn. Syst. Ser. B **12**(2), 401 (2009)

60. Li, J., Kuang, Y.: Analysis of a model of the glucose-insulin regulatory system with two delays. SIAM J. Appl. Math. **67**(3), 757–776 (2007)
61. Li, J., Kuang, Y.: Systemically modeling the dynamics of plasma insulin in subcutaneous injection of insulin analogues for type 1 diabetes. Math. Biosci. Eng. MBE **6**(1), 41–58 (2009)
62. Li, J., Kuang, Y., Li, B.: Analysis of ivgtt glucose-insulin interaction models with time delay. Discrete Contin. Dyn. Syst. Ser. B **1**(1), 103–124 (2001)
63. Li, J., Kuang, Y., Mason, C.C.: Modeling the glucose-insulin regulatory system and ultradian insulin secretory oscillations with two explicit time delays. J. Theor. Biol. **242**(3), 722–735 (2006)
64. Li, J., Wang, M., De Gaetano, A., Palumbo, P., Panunzi, S.: The range of time delay and the global stability of the equilibrium for an ivgtt model. Math. Biosci. **235**(2), 128–137 (2012)
65. Makroglou, A., Karaoustas, I., Li, J., Kuang, Y.: Delay differential equation models in diabetes modeling. Theor. Biol. Med, Model (2009)
66. Makroglou, A., Li, J., Kuang, Y.: Mathematical models and software tools for the glucose-insulin regulatory system and diabetes: an overview. Appl. Numer. Math. **56**(3–4), 559–573 (2006)
67. Mari, A.: Mathematical modeling in glucose metabolism and insulin secretion. Curr. Opin. Clin. Nutr. Metab. Care **5**(5), 495–501 (2002)
68. Mohammed, I.I., Adamu, I.I., Barka, S.J.: Mathematical model for the dynamics of glucose, insulin and β-cell mass under the effect of trauma, excitement and stress. Model. Numer. Simul. Mater. Sci. **9**(4), 71–96 (2019)
69. Moscardó, V., Rossetti, P., Ampudia-Blasco, F.J., Bondia, J.: Modellig of adrenaline counterregulatory action during hypoglycaemia in type 1 diabetes. In: 2016 IEEE Conference on Control Applications (CCA), pp. 430–435. IEEE (2016)
70. Mukhopadhyay, A., De Gaetano, A., Arino, O.: Modeling the intra-venous glucose tolerance test: a global study for a single-distributed-delay model. Discrete Contin. Dyn. Syst. Ser. B **4**(2), 407–418 (2004)
71. Nathan, D.M., Kuenen, J., Borg, R., Zheng, H., Schoenfeld, D., Heine, R.J.: Translating the a1c assay into estimated average glucose values. Diabetes Care **31**(8), 1473–1478 (2008)
72. Nucci, G., Cobelli, C.: Models of subcutaneous insulin kinetics. a critical review. Comput. Methods Progr. Biomed. **62**(3), 249–257 (2000)
73. Paidipati, K.K., Padi, T.R.: Bivariate stochastic modelling of glucose and insulin regulatory system among type-2 diabetes mellitus patients. Bull. Math. Sci. Appl. ISSN **2278**, 9634 (2015)
74. Palumbo, P., Panunzi, S., De Gaetano, A.: Qualitative behavior of a family of delay-differential models of the glucose-insulin system. Discrete Contin. Dyn. Syst. Ser. B **7**(2), 399–424 (2007)
75. Prager, R., Wallace, P., Olefsky, J.M., et al.: In vivo kinetics of insulin action on peripheral glucose disposal and hepatic glucose output in normal and obese subjects. J. Clin. Invest. **78**(2), 472–481 (1986)
76. Rafacho, A., Ortsäter, H., Nadal, A., Quesada, I.: Glucocorticoid treatment and endocrine pancreas function: implications for glucose homeostasis, insulin resistance and diabetes. J. Endocrinol. **223**(3), R49–R62 (2014). https://doi.org/10.1530/joe-14-0373
77. Rohlfing, C.L., Wiedmeyer, H.M., Little, R.R., England, J.D., Tennill, A., Goldstein, D.E.: Defining the relationship between plasma glucose and hba1c: analysis of glucose profiles and hba1c in the diabetes control and complications trial. Diab. Care **25**(2), 275–278 (2002)
78. Roy, A., Parker, R.S.: Dynamic modeling of free fatty acid, glucose, and insulin: an extended "minimal model". Diab. Technol. Ther. **8**(6), 617–626 (2006)
79. Saber, S., Alzahrani, S.: Hopf bifurcation on fractional ordered glucose-insulin system with time-delay. Albaha Univ. J. Basic Appl. Sci. **3**(2), 27–34 (2019)
80. Saber, S., Alzahrani, S.: Stability analysis of a fractional order delayed glucose-insulin model. Albaha Univ. J. Basic Appl. Sci. **3**(2), 7–14 (2019)
81. Saber, S., Bashier, E., Alzahrani, S., Noaman, I.: A mathematical model of glucose-insulin interaction with time delay. J. Appl. Computat. Math. **7**(416), 2 (2018)

82. Sarikaa, W., Lenburya, Y., Kumnungkitb, K., Kunphasuruangc, W.: Modelling glucose-insulin feedback signal interchanges involving β-cells with delays. Sci. Asia **34**, 77–86 (2008)
83. Segre, G., Turco, G., Vercellone, G.: Modeling blood glucose and insulin kinetics in normal, diabetic and obese subjects. Diabetes **22**(2), 94–103 (1973)
84. Shabestari, P.S., Rajagopal, K., Safarbali, B., Jafari, S., Duraisamy, P.: A novel approach to numerical modeling of metabolic system: investigation of chaotic behavior in diabetes mellitus. Complexity **2018** (2018)
85. Shimoda, S., Nishida, K., Sakakida, M., Konno, Y., Ichinose, K., Uehara, M., Nowak, T., Shichiri, M.: Closed-loop subcutaneous insulin infusion algorithm with a short-acting insulin analog for long-term clinical application of a wearable artificial endocrine pancreas. Front. Med. Biol. Eng. Int. J. Jpn. Soc. Med. Electron. Biol. Eng. **8**(3), 197–211 (1997)
86. Smith, J.M., Maas, J.A., Garnsworthy, P.C., Owen, M.R., Coombes, S., Pillay, T.S., Barrett, D.A., Symonds, M.E.: Mathematical modeling of glucose homeostasis and its relationship with energy balance and body fat. Obesity **17**(4), 632–639 (2009)
87. Srinivasan, R., Kadish, A.H., Sridhar, R.: A mathematical model for the control mechanism of free fatty acid-glucose metabolism in normal humans. Comput. Biomed. Res. **3**(2), 146–165 (1970)
88. Sturis, J., Polonsky, K.S., Mosekilde, E., Van Cauter, E.: Computer model for mechanisms underlying ultradian oscillations of insulin and glucose. Am. J. Physiol.-Endocrinol. Metab. **260**(5), E801–E809 (1991)
89. Sundell, J., Knuuti, J.: Insulin and myocardial blood flow. Cardiovasc. Res. **57**(2), 312–319 (2003)
90. Tahara, Y., Shima, K.: The response of ghb to stepwise plasma glucose change over time in diabetic patients. Diab. Care **16**(9), 1313–1314 (1993)
91. Temsch, W., Luger, A., Riedl, M.: Hba1c values calculated from blood glucose levels using truncated fourier series and implementation in standard sql database language. Methods Inf. Med. **47**(04), 346–355 (2008)
92. Toffolo, G., Bergman, R.N., Finegood, D.T., Bowden, C.R., Cobelli, C.: Quantitative estimation of beta cell sensitivity to glucose in the intact organism: a minimal model of insulin kinetics in the dog. Diabetes **29**(12), 979–990 (1980)
93. Tolić, I.M., Mosekilde, E., Sturis, J.: Modeling the insulin-glucose feedback system: the significance of pulsatile insulin secretion. J. Theor. Biol. **207**(3), 361–375 (2000)
94. Topp, B., Promislow, K., Devries, G., Miura, R.M., T FINEGOOD, D.: A model of β-cell mass, insulin, and glucose kinetics: pathways to diabetes. J. Theor. Biol. **206**(4), 605–619 (2000)
95. Treviño, G.: On the weighted-average relationship between plasma glucose and hba1c. Diab. Care **29**(2), 466–466 (2006)
96. Vosoughi, R., Goghari, Z.S., Jafari, A.H.: Modelling system of two insulin-glucose delays to achieve the dynamics of glucose changes. J. Biomed. Phys. Eng. **12**(2), 189–204 (2022)
97. Wang, H., Li, J., Kuang, Y.: Mathematical modeling and qualitative analysis of insulin therapies. Math. Biosci. **210**(1), 17–33 (2007). https://doi.org/10.1016/j.mbs.2007.05.008
98. WHO: Diabetes. http://www.who.int/news-room/factsheets/detail/diabetes
99. Yang, B., Li, J., Haller, M.J., Schatz, D.A., Rong, L.: The progression of secondary diabetes: a review of modeling studies. Front. Endocrinol. **13** (2022)
100. Zavala, E., Gil-Gómez, C.A., Wedgwood, K.C., Burgess, R., Tsaneva-Atanasova, K., Herrera-Valdez, M.A.: Dynamic modulation of glucose utilisation by glucocorticoid rhythms in health and disease. BioRxiv, pp. 2020–02 (2020)

A Comprehensive Review of Mathematical and Data-Driven Models in Glucose Homeostasis and Diabetes Pathways

Hannah Al Ali, Alireza Daneshkhah, Abdesslam Boutayeb, Nora Merabet, and Zindoga Mukandavire

Abstract There are numerous mathematical models of the glucose homeostasis system, serving as vital tools in understanding the pathways and treatment strategies for diabetes. This paper provides an extensive review of these models, categorising them into clinical and non-clinical frameworks. We explore various approaches, including minimal models, minimal control networks, and the more recent data-driven models. Through our review, we highlight the strengths and limitations of each category, identifying significant gaps in the existing literature. A trend is the increasing importance of data-driven, machine learning methods, which capture the complex, nonlinear dynamics of glucose regulation more effectively than traditional mathematical models. These advanced methods offer the potential for more personalised and adaptive treatment strategies, reflecting the variability in patient responses and disease progression. Furthermore, machine learning models can continuously improve and adapt with the influx of new data, making them highly dynamic and responsive to changes over time. This adaptability contrasts with the static nature of many traditional models, allowing for real-time adjustments in treatment plans. By leveraging diverse heterogeneous datasets and sophisticated algorithms, machine

H. Al Ali · A. Daneshkhah · A. Boutayeb (✉)
Faculty of Mathematics and Data Science, Emirates Aviation University, Dubai, United Arab Emirates
e-mail: x.boutayeb@gmail.com

H. Al Ali
e-mail: hannah.alali@emirates.com

A. Daneshkhah
e-mail: alireza.daneshkhah@emirates.com

N. Merabet · Z. Mukandavire
Institute of Applied Research and Technology, Emirates Aviation University, Dubai, United Arab Emirates
e-mail: noramerabet@ymail.com

Z. Mukandavire
e-mail: zindoga.mukandavire@emirates.com

A. Boutayeb and M. E. N. Lamlili (eds.), *Computational Mathematics and Modelling for Diabetes*, Industrial and Applied Mathematics,
https://doi.org/10.1007/978-981-96-1925-2_3

learning approaches can uncover hidden patterns and relationships that might be overlooked by conventional models. Our findings emphasise the need for further research and development in this area, aiming to replace traditional mathematical models with more sophisticated, data-driven approaches. By addressing these gaps, we contribute to the ongoing efforts to improve diabetes management and treatment outcomes, leading to more accurate and individualised treatment interventions.

1 Introduction

Understanding Glucose Homeostasis

Before conducting a comprehensive literature review of data-driven approaches for modelling the glucose homeostasis system and its pathways to diabetes, it is essential to thoroughly understand the biological processes involved in glucose homeostasis. These processes are the foundation upon which any accurate and effective system model is built. Therefore, we will start with an in-depth review of the critical biological mechanisms that regulate glucose levels in the body. By doing so, we can ensure a solid grounding in the relevant biological material, enabling us to better understand the complexities and challenges of accurately modelling glucose homeostasis and diabetes.

Diabetes is primarily classified into two major types: Type 1, which requires insulin for management, and Type 2, which is typically managed without insulin. However, the International Diabetes Federation (IDF) reports that, in addition to Type 1, Type 2, and gestational diabetes, which are the most prevalent forms, approximately 1.5–2% of individuals are affected by other less common forms of diabetes (Table 1) [53]. Shared symptoms across both Type 1 and Type 2 diabetes include frequent thirst, excessive urination, unexplained weight loss, increased appetite, blurred vision, and extreme fatigue [31]. When left untreated, diabetes can result in serious health complications such as cardiovascular disease, kidney failure, nerve damage, diabetic comas, and potentially death [27]. It should be noted that the rare diabetes types mentioned in Table 1 fall outside the scope of this paper, and readers can refer to [53] for more information on these rarer forms.

Insulin and Glucagon Regulation

In individuals without diabetes, the body regulates glucose levels through the release of insulin from β-cells in response to elevated blood sugar levels following meals, while physical activity typically leads to a decrease in blood glucose. This interplay between insulin and glucagon is essential for maintaining balance. When blood sugar levels increase (so-called hyperglycemia), it prompts the β-cells to release insulin, facilitating glucose uptake and further stimulating insulin secretion. Conversely, when blood glucose levels fall (hypoglycemia), insulin secretion is switched off [31]. On the other hand, when blood sugar levels decrease, which is known as "hypoglycemia", the secretion of insulin is halted. At the same time, α-cells release

Table 1 Rare forms of diabetes [53]

Rare forms of diabetes	Definition and characteristics
MODY (Maturity-Onset Diabetes of the Young)	A rare genetic form of diabetes which develops as a result of gene mutations before age 25 years
LADA (Latent Auto-immune Diabetes in Adults)	A form of Type 1 diabetes affecting adults and progressing more slowly than Type 1 diabetes
Secondary diabetes	It develops after acquiring other diseases, conditions or medications that destruct beta-cells and/or induces insulin resistance
Neonatal diabetes	This rare form of diabetes resembles Type 1 diabetes in children older than six months but develops during the first six months of life
Diabetes insipidus	This condition arises when the kidneys are unable to retain water, leading to excessive thirst and frequent urination, though it does not impact blood glucose levels
Alström Syndrome	A rare genetic disorder, currently incurable, that can lead to the development of Type 2 diabetes
Steroid-induced diabetes	This rare form is more prevalent in individuals with a higher risk of Type 2 diabetes and can also develop in those taking steroids

glucagon, a hormone that signals the liver to release glucose into the bloodstream. This mechanism is crucial for maintaining blood glucose homeostasis, with β-cells being essential to this process and in the progression of diabetes. These specialised cells, found in the pancreatic islets of Langerhans, are responsible for the production, storage, and secretion of insulin.

Type 1 Diabetes

Type 1 diabetes is an autoimmune disease, characterised by the immune system's attack on β-cells [111], preventing them from secreting insulin [27]. The lack of β-cells prevents the regulation of blood glucose levels [31]. Prolonged high blood glucose levels over weeks and months due to insufficient insulin can result in various complications. For individuals with Type 1 diabetes, daily insulin injections are necessary for management [111].

Type 2 Diabetes

Type 2 diabetes is a polygenic disease [31]. Overweight individuals are more likely to develop diabetes, with 80% of Type 2 diabetes patients being overweight [27]. This is due to insufficient insulin secretion from β-cells and insulin resistance, where the body produces insulin but cannot effectively use it [2, 89]. The body compensates by producing more insulin [111], but when insulin resistance occurs, hyperglycemia develops [31]. Persistent hyperglycemia can lead to β-cell burnout [111]. This condition is not solely caused by a lack of insulin; impaired glucagon secretion from α-cells also plays a significant role [2, 3, 89]. Many studies suggest these issues have a strong genetic component, exacerbated by factors like obesity, lack of physical activity, and overeating [2, 89]. Current treatments for Type 2 diabetes include a balanced diet and exercise, oral medications, and insulin injections if needed [31].

Current Modelling Approaches

There are many drugs available to manage diabetes symptoms, but no reliable method exists to replace β-cells and cure the disease [94]. To prevent or cure diabetes, a validated mathematical model is needed to understand the dynamics of the blood glucose regulatory system.

In this review, we classify diabetes models into two main categories: clinical and non-clinical, according to their intended use. Our emphasis is on minimal models and minimal control networks, synthesising existing research to uncover researh gaps and identify their drawbacks within the field. This is particularly important in regions like the Middle East, where the prevalence of diabetes is increasing at an alarming rate. The Gulf Cooperation Council (GCC), which includes the United Arab Emirates (UAE), has one of the highest prevalences of diabetes [1, 8, 11]. The UAE had a prevalence of 16.4% reported in 2021 in comparison with 9.8% globally [73]. This alarming statistic highlights the urgent need for effective modelling and intervention strategies in this region.

Research Methodology

To ensure a thorough and comprehensive review of the literature on diabetes modelling, we employed a meticulous search strategy across multiple databases. We began by searching PubMed and Google Scholar, two of the most extensive and widely respected repositories of scientific literature. Our search was guided by carefully selected keywords related to diabetes modelling, including terms such as "glucose homeostasis," "diabetes pathways," "minimal models," "minimal control networks," "data-driven models," and "machine learning in diabetes." To refine our search and capture the breadth of relevant studies, we employed Boolean operators (AND, OR, NOT) to combine keywords and phrases, ensuring comprehensive coverage of the topic. Additionally, we utilised advanced search filters to limit our results to peer-reviewed articles, reviews, and meta-analyses published within the last four decades, thereby focusing on the most current and impactful research.

We also reviewed references from key papers to identify additional relevant studies that may have been missed in the initial search. This backward snowballing technique allowed us to trace the development and evolution of modelling approaches over time. Furthermore, we examined previous consolidated reviews on diabetes modelling to identify recurring themes, gaps in the literature, and emerging trends. This comprehensive approach enabled us to map the landscape of diabetes modelling thoroughly and systematically, ensuring that our review is both exhaustive and up-to-date. Our rigorous search strategy underscores our commitment to providing a detailed and insightful overview of the state-of-the-art in diabetes modelling, highlighting the strengths and limitations of existing approaches and paving the way for future research in this critical area.

Structure of the Review

In this chapter, we present a comprehensive review of mathematical and data-driven models in the context of glucose homeostasis and diabetes. We begin by exploring clinical models, including the widely used minimal model, in Sect. 2. In Sect. 3, we examine non-clinical models, focusing on the complex dynamics of minimal control networks and their role in understanding glucagon dynamics. Section 4 addresses the emerging field of machine learning models, with a particular emphasis on their application to the development of artificial pancreas systems and the most effective data-driven approaches for this purpose and modelling glucose homeostasis and its pathways to diabetes. Finally, we offer concluding remarks in Sect. 5, summarising key findings and highlighting avenues for future research in the field of diabetes modelling.

2 Clinical Models

A clinical model is designed to incorporate the stages of a disease and make predictions about its progression. These models play a vital role in assessing diagnostic tests, creating glucose control systems, comprehending disease progression, and forecasting the potential risk of complication [2, 89]. For instance, Glucose Tolerance Tests (GTT) are used to assess diabetic and pre-diabetic conditions in individuals, offering assessments of insulin sensitivity, glucose effectiveness, insulin secretion, and β-cell functionality [2, 89]. As a result, initial models concentrated on diagnostic evaluations to create insulin-glucose system models, including those outlined in Bolie [24], Gao [54], Topp [130], Jain [75], and Palumbo [102].

2.1 Minimal Models

A prevalent example of a clinical model is the minimal model, originally designed to analyse clinical data effectively. The basis for many minimal models was suggested in [24], who introduced a straightforward linear model characterised by a set of ordinary differential equations (ODEs) [27]. This foundational work paved the way for the official minimal model later established by [129]. This progression of mathematical models significantly advanced the understanding of insulin-glucose interactions, marking a pivotal moment in modelling by creating a physiological framework to describe glucose disappearance following the Intravenous Glucose Tolerance Test (IVGTT) [2, 89].

The IVGTT bypasses the digestive system, allowing for a controlled study of insulin secretion and glucose uptake dynamics. Typically, after the glucose injection, blood samples are taken at various intervals to measure glucose and insulin levels.

This test is particularly useful in research settings for detailed kinetic studies of glucose and insulin dynamics because it eliminates variables related to gastrointestinal absorption [120].

In contrast, the oral GTT requires the patient to consume a glucose-rich beverage orally. Blood samples are then collected at specific intervals over the next 2–3 h to measure blood glucose levels. The GTT mimics the natural process of glucose entering the bloodstream through digestion and is commonly used in clinical settings to diagnose diabetes and other disorders of glucose metabolism.

Both these tests provide valuable insights into the body's glucose regulation mechanisms but differ in their approach. The IVGTT allows for more precise control and measurement of insulin response to glucose directly introduced into the bloodstream, while the GTT provides a more comprehensive picture of how the body handles glucose in a typical dietary intake scenario. By employing these tests, researchers and clinicians can gain a deeper understanding of glucose homeostasis and the factors that contribute to diabetes and its management. Both these models [24, 89, 129] are pivotal and revolutionary in modelling the glucose homeostasis system [3].

The minimal model is fundamentally structured around two essential components: the glucose minimal model, which focuses on the uptake of glucose, and the insulin minimal model, which pertains to insulin release [129]. Its straightforward design and limited number of parameters allow for efficient evaluation, making it a fundamental part for a wide array of research studies [2, 4, 5, 27, 75, 89, 102, 130]. Introduced by [129], this model has seen numerous adaptations aimed at rectifying its initial deficiencies [2, 27, 89]. Among its notable shortcomings are the tendency to overestimate glucose effectiveness—the ability of glucose to promote its own uptake and reduce internal production when insulin levels are low—and to underestimate insulin sensitivity, which refers to insulin's role in helping glucose enter cells and limiting its production [2, 5, 89].

The minimal model for diabetes consists of two key components: the glucose minimal model, which addresses glucose uptake, and the insulin minimal model, which focuses on insulin release [129]. Its straightforward structure and essential parameters make it an accessible and foundational framework for various studies [2, 4, 5, 27, 75, 89, 102, 130]. Since its inception by Bergman and his team [129], the model has undergone numerous adaptations to address its initial shortcomings [2, 27, 89]. Notably, it tends to overestimate glucose effectiveness—defined as glucose's ability to enhance uptake and inhibit endogenous production at low insulin levels—while underestimating insulin sensitivity, which is crucial for promoting glucose uptake and limiting its production [2, 5, 89].

Extensions to this model have underscored its transformative impact on diabetes research and modelling. For instance, additional compartments were integrated to represent glucose kinetics during intravenous glucose tolerance tests (IVGTT), facilitating more accurate estimates of insulin sensitivity and glucose effectiveness. However, the costs and technical requirements associated with IVGTT limit its practicality for widespread use in the general population [2, 89]. Further developments included the incorporation of data from healthy subjects, leading to the emergence of a new variant of the minimal model developed by Topp and colleagues, which is another significant model relevant to our research [75, 102, 130].

A deep understanding of the dynamics underlying diabetes is essential for effective prevention and treatment strategies. Research indicates that abnormalities in β-cells play a critical role in both types of diabetes, emphasizing their importance in comprehending glucose regulatory system dynamics and the pathways to the disease [52, 60, 127]. This insight prompted the inclusion of β-cell dynamics in Topp and colleagues' model, an enhancement of the original minimal model established by Bergman [21, 75, 102, 130]. Their model is represented by a system of three non-linear ordinary differential equations (ODEs), each depicting a key element of the glucose homeostasis mechanism: insulin, glucose, and β-cell dynamics.

Historically, minimal model-based analyses concentrated on either glucose or insulin dynamics as standalone variables [75, 102, 130]. The innovative structure of Topp's model allows for a comprehensive examination of the glucose regulatory system's normal behavior and its deviations leading to diabetes [75, 102, 130]. Specifically, it assesses the β-cell response and the impacts of reduced β-cell mass under various physiological conditions [2, 3, 89]. This framework also enables the identification of factors contributing to feedback loop deficiencies by providing a platform for parameter manipulation and result analysis [75, 102, 130].

Blood glucose homeostasis is maintained via feedback mechanisms that rely on two negative feedback loops: a short-term and a long-term loop [75, 102, 130]. The short-term loop activates in response to mild hyperglycemia, prompting β-cells to release insulin, which subsequently lowers blood glucose by diminishing production and enhancing uptake [20, 54].

The long-term feedback loop, conversely, is initiated during chronic hyperglycemia, resulting in an increase in β-cell mass due to altered replication and death rates of these cells [26, 45, 68, 72, 125, 141]. The model illustrates how these two feedback systems regulate blood glucose levels. In Type 2 diabetes, defects in both short-term and long-term feedback mechanisms, along with chronic negative feedback loops, are common [75, 102, 130].

It is also noteworthy that such defects can be present in non-diabetic individuals, particularly in those with obesity or during periods of aging, puberty, and pregnancy [75, 102, 130]. The model's behaviors align well with established biological processes, forming a strong foundation for further investigation. A critical observation was that glucose and insulin dynamics occurred at a faster rate than β-cell dynamics, prompting adjustments to the model's time scale from minutes to days for improved alignment with observed trends.

Further analyses indicated that mild hyperglycemia can lead to an increase in β-cell mass, establishing a negative feedback loop. However, severe hyperglycemia results in reduced β-cell mass and initiates a positive feedback loop. The model identified three pathways leading to diabetes in the context of prolonged hyperglycemia [75, 102, 130].

The model's limitations include the exclusion of several known physiological factors, such as the effects of hyperglycemia on neogenesis, insulin sensitivity, and β-cell heterogeneity. Additionally, it does not consider the influences of insulin and incretin hormones on β-cell mass dynamics, nor the potential adjustments in insulin clearance rates in response to insulin resistance and elevated free fatty acids. These omissions may be attributed to the insufficient experimental quantification of these biological processes, leading to reliance on estimated parameter values for modelling [3].

Primary analyses and simulations suggested that while these physiological responses impact the quantitative aspects of the model, they do not significantly alter its qualitative behavior [75, 102, 130]. Overall, the model lays a theoretical foundation for the indirect measurement of β-cell mass in vivo, eliminating the need for invasive procedures like pancreatic removal [75, 102, 130]. It also facilitates longitudinal investigations into β-cell mass dynamics without complex interventions, enhancing our understanding and quantification of these dynamics in individuals [75, 102, 130].

Type 2 diabetes is characterised by elevated blood glucose levels [66], leading to chronic insulin resistance and significant reductions in β-cell mass or glucose toxicity [66]. Individuals with Type 2 diabetes typically exhibit approximately 50% less β-cell mass compared to their non-diabetic counterparts [138]. This loss in β-cell mass further contributes to diminished insulin secretion, exacerbated by insulin resistance [130]. Building on Topp's work, Hernandez and colleagues [66] recognised the critical role of accurate modelling equations and introduced insulin receptor dynamics as a novel parameter. This extension accounts for the chronic down-regulation of insulin receptors on muscle cell surfaces, providing a more comprehensive understanding of receptor behavior and enhancing existing modelling frameworks [130].

It is crucial to recognise that insulin resistance can manifest as defects in either insulin secretion, insulin action, or both [66]. The decision to incorporate dynamics related to insulin receptors was informed by several key findings [16, 30, 51, 71, 77, 90, 91, 95, 121]. Firstly, as plasma insulin concentration increases, the rate of insulin clearance plateaus, which is closely associated with insulin receptors found on the surface of muscle cells. These receptors bind to insulin molecules, enabling the uptake and transport of glucose [66]. Once this process is complete, the receptors are recycled back to the cell membrane surface [66]. Furthermore, the inclusion of the insulin receptor parameter is supported by evidence of insulin resistance in both skeletal muscle and adipose tissue. Since skeletal muscle is responsible for approximately 75% of total body glucose disposal, insulin-resistant individuals experience reduced muscle glucose uptake, contributing significantly to the overall decline in whole-body glucose disposal [66].

Before the incorporation of insulin receptors, the model presented by [75, 102, 130] delineated two stable physiological points alongside a saddle point. However, one of these points was quantitatively untenable, with the β-cell value reaching an exceptionally high and biologically implausible level. Subsequently, adjustments were made by [66] to certain parameters to achieve a more realistic physiological equilibrium. In a manner comparable to the model outlined by [75, 102, 130], the findings of [66] identified three equilibrium points when assessing average health metrics, although these points were deemed to be more physiologically realistic. While the results contribute significantly to the understanding established by [75, 102, 130], there is still an opportunity for further enhancement, such as setting a maximum limit for β-cell mass. This modification is necessary since simulations conducted at various levels indicate that the β-cell masses produced are not always plausible. For example, a computed β-cell mass of 2179.38 mg is nearly impossible within physiological norms, which typically fall between 600–1000 mg [27]. Such an aberration would signify a diabetic condition prevailing due to insufficient insulin production to counter the elevation in blood glucose [66]. Furthermore, other enhancements could be explored, including the incorporation of additional parameters such as α-cells and δ-cells, to yield even more accurate equilibrium points and delineate disease pathways.

A common approach to preventing or managing Type 2 diabetes involves increasing physical activity and improving diet, highlighting the potential link between obesity and the onset of diabetes. This connection likely stems from the relationship between obesity and insulin resistance, making obesity a significant risk factor for the development of diabetes [27]. Building on this significant discovery, Boutayeb et al. [27] advanced the model developed by [130], by incorporating genetic factors influencing β-cell dynamics. Their work highlights the crucial role of genetics in the interplay between obesity and diabetes [27, 28]. This refined system is characterised by three differential equations (DEs) and enhances earlier models by introducing a genetic parameter, ϵ, which specifically quantifies the genetic susceptibility to diabetes. The DEs governing the model are as follows:

$$\frac{dG}{dt} = a - bG(t) - \frac{cI(t)G(t)}{(\alpha G(t) + 1)} \tag{1}$$

$$\frac{dI}{dt} = \frac{d\beta(t)G(t)^2}{(e + G(t)^2)} - fI(t) \tag{2}$$

$$\frac{d\beta}{dt} = (1 - \epsilon)r\beta(t)\left(1 - \frac{\beta(t)}{K}\right) + \epsilon(-g + hG(t) - iG(t)^2)\beta(t), \tag{3}$$

where $\epsilon = 1$ represents a predisposition to diabetes, and $\epsilon = 0$ signifies a lack of genetic predisposition. This parameter, previously overlooked, plays a crucial role in comprehending the dynamics of the glucose regulatory system [27].

The results of this model yield similar findings to those by [75, 102, 130] when $\epsilon = 1$, using parameters corresponding to healthy individuals. However, in the absence of genetic predisposition ($\epsilon = 0$), the model identifies only two pathways to the disease. These results represent an improvement over those by [75, 102, 130] as they align more closely with actual clinical observations, underscoring the significance of genetic predisposition in influencing disease dynamics. The findings also underscore the importance of incorporating the predisposition effect parameter to fully capture the various pathways leading to diabetes [27].

It is crucial to recognise that despite the extensive research on minimal models, our comprehension of the dynamics of the glucose regulatory system and the disease is still incomplete. Numerous parameters and factors have yet to be explored within current models, which could potentially yield new insights and enhance our comprehension of the disease. This ongoing exploration is why many researchers continue to focus on refining and expanding minimal models.

3 Non-clinical Models

Non-clinical models play an essential role in elucidating biological systems prior to their implementation in clinical environments. These models tend to be more intricate, offering comprehensive insights that are vital for clinical applications. A prominent example of a non-clinical model is the minimal control network, which is instrumental in deciphering the dynamics of glucagon.

Glucose Homeostasis and Glucagon Dynamics

Glucose homeostasis is maintained through the coordinated actions of several hormones, primarily insulin and glucagon. While insulin lowers blood glucose levels, glucagon raises them, ensuring a balance that is crucial for normal bodily functions. Understanding the dynamics of glucagon is essential for maintaining this balance and preventing the extremes of hyperglycemia and hypoglycemia, both of which are characteristic of diabetes. The minimal control network model helps in comprehending these dynamics, offering insights into how glucagon affects glucose regulation.

3.1 Hormones Affecting Glucose Homeostasis

Growth hormone (GH), an essential hormone involved in glucose counter-regulation, is released by somatotropic cells within the pituitary gland [67, 70, 76, 88]. When hypoglycemia occurs, GH levels rise [67, 70, 88], stimulating glucose production through gluconeogenesis and glycogenolysis while decreasing glucose uptake in fat tissues [76, 100]. GH interacts with insulin by promoting protein synthesis and enhancing fat breakdown to supply energy for tissue growth, which leads to an increase in free fatty acids (FFA) released from adipose tissues [76]. This process can result in insulin resistance, as GH inhibits insulin receptor activity, a significant challenge in managing diabetes [67, 70, 88]. Additionally, GH stimulates the production of insulin-like growth factor (IGF-I), whose receptors closely resemble those of insulin. IGF-I and insulin can bind to each other's receptors, potentially disrupting glucose homeostasis [67, 70, 76, 88, 100]. Elevated FFA levels can further contribute to β-cell toxicity, leading to an increased rate of β-cell death [76, 100]. The role of GH within the glucose regulatory system is depicted in Fig. 1.

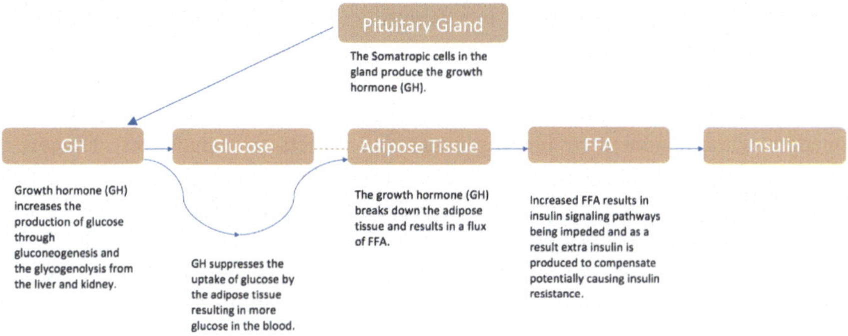

Fig. 1 A schematic flowchart illustrating the role of GH in maintaining glucose homeostasis (adapted from [3])

3.2 *Minimal Control Networks*

The Role of α-Cells and Glucagon

Maintaining blood glucose homeostasis involves glucose counter-regulation, in which α-cells play a critical role. These α-cells, which secrete glucagon, are essential in this process; without their regulation, diabetes would be more prevalent [2, 89]. Despite their importance, research on the involvement of α-cells in diabetes remains limited. The main function of α-cells is to regulate blood glucose levels and prevent chronic hypoglycemia by working in conjunction with β-cells. While glucagon, the hormone released by α-cells, counters the effects of insulin from β-cells, studies on glucagon's role are notably underrepresented in comparison to the focus on insulin and β-cell function [2, 89].

Development and Components of Minimal Control Network Models

Acknowledging the significance of α-cells and glucagon in the regulation of glucose, Farhy et al. introduced a minimal control network model that incorporates α-cells as a new parameter [50]. This model underscores the importance of delayed feedback from α-cells in the counter-regulation of glucagon and consists entirely of clinically measurable components. The previous complex network has been simplified into a more accessible model that elucidates glucose counter-regulation in diabetic patients. The interactions between β-cells, α-cells, δ-cells, and blood glucose are represented through a system of non-linear ordinary differential equations (ODEs).

Such models aim to clarify the role of delayed feedback from α-cells in the regulation of glucagon using measurable components [2, 89]. Previously, glucose counter-regulation was described using a complex network, but this has been replaced by a minimal control model that addresses issues encountered in diabetic patients [50]. The model represents a pancreatic network that integrates interactions between β-cells, α-cells, δ-cells, and blood glucose [50]. Specifically, the model highlights the inhibition of α-cells by β-cells, the inhibition of α-cells by δ-cells, the stimulation of δ-cells by α-cells, and the stimulation of β- and δ-cells by glucose, alongside the inhibition of α-cells by glucose. It is noteworthy that δ-cells produce somatostatin, which was included implicitly in the original complex model [109] and continues to be included in this simplified version [50]. The model is formulated using a system of non-linear ODEs that approximate the rate of change in glucagon and somatostatin concentrations, regulated by blood glucose and switch-off signals [50]. However, as somatostatin cannot be measured accurately in humans, parameters related to somatostatin are estimated, limiting the model's use in clinical studies. This limitation hinders validation of the glucose counter-regulation control axis in humans [50]. To address this issue, Farhy and colleagues proposed a less complex model in which somatostatin is included implicitly, simplifying the network further.

Advantages, Limitations, and Improvements

Both models demonstrated that reduced insulin secretion results in delays and decreases in glucose counter-regulation. This condition results in delays and decreases in glucose counter-regulation. The findings also confirmed that this new model is adequate to replace the older, more complex network, as it produces comparable results [50]. A significant advantage of the new model is that all components explicitly involved can be measured clinically. As a result, abnormalities in glucose counter-regulation can be studied by applying the model to the human pancreas [50].

However, limitations remain, such as challenges in evaluating system behavior, as the specific contributions of each interaction are not clearly defined. Additionally, the model describes sequential relationships, but the individual approach with feedback is neither optional nor fully representative of the system's dynamics. Improvements are necessary, particularly in the form of time-varying interactions of multiple components, including the ability to simulate oscillations and rebound responses to switch-off signals. Previous studies [3, 22, 23, 63, 107, 108, 110, 122, 123] have shown that insulin secretion by β-cells is oscillatory, which could indicate that the same might hold true for α-cells [2, 89]. This led to the development of a nonlinear ODE model demonstrating ultradian oscillations in insulin secretion, aiming to provide insights into its impact on glucose utilization and production [38, 124].

Another strength of the minimal control network model is its ability to simulate the dynamics of the entire system rather than studying components in isolation, which would provide little information on network behavior. By integrating key components, this model is essential for understanding the interplay between insulin, glucose, and α-cells within the system [50]. It also defines the system's components more clearly and suggests an auto-feedback mechanism for α-cells, which play a crucial role in glucose counter-regulation.

The modelling of somatostatin raises the question of why growth hormone was excluded from existing diabetes models. Somatostatin, a growth-inhibiting hormone, is incorporated into the model by Farhy et al. [50]. Growth hormone, secreted by somatotropic cells in the pituitary gland, was not included despite its potential importance, as changes in somatostatin affect growth hormone concentrations. Although the model indirectly accounts for growth hormone through somatostatin [50], explicit inclusion of growth hormone would provide a more comprehensive understanding of the system.

Whole Body Dynamics and Multi-scale Modelling

The whole-body dynamics feature of the minimal control network has inspired the development of other models that adopt multi-scale, hierarchical, and compartmental modelling approaches [2, 89]. These emerging methodologies are increasingly significant, as they hold potential for advancing control and treatment strategies for diabetes [2, 89]. modelling studies that explore the mechanisms leading to diabetes complications, while identifying intervention points for self-management, could prove life-changing for many [2, 89]. Accurate models are essential for tracing the pathways that lead to diabetes, offering ways to prevent or manage the disease effectively. In this context, a model that can represent the system accurately and provide improved predictions is crucial.

3.3 *Applications of Minimal Control Networks*

Artificial Pancreas and Automated Glucose Controllers

Mathematical models capable of accurately describing the glucose homeostasis system are essential for devices like artificial pancreases, which offer improved diabetes treatments. An artificial pancreas consists of a blood glucose sensor or continuous glucose monitor (CGM), an insulin pump, and a control algorithm to administer the correct insulin dose [94]. Currently, subcutaneous insulin injections or insulin pumps are the most common methods for managing diabetes. For mathematical models to be integrated into artificial pancreas systems, evaluating β-cell function and insulin secretion is crucial to emulate natural processes. A maximal model developed by Grodsky [58, 59] provides a detailed view of the pancreatic insulin response to glucose stimuli [2, 89]. While maximal models focus on cellular processes, making them suitable for computer simulations, they are not widely applicable in clinical settings due to the invasive and complex nature of cellular quantification.

Control Algorithms and Model-Free Methods

Another important application of mathematical models is in automated glucose controllers. These systems rely on models to link input and output data via mathematical equations to predict plasma glucose levels automatically [2, 89]. Several approaches, including algorithms, stochastic methods, and artificial neural networks, have been utilised to develop validated models [2, 89]. Hovorka and colleagues [69] created a nonlinear predictive control model that maintains normal glucose levels during fasting in individuals with Type 1 diabetes, using previous glucose kinetics models alongside Bayesian parameter estimation techniques. Currently, insulin-glucose dynamic models for Type 1 diabetes are available, which can be used to design feedback control algorithms for artificial pancreases and validated with clinical data [2, 89]. Additionally, model-free methods, such as intelligent proportional controllers, can be employed for artificial pancreas control [94]. Proportional-integral-derivative (PID) control and model-based predictive control are commonly used, but PID tuning is often difficult and sensitive to disturbances like meals. A mathematical model, along with optimization procedures, is required for successful tuning [94].

Practical Applications and Future Directions

There is an urgent need for a wearable, automated, bihormonal bionic pancreas that can improve mean glycemic levels and reduce hypoglycaemic episodes. Model-free control methods have been applied successfully, bringing us closer to realizing an artificial pancreas [94]. Control algorithms offer several advantages: tuning intelligent controllers is straightforward, no complex modelling is required, and they can be easily implemented on inexpensive programmable devices [94]. Unlike model-based algorithms, which require detailed physiological studies, model-free algorithms rely on readily available patient data [94]. Additionally, these systems are not dependent on individual patient characteristics, meaning they do not require parameter adjustments when applied to different patients [94].

Although these methods have shown promising results in silico, further testing is needed. Notably, they have yet to be validated on an FDA-approved simulator, such as the UVa simulator, and have only been tested on small cohorts of patients [94]. The FDA-approved simulator could serve as a substitute for animal trials, but its application to unstable diabetic individuals remains a future goal.

Subcutaneous insulin injections and insulin pumps remain the most widely used methods for diabetes management today. However, a key challenge with these methods is the variability in insulin absorption kinetics during external administration, which can complicate effective glucose regulation [3]. To better understand and predict this variability, researchers have developed models that explore the relationship between different insulin polymers and their absorption dynamics. These models have been pivotal in advancing quantitative approaches to insulin absorption. For instance, a comprehensive model was introduced that effectively illustrates how external insulin infusion, whether through injections or pumps, can replicate the body's natural pancreatic insulin secretion for both Type 1 and Type 2 diabetes [2, 3, 89, 135]. Such models have become essential tools in computational simulators, offering valuable insights into the interplay between external insulin administration, meal intake, and subsequent glucose response, thus aiding in the development of more precise diabetes management strategies.

Diabetes can lead to a range of complications, such as impaired wound healing, retinopathy, foot ulcers, and cardiovascular disease. While obesity is a major risk factor for Type 2 diabetes rather than a direct complication, several models have been developed to better predict and understand these related conditions [2, 89]. In addition, population-based risk models have been designed to describe complications like diabetic retinopathy, nephropathy, and coronary heart disease. However, the complexity of these models has limited their widespread application. Typically, these models assume that insulin-filled granules are stored in β-cells and released in phases in response to specific glucose thresholds, aligning with the first and second phases of insulin secretion. Over time, these models have been refined to integrate newer findings on β-cell functionality [64, 65, 112].

Many variations of diabetes models have been developed over the years. The minimal model, introduced by Bergman and colleagues [129], has undergone numerous adaptations to address its initial limitations. This minimal model, alongside the minimal control network, remains vital for accurately representing glucose counter-regulation dynamics. Unlike previous models that assumed a constant disease state, the progressive nature of diabetes is now recognised. Greater focus on understanding glucose-insulin dynamics is essential for developing new anti-diabetic treatments capable of altering disease progression [2, 89, 96].

In summary, the literature reveals certain gaps in diabetes research, with opportunities for further development of current models. For instance, limited attention has been given to α-cells and glucagon in comparison to β-cells, and more research is needed on the effects of growth hormone. Additionally, models addressing obesity as a factor in diabetes complications are scarce, often due to a lack of experimental data [2, 89]. Incorporating the oscillatory nature of β-cells could also enhance current models [104]. Finally, incorporating delay equations may offer valuable flexibility to the next generation of diabetes models.

4 Machine Learning Models

An artificial pancreas is in high demand for managing Type 1 diabetes. To create such a device, a model capable of accurately predicting and responding to changes in blood glucose levels influenced by daily factors like physical activity, food intake, and other variables is essential. Gaining insights into these effects can help refine treatment approaches, ultimately reducing the risk of diabetes-related complications. The development of predictive control devices, such as artificial pancreases, continues to evolve, with the aim of automatically regulating blood glucose fluctuations in individuals with diabetes [10, 140]. However, creating these devices requires a model that can precisely describe the interactions between glucose, insulin, and food intake [10, 140]. Therefore, realistic mathematical models of the glucose regulatory system are crucial for improving the diagnostic, management, and predictive capabilities of these devices. Developing such models remains a significant challenge, largely due to limited access to high-quality data, the high cost of data collection equipment, and the complexity of the underlying system dynamics [10, 140]. These challenges often result in models with many parameters, which, while accurate, become too complex for practical implementation. Data-driven methods offer a potential solution, which will be discussed later in this review.

The primary goal of an artificial pancreas is to automate the regulation of blood glucose levels, providing a reliable treatment option for individuals with Type 1 diabetes (and some Type 2 patients undergoing intensive insulin therapy) [33, 46]. By simplifying treatment, an artificial pancreas can greatly improve the daily life of a person with Type 1 diabetes, offering a more accurate and efficient alternative. Traditionally, patients manage their condition with subcutaneous insulin injections, but this requires constant vigilance from the patient, leading to potential dosing errors. Another common approach involves the use of insulin pumps combined with continuous glucose monitors; however, even these systems are not without issues. For instance, bolus doses (high doses of insulin taken with meals) are determined by the patient, leaving room for incorrect dosing [81, 103].

The need for more effective blood glucose control in individuals with Type 1 diabetes led to the development of closed-loop control devices, also referred to as artificial pancreases. These devices are designed to regulate blood glucose levels and maintain homeostasis. A fully automated artificial pancreas would ease the management of the disease, reduce reliance on patient input, and improve the quality of life for people with Type 1 diabetes [131].

4.1 Artificial Pancreas

An artificial pancreas system generally comprises three essential elements: a blood glucose sensor or continuous glucose monitoring (CGM) device, an insulin pump, and a control algorithm responsible for managing insulin delivery [39, 94]. These devices are worn externally by the patient [131]. Although most current artificial pancreas systems are only semi-automated, the goal is to develop fully automated, or closed-loop, systems. The most commonly used devices for managing Type 1 diabetes include CGMs and insulin pumps that administer continuous subcutaneous insulin infusion (CSII) [131]. There are two main types of semi-automated artificial pancreas devices: low-glucose suspend and predictive glucose suspend systems. Research suggests that artificial pancreas systems offer advantages over standard CSII therapy, such as reduced hypoglycemia frequency, better nocturnal control, and more time spent within target glucose levels [131]. Currently, the most widely accepted method for insulin therapy is CSII.

The first commercial artificial pancreas, the Biostator, was introduced in 1974 [105], followed by the Nikkiso STG-22, still used in Japan today, with the updated STG-55 model [9, 39, 116]. However, these early devices relied on intravenous glucose sensing and insulin delivery, which is not practical for outpatient settings [39]. This challenge led to the exploration of subcutaneous methods. Early artificial pancreas systems faced issues in outpatient use, but later developments improved the feasibility of subcutaneous insulin delivery [83, 106, 126, 131]. Subcutaneous insulin delivery was demonstrated to be effective in continuous insulin administration [39]. Subsequently, a wearable artificial pancreas prototype was developed [114, 119], which was further refined by researchers [3, 62, 118]. An implantable artificial pancreas was also introduced, utilizing intravenous glucose sensing and intraperitoneal insulin infusion [12, 82], though it was eventually discontinued due to the invasive surgical procedures required [39].

A critical component of artificial pancreas systems is the control algorithm [131]. Early models used proportional-integral-derivative (PID) controllers, but these have limitations, particularly with time delays in subcutaneous glucose sensing and insulin delivery [39]. More recent advancements have introduced model-predictive control

(MPC) algorithms, which address some of these limitations by incorporating models of the individual's metabolic system [39]. Many MPC algorithms are based on the minimal model introduced by Bergman and Cobelli [129], though challenges remain in developing fully automated systems.

Certainly! Here's a restructured and slightly expanded version of the paragraph:

The evolution of closed-loop artificial pancreas systems began as early as the 1960s with the introduction of intravenous (IV) closed-loop control systems. By the 1970s and 1980s, significant advancements were made, including the development of the Biostator, an early artificial pancreas device, as well as the autosyringe, and the first utilization of continuous subcutaneous insulin infusion (CSII) for insulin delivery. In 1979, Bergman and Cobelli introduced the minimal model of glucose kinetics, laying the groundwork for future advancements in diabetes management [19].

The late 1990s marked a breakthrough in subcutaneous glucose sensing with the introduction of the MiniMed continuous glucose monitoring system (CGMS). Building on these advancements, the European Commission funded the ADICOL project in 2000, which demonstrated the feasibility of using model predictive control (MPC) strategies in closed-loop insulin delivery systems. Further support for artificial pancreas research came in 2006 when the Juvenile Diabetes Research Foundation International (JDRF) launched the Artificial Pancreas Project, funding research centers dedicated to developing these systems. A major regulatory milestone was achieved when the FDA approved the Padova Type 1 simulator, developed by the University of Virginia (UVA), as a substitute for animal trials in closed-loop research [3].

In 2009, the National Institutes of Health (NIH) funded additional projects focused on artificial pancreas development, followed by the launch of the AP@Home project by the European Commission in 2010 [3]. Today, artificial pancreas systems are being actively trialed and implemented in patients, offering promising results. However, further clinical trials and research are still necessary to tackle the challenges of cost-effectiveness, accessibility, and widespread adoption [13, 85, 93, 98, 115].

Continuous glucose monitors (CGMs) play a crucial role in providing input data to the control algorithms used in artificial pancreas systems [39]. Early versions of CGMs faced performance issues, particularly during hypoglycemic events [36, 55, 79], but these devices have seen substantial improvements since their initial introduction [39]. Despite their advancements, CGMs still encounter certain challenges that need to be addressed, such as time lag, random noise, and the need for calibration.

Most CGMs measure glucose levels in the interstitial fluid. However, during periods of rapid glucose fluctuations—such as after meals or during hypoglycemic episodes—the interstitial glucose levels may differ from blood glucose levels, requiring additional blood samples for accurate calibration [39]. To mitigate this issue, frequent calibrations are often necessary, typically involving multiple daily blood glucose measurements. Studies have shown that the accuracy of CGMs improves when calibrations are performed during stable glucose periods [34, 39].

The time lag between blood glucose and interstitial glucose is another significant issue, as it affects how quickly the CGM can provide accurate readings. This lag, combined with the time required for sensor processing, directly impacts the monitor's accuracy [14, 39, 74, 80, 117]. Additionally, CGMs are prone to errors caused by random noise, which necessitates the use of filtering and denoising techniques to improve data reliability [35, 37, 56, 117].

Overall, the performance of CGMs depends heavily on regular calibrations through self-monitoring of blood glucose, as well as the application of filtering techniques to reduce noise [39, 43, 57]. These limitations are some of the reasons why current artificial pancreas systems are not yet fully closed-loop or completely automated.

In the early stages of insulin delivery (1972–1990s), the method used was intravenous, and later, it was adapted for implantable devices [39]. However, this approach encountered complications such as blood clotting and issues with catheters, which limited its effectiveness and ultimately led to discontinuation of intravenous insulin delivery [39]. Since the 1990s, the standard method for insulin pump therapy has become continuous subcutaneous insulin infusion (CSII) [3]. Nevertheless, a significant challenge with subcutaneous delivery remains: the delay in insulin absorption, which can cause insulin peaks to occur later than needed, making glucose control more difficult [39]. Additionally, varying absorption rates depending on the delivery site introduce the risk of under-delivery of insulin [39, 43, 57, 97].

Another key aspect of modern insulin pumps is their reliance on user input. Users are required to estimate carbohydrate intake and either approve or adjust the system's recommended bolus dose [33, 46]. Although insulin pumps have improved glucose regulation compared to earlier models, none can yet accurately deliver bolus doses without user intervention [33, 46].

It's also worth noting that most insulin pumps do not automatically adjust the predicted bolus dosage [33, 46]. This highlights the need for further research to assess how these devices affect users from both physiological and psychological perspectives in order to enhance their design [33, 46]. A model developed by [33] used an "Eat, Trust, and Correct" framework alongside a clustering algorithm to analyse user behavior patterns. This model provided valuable insights into how behavioral shifts may impact blood glucose control, helping to refine pump functionality [33, 46].

Artificial pancreas systems are categorised into six generations based on the degree of automation. The first generation includes systems that are not closed-loop, with features like glucose insulin shutoff when levels are dangerously low, and hypo/hyperglycemia minimisers. The second generation encompasses automated basal systems, hybrid closed-loop systems, and fully closed-loop insulin delivery. The third generation includes fully automated, multihormonal closed-loop devices. Currently, artificial pancreas technology has reached stages 5 and 6, where insulin delivery is either fully closed-loop or involves multihormonal control [3, 131].

In their analysis of artificial pancreas systems under development, Trevitt et al. [131] classified these systems into four distinct categories based on key characteristics: level of automation, hormonal regulation, control algorithms, and strategies for glycemic control. They identified a total of 18 closed-loop systems in clinical trials, with six of these being developed by commercial companies, some of which had projected launch timelines [131]. By 2021, several fully automated artificial pancreas systems had been introduced in select countries, though global availability remains limited [17, 25, 84, 86, 92, 136, 139]. For example, the Inreda bihormonal artificial pancreas, while accessible to residents of the Netherlands, has not yet been made available for general consumer purchase [25]. Meanwhile, the Minimed 770 g system has been launched in specific regions and is available for purchase in select countries [92].

The findings in [131] indicated that most of the systems under development—specifically twelve—were classified as second-generation automation. Three systems were labeled as second/third generation, while the remaining three were categorised as third-generation, representing fully automated systems.

Artificial pancreas systems can use either of two primary hormonal approaches: a single-hormone method (insulin alone) or a multihormonal approach [39, 131].

The multihormonal approach aims to more closely replicate the body's natural regulation of blood glucose by incorporating glucagon alongside insulin, although other hormones could be used as alternatives [131]. This approach offers tighter glycemic control. However, there are practical challenges, particularly because glucagon is unstable in solution and must be replaced every eight hours. This creates the need for commercial dual infusion pumps to deliver both hormones effectively [131]. According to the review by [131], the majority of systems—twelve—employed a single-hormone approach, while six used the multihormonal method. The relatively lower adoption of the multihormonal approach may be due to the increased complexity compared to the single-hormone method, but the benefits appear to outweigh the challenges.

The main algorithms used in artificial pancreas system controllers include proportional-integral-derivative (PID), fuzzy logic, model-predictive control (MPC), and bio-inspired algorithms. Most systems currently rely on either PID or MPC algorithms, which have proven effective for artificial pancreas systems and remain relevant for future improvements [131]. While fewer systems utilise fuzzy logic or bio-inspired algorithms, their potential to enhance the performance of artificial pancreas systems suggests that further research and development could lead to their wider adoption [86, 131].

According to the review by [131], 61% of artificial pancreas systems followed a "treat to target" strategy, aiming to maintain blood glucose levels as close as possible to a specific target value. Meanwhile, 22% of the systems used a "treat to range" strategy, which focuses on keeping blood glucose within a personalised range, and 17% utilised a combination of both strategies. The "treat to target" approach has been extensively studied, whereas the "treat to range" strategy has received less attention. Given that the latter is personalised to the individual, it may be worth exploring further in the future, as personalised ranges tailored to each person's unique needs could offer more effective long-term management of diabetes.

Artificial pancreas systems have been tested primarily in inpatient settings and for short durations [18, 84, 86, 131]. Overall, these trials have demonstrated that the systems are safe for short-term use and have shown encouraging results across various clinical and outpatient environments. However, longer-term studies are necessary to fully evaluate the cost and clinical effectiveness of artificial pancreas systems before they can be commercialised and distributed globally [18, 84, 86]. Ultimately, the goal of these systems is to improve the quality of life and management of diabetes for individuals with Type 1 diabetes, so it is essential that their design and functionality meet users' expectations [84, 86, 131].

Determining the correct insulin dosage to maintain optimal blood glucose levels during meals remains one of the main challenges preventing the success of a fully automated closed-loop control system. Trials conducted in 2011 revealed that all artificial pancreas systems exhibited a significant postprandial glucose spike, exceeding normal ranges, indicating a critical issue to be addressed in future models [18, 39]. The choice of algorithm plays a crucial role in managing the delays and inaccuracies associated with glucose sensing and insulin delivery systems [39]. Both excessively slow and overly fast responses can destabilise the system, making it essential to fine-tune the timing of insulin delivery [39].

For a fully automated closed-loop control to succeed, the control algorithm must strike a balance in responsiveness, effectively regulating blood glucose levels throughout both day and night. The initial proportional–integral–derivative (PID) algorithms in closed-loop systems were reactive and thus limited in their capabilities [39]. To accommodate individual differences, the algorithm must also be adaptable, with customizable controllers tailored to each patient. Model predictive control (MPC) is a promising solution, offering the ability to personalise and learn from the patient's data [39]. This is particularly important since patients' meal times are often routine, and such information can enhance treatment strategies [18, 39, 86]. Another alternative proposed by [94] is a simpler, model-free method, which reduces computational complexity and costs. This approach has been tested in silico and has shown long-term accuracy with clinical data, improving post-meal glucose peaks and overall average glucose levels, a critical challenge for Type 1 diabetes management [94].

One major hurdle to achieving fully automated artificial pancreas systems is the lack of real-time insulin sensors. Accurately monitoring plasma insulin concentration is critical for managing glycemic variability, and without it, the effectiveness of these systems is compromised [15, 44, 134]. To address this, Avila et al. [15] proposed using Gaussian-process-Bayesian filtering methods for model-based insulin estimation, which introduces uncertainty quantification into the prediction and filtering process. This differs from traditional parametric filtering techniques by explicitly accounting for uncertainty during estimation[15, 134]. However, challenges remain, such as the time delay between subcutaneous insulin injections and their impact on blood glucose levels—a persistent issue across all approaches [15, 134]. Additionally, the method requires patients to count carbohydrates, adding an inconvenience factor [15, 134].

Another potential solution involves combining reinforcement learning with GP approximations, as proposed by De et al. [44]. This approach is designed to address issues such as sensitivity, stability, calibration, and the time lag between blood glucose and interstitial glucose levels. While it has shown promise in managing individual variability due to differences in lifestyle and metabolic responses, it still faces technical challenges, such as errors related to the infusion pump and wireless signal reliability, both of which must be addressed to improve the robustness of artificial pancreas systems [44]. Additionally, there is a trade-off between robustness and optimal performance—enhancing one often compromises the other [44]. Finally, clinical trials and validations are required before these control algorithms can be widely implemented in artificial pancreas systems [44].

In summary, early closed-loop systems have reduced the need for manual carbohydrate counting by relying on glucose sensor readings. However, this method leads to prolonged postprandial hyperglycemia [48, 137]. While current closed-loop systems are more effective at managing diabetes than insulin pump therapy, they still require users to input meal information, such as carbohydrate count or meal size, which means they are not yet fully automated [47, 113]. A fully closed-loop system using Fiasp plus pramlintide has been developed that does not require complete carbohydrate counting, but it still exhibited post-meal hyperglycemia for the first two hours [132]. Currently approved commercial closed-loop systems [17, 18, 25, 84, 86, 92, 136, 139] are only available in certain countries as they await further long-term outpatient trials before global approval.

Additional advancements are required to create a fully automated closed-loop system that eliminates hyperglycemia episodes and requires no user involvement.

4.2 Data Driven Models

Mathematical models are typically employed to capture the complex dynamics of the glucose regulatory system by formulating systems of differential equations (DEs) that describe real-world processes. However, an alternative approach is to use data-driven methods. A vast amount of data is being collected from diabetes patients, particularly through continuous glucose monitors, but much of this information remains underutilised. As the volume of data continues to grow, data-driven approaches are increasingly necessary. These methods have the potential to uncover new insights and yield valuable results related to diabetes management. Techniques such as data mining allow researchers to analyse large datasets, generating new hypotheses and enabling earlier detection of diseases.

Another data-driven strategy for understanding glucose homeostasis involves converting a physiology-based model of blood glucose dynamics into a Bayesian network, accounting for measurement error and variability in physiological processes [49]. This approach led to the development of the stochastic model by [49]. However, current models using this method face limitations, as much of the data is gathered from healthy individuals who participate in controlled, simplified exercise protocols that do not accurately represent the experiences of active diabetic individuals. As a result, there is uncertainty about how these models perform when applied to free-living data, where individuals are not confined to specific protocols. To address this issue, [49] introduced a Bayesian network to offer a more accurate depiction of the physiological process. One specific limitation is that model performance tends to degrade when tested with free-living data [49]. This may stem from the assumption that post-exercise glucose levels return to basal conditions, an assumption disproven by a Diabetes UK study [133]. Consequently, the model produced suboptimal results, particularly with glycogenolysis and insulin clearance roles becoming disproportionately emphasised. Additionally, the short-term nature of the exercise trials did not accurately reflect the day-to-day activity patterns of individuals.

The study by [49] sought to address the limitations of their previous exercise model by making key adjustments. These included removing the rate of decline in glycogenolysis and insulin clearance during activity and incorporating renal clearance of blood glucose. This led to a physiologically viable output that better handled free-living data [49]. One advantage of Bayesian methods is their ability to incorporate prior information about parameters, allowing models to be updated as new data becomes available. To validate the modified Bayesian network, [49] used Markov Chain Monte Carlo (MCMC) methods for parameter fitting and blood glucose concentration predictions. The model's suitability was tested through simulations where fixed parameter values generated blood glucose profiles, and MCMC estimates were obtained for the known parameters [49]. The results demonstrated good predictive performance in some scenarios, with credible intervals that included true values and no evidence of prediction bias [49]. However, the intervals used were extremely wide, ranging from 3 to 2980 mg/dl, which is not realistic for biological systems. Such a wide range suggests the possibility of either hyperglycemia or hypoglycemia, and a blood glucose concentration as high as 2980 mg/dl is physiologically unfeasible [49]. Despite this significant issue, the model offers a new approach for analyzing diabetes and could play a role in the development of artificial pancreas systems [49].

Data-driven methods have demonstrated their utility, but they often fall short by treating every poor fit as noise, without adequately accounting for stochasticity and uncertainty in the data. This oversight can significantly alter the deterministic dynamics, making such models less representative of the real system. The study by [10, 140] proposed a minimal-order model with a limited number of parameters while still capturing nonlinearity and stochasticity. It is well-known that blood glucose responses to food intake demonstrate significant nonlinear behavior, highlighting the necessity of incorporating nonlinear terms into the modelling equations. This nonlinear stochastic model plays an important role in supporting control strategies for managing blood glucose in clinical environments [3]. The stochastic elements in this model arise from two key sources: measurement errors introduced by the monitoring device and dynamic intrinsic noise generated by factors like physical activity and stress [10, 140].

This model marks a significant improvement over previous approaches, which often attributed poor fits solely to noise without properly accounting for its underlying causes. The model by [10, 140] was developed as four models with two functions, each containing no more than three polynomial terms. This constraint was justified, as using more polynomials could increase the risk of overfitting and lead to system instability without improving parameter estimation. To identify the best-fitting model and estimate both deterministic and stochastic parameters, a Bayesian learning algorithm was applied [10, 140]. The algorithm successfully detected the stochastic nature of glucose dynamics, confirming the appropriateness of the Bayesian approach [10, 140]. Moreover, it effectively distinguishes between measurement and dynamic noise. In addition to being more flexible and robust to model misspecification, this algorithm estimates model parameters in an ideal form, rather than as fixed values [10, 140], addressing the uncertainties around parameter estimates.

Overall, the model by [10, 140] effectively describes the dynamics of the glucose regulatory system in response to food intake. It holds promise for clinical applications, particularly in early diagnosis and evaluating drug effects on glucose response stability. Additionally, it offers deterministic parameters for diabetes patients, aiding in control strategies. The model's compatibility with both phenomenological and physiological models further enhances its utility for parameter interpretation. However, the study had a small sample size, and a larger dataset would be needed to confirm and validate the observed relationships. The Bayesian approach is computationally efficient and allows for sensitivity analysis with fewer model runs compared to traditional Monte Carlo methods [99]. Moreover, this approach can calculate all measures of interest from a single set of runs [99].

4.3 Bayesian Approaches for Analysing Type 1 Diabetes Experimental Data

Mismanagement of diabetes can lead to elevated blood glucose levels, resulting in severe secondary diseases and potentially death [61]. Understanding Type 1 diabetes and monitoring blood glucose concentration at critical time points are crucial for effective treatment and management. As described above, Mathematical models play a pivotal role in elucidating the complex pathways and critical thresholds necessary to maintain glucose homeostasis. These models are fundamental components in the algorithms used in continuous glucose monitoring devices, insulin pumps, and artificial pancreas systems [78, 135]. Accurately selecting the best mathematical model based on the collected experimental data is indispensable for predicting blood glucose levels and determining the appropriate insulin dosage, underscoring the importance of rigorous model comparison and selection in establishing an effective glucose-insulin regulatory system.

Bayesian machine learning methods are essential for selecting the best model out of various mathematical models developed for modelling glucose dynamics and refining treatment strategies in diabetes. There could be different diabetes models, including those with and without β-cells, while incorporating variables like subcutaneous insulin injections. As a result, comparing these models to the experimental data is crucial to select the most suitable one. Al Ali et al. [7] proposed a methodology by benefiting various model selection creteria including Akaike Information Criteria (AIC) and Bayesian Information Criteria (BIC) values to conduct a precise benchmark among the plausible developed mathematical models. Their findings reveal that the model excluding β-cells demonstrated superior performance in severe cases, as evidenced by significant differences in AIC and BIC values. These findings emphasise the critical role of Bayesian machine learning in predicting blood glucose levels and managing diabetes, especially in advancing automated insulin delivery systems [6, 7, 135].

4.4 Gaussian Processes in Glucose Regulation

Insulin sensitivity plays a crucial role in the glucose regulatory system, but it is difficult to model and control because it varies in a circadian pattern [101]. A potential solution to this challenge involves incorporating a time-varying feature, as proposed by [101]. This approach uses a Gaussian process (GP), a data-driven method, to predict future insulin sensitivity based on past data, which is continuously updated [101]. This technique enables more personalised control for individuals, leading to more precise blood glucose regulation. The system combines model predictive controllers with an Unscented Kalman Filter (UKF) and a GP to capture the effects of circadian rhythms, which are then integrated into a closed-loop controller [101]. In contrast, an earlier model by [128] relied solely on model predictive control, making the approach by [101] an improvement.

The GP model predictive controller significantly improves accuracy by minimizing tracking errors and providing more precise blood glucose regulation during fasting, outperforming traditional model predictive control methods. It is especially effective during mealtimes, which are typically the most challenging periods for glucose control, helping to prevent hyperglycemic events [101]. However, the model by [101] does have limitations, as it struggles to maintain blood glucose levels within the "tight" zone during meals or when meals are skipped [84, 101]. The "tight" zone refers to maintaining blood glucose levels within the ideal range typically observed in non-diabetic individuals, a state known as glucose homeostasis [29].

Unlike conventional model predictive control systems, this approach offers flexibility in meal timing and portion sizes, making it a promising candidate for future models [101].

4.5 Enhancing Type 1 Diabetes Models with Sensitivity Analysis Techniques

As discussed above, mathematical modelling plays a pivotal role in understanding insulin dynamics and designing treatments for Type 1 diabetes [6, 87]. Various models, including depot injections, compartmental, and systemic models, have been developed to capture insulin dynamics, typically assuming inverse proportional absorption [87]. However, these models often overlook critical biological factors such as growth hormone dynamics and fail to fully represent the unique pathways of Type 1 diabetes, including scenarios with zero insulin production [5, 6]. To address these limitations, Al Ali et al. [6] propose a simplified Type 1 diabetes model with an insulin bolus injection component, aiming to identify critical parameters for insulin management. This approach emphasises considering the entire glucose-insulin-growth hormone system rather than focusing solely on insulin.

Despite the utility of these models in clarifying disease pathways and dynamic behaviours of Type 1 diabetes, understanding its pathological and manageable equilibrium states at different model thresholds remains challenging. Efficient sensitivity analysis (SA) approaches are therefore indispensable to simplify the model and pinpoint key parameters at significant time points. Al Ali et al. [6] explore several SA methods, including partial rank correlation coefficient (PRCC) and probabilistic SA methods, including Sobol indices, to tackle these challenges. The computational demands of probabilistic SA methods are addressed using a GP emulator algorithm, enhancing the efficiency of evaluating these SA measures.

The comprehensive evaluation conducted by [6], focusing on model parameters during the glucose homeostasis system's return to equilibrium post food uptake, highlights the significant influence of insulin clearance and absorption rates on model response variables. The PRCC method highlights the importance of the subcutaneous bolus injection term, while probabilistic methods, particularly the Sobol' method, complement these findings by revealing complex parameter relationships. Importantly, both PRCC and probabilistic SA methods consistently identify influential parameters across different time points, demonstrating their robustness in characterising the dynamics of Type 1 diabetes models.

Sobol's variance-based indices prove computationally efficient for complex models, emphasising the inadequacy of classical SA approaches in comprehending model dynamics and parameter interactions. Variations in sensitivity analysis outcomes across studies indicate the sensitivity of methodological selection to model complexity and research objectives [7, 32, 40–42]. Comparative findings with previous studies [6] highlight the complementary strengths of PRCC and Sobol methods; PRCC identifies influential parameters, including those defining the threshold, , while Sobol method provides insights into parameter interactions. These insights are crucial for developing robust mathematical algorithms for artificial pancreas systems, capable of managing zero-insulin states common in Type 1 diabetes, thereby enhancing treatment efficacy and patient outcomes.

5 Conclusion

Current mathematical models for diabetes face several challenges that limit their predictive power, and their mathematical properties have not been thoroughly analysed. Key issues with existing models include:

- The omission of critical biological processes, such as the role of growth hormone in glucose regulation.

- Limited mathematical examination of dynamic properties and parameter spaces, which, if explored, could lead to better understanding of diabetes progression and management.
- A lack of integration with machine learning techniques, which are essential for exploring parameter spaces, enhancing model predictions, and improving strategies for managing or treating diabetes.

Future advancements in this field will depend heavily on the use of computer simulations. These simulations can reduce the need for time-consuming and expensive animal testing by providing in silico alternatives, accelerating progress toward clinical trials. This is particularly important for developing a commercially viable, fully automated closed-loop device, such as an artificial pancreas. Additional research is required to enhance the long-term predictive capacity of models for high-risk diabetic patients, especially in outpatient environments. Furthermore, artificial pancreas systems need improvements in wearability, user-friendliness, and their ability to handle unexpected situations. The ultimate goal is to create a digital controller that provides automatic regulation, removing the need for patients or specialists to make manual adjustments. This would streamline disease management, making it faster, more efficient, and easier for individuals with Type 1 diabetes.

In our review, we systematically evaluated both clinical and non-clinical models, including those driven by machine learning, and a range of mathematical modelling techniques, such as the minimal model. We highlighted several gaps in the literature and pinpointed areas where improvements are needed. A common flaw in many models is the assumption of a constant disease state, which is not realistic for diabetes. Incorporating disease progression into models is essential for studying the long-term effects of anti-diabetic treatments across different stages of the disease. To address these limitations, it is crucial to identify non-clinical and genetic factors linked to complications from prolonged hyperglycemia and integrate them into existing models. Doing so would allow for a more comprehensive approach, enhancing models like those developed by Boutayeb and colleagues [27] by introducing new variables or parameters to reflect these additional factors.

References

1. Abdulaziz Al Dawish, M., Alwin Robert, A., Braham, R., Abdallah Al Hayek, A., Al Saeed, A., Ahmed Ahmed, R., Sulaiman Al Sabaan, F.: Diabetes mellitus in Saudi Arabia: a review of the recent literature. Current Diab. Rev. **12**(4), 359–368 (2016)
2. Ajmera, I., Swat, M., Laibe, C., Le Novere, N., Chelliah, V.: The impact of mathematical modeling on the understanding of diabetes and related complications. CPT: Pharmacomet. Syst. Pharmacol. **2**(7), 1–14 (2013)
3. Al Ali, H.: Mathematical and data-driven modelling of glucose homeostasis system to understand diabetes pathways. Ph.D. thesis, Coventry University (2022)
4. Al Ali, H., Boutayeb, W., Boutayeb, A., Merabet, N.: A mathematical model for type 1 diabetes, on the effect of growth hormone. In: 2019 8th International Conference on Modeling Simulation and Applied Optimization (ICMSAO), pp. 1–4. IEEE (2019)

5. Al Ali, H., Boutayeb, W., Boutayeb, A., Merabet, N.: A mathematical model on the effect of growth hormone on glucose homeostasis. Revue Africaine de Recherche en Informatique et Mathématiques Appliquées **30** (2019)
6. Al Ali, H., Daneshkhah, A., Boutayeb, A., Malunguza, N.J., Mukandavire, Z.: Exploring dynamical properties of a type 1 diabetes model using sensitivity approaches. Math. Comput. Simul. **201**, 324–342 (2022)
7. Al Ali, H., Daneshkhah, A., Boutayeb, A., Mukandavire, Z.: Examining type 1 diabetes mathematical models using experimental data. Int. J. Environ. Res. Public Health **19**(2), 737 (2022)
8. Al Awadi, F., Hassanein, M., Hussain, H.Y., Mohammed, H., Ibrahim, G., Khater, A., Suliman, E.: Prevalence of diabetes and associated health risk factors among adults in Dubai, United Arab Emirates: results from Dubai household survey 2019. Dubai Diab. Endocrinol. J. **26**(4), 164–173 (2021)
9. Alikhani, G., Balochian, S.: Robust stability control for nonlinear time varying delay fractional order practical systems and application in glucose-insulin system. Comput. Methods Biomech. Biomed. Eng. **26**(15), 1796–1805 (2023)
10. Aljamaan, I., Al-Naib, I.: Prediction of blood glucose level using nonlinear system identification approach. IEEE Access **10**, 1936–1945 (2021)
11. Aljulifi, M.Z.: Prevalence and reasons of increased type 2 diabetes in gulf cooperation council countries. Saudi Med. J. **42**(5), 481 (2021)
12. Alshalalfah, A.L., Hamad, G.B., Mohamed, O.A.: Towards safe and robust closed-loop artificial pancreas using improved pid-based control strategies. IEEE Trans. Circuits Syst. I: Regul. Pap. **68**(8), 3147–3157 (2021)
13. Åm, M.K., Teigen, I.A., Riaz, M., Fougner, A.L., Christiansen, S.C., Carlsen, S.M.: The artificial pancreas: two alternative approaches to achieve a fully closed-loop system with optimal glucose control. J. Endocrinol. Invest. **47**(3), 513–521 (2024)
14. Aussedat, B., Dupire-Angel, M., Gifford, R., Klein, J., Wilson, G., Reach, G.: Interstitial glucose concentration and glycemia: implications for continuous subcutaneous glucose monitoring. Am. J. Physiol.-Endocrinol. Metab. **278**(4), E716–E728 (2000)
15. Avila, L.O., De Paula, M., Martinez, E.C., Errecalde, M.L.: Robust insulin estimation under glycemic variability using Bayesian filtering and gaussian process models. Biomed. Signal Process. Control **42**, 63–72 (2018)
16. Backer, J.M., Kahn, C.R., White, M.: Tyrosine phosphorylation of the insulin receptor during insulin-stimulated internalization in rat hepatoma cells. J. Biol. Chem. **264**(3), 1694–1701 (1989)
17. Beck, R.W., Kanapka, L.G., Breton, M.D., Brown, S.A., Wadwa, R.P., Buckingham, B.A., Kollman, C., Kovatchev, B.: A meta-analysis of randomized trial outcomes for the t: slim x2 insulin pump with control-iq technology in youth and adults from age 2 to 72. Diab. Technol. Ther. **25**(5), 329–342 (2023)
18. Bekiari, E., Kitsios, K., Thabit, H., Tauschmann, M., Athanasiadou, E., Karagiannis, T., Haidich, A.B., Hovorka, R., Tsapas, A.: Artificial pancreas treatment for outpatients with type 1 diabetes: systematic review and meta-analysis. bmj **361** (2018)
19. Bergman, R., Bortolan, G., Cobelli, C., Toffolo, G.: Identification of a minimal model of glucose disappearance for estimating insulin sensitivity. IFAC Proc. Vol. **12**(8), 883–890 (1979)
20. Bergman, R.N., Finegood, D.T., Ader, M.: Assessment of insulin sensitivity in vivo. Endocr. Rev. **6**(1), 45–86 (1985)
21. Bergman, R.N., Phillips, L.S., Cobelli, C., et al.: Physiologic evaluation of factors controlling glucose tolerance in man: measurement of insulin sensitivity and beta-cell glucose sensitivity from the response to intravenous glucose. J. Clin. Invest. **68**(6), 1456–1467 (1981)
22. Bergsten, P., Grapengiesser, E., Gylfe, E., Tengholm, A., Hellman, B.: Synchronous oscillations of cytoplasmic ca2+ and insulin release in glucose-stimulated pancreatic islets. J. Biol. Chem. **269**(12), 8749–8753 (1994)

23. Bergsten, P., Hellman, B.: Glucose-induced cycles of insulin release can be resolved into distinct periods of secretory activity. Biochem. Biophys. Res. Commun. **192**(3), 1182–1188 (1993)
24. Bolie, V.W.: Coefficients of normal blood glucose regulation. J. Appl. Physiol. **16**(5), 783–788 (1961)
25. van Bon, A., Blauw, H., Jansen, T., Laverman, G., Urgert, T., Geessink-Mennink, J., Mulder, A., Out, M., Veldman, R.G., Onvlee, A., et al.: Bihormonal fully closed-loop system for the treatment of type 1 diabetes: a real-world multicentre, prospective, single-arm trial in the Netherlands. Lancet Digit. Health **6**(4), e272–e280 (2024)
26. Bonner-Weir, S., Deery, D., Leahy, J.L., Weir, G.C.: Compensatory growth of pancreatic β-cells in adult rats after short-term glucose infusion. Diabetes **38**(1), 49–53 (1989)
27. Boutayeb, W., Lamlili, M.E., Boutayeb, A., Derouich, M.: Mathematical modelling and simulation of β-cell mass, insulin and glucose dynamics: effect of genetic predisposition to diabetes. J. Biomed. Sci. Eng. **2014** (2014)
28. Boutayeb, W., Lamlili, M.E., Boutayeb, A., Derouich, M.: The impact of obesity on predisposed people to type 2 diabetes: mathematical model. In: Bioinformatics and Biomedical Engineering: Third International Conference, IWBBIO 2015, Granada, Spain, April 15–17, 2015, Proceedings, Part I 3, pp. 613–622. Springer (2015)
29. Buppajarntham, S., Junpaparp, P., Salameh, R., Anastasopoulou, C., Eric, B.: Insulin: Reference range, interpretation, collection and panels (2019)
30. Burgess, J.W., Wada, I., Ling, N., Khan, M., Bergeron, J., Posner, B.: Decrease in beta-subunit phosphotyrosine correlates with internalization and activation of the endosomal insulin receptor kinase. J. Biol. Chem. **267**(14), 10077–10086 (1992)
31. Cantley, J., Ashcroft, F.M.: Q&a: insulin secretion and type 2 diabetes: why do β-cells fail? BMC Biol. **13**, 1–7 (2015)
32. Chatrabgoun, O., Esmaeilbeigi, M., Cheraghi, M., Daneshkhah, A.: Stable likelihood computation for machine learning of linear differential operators with gaussian processes. Int. J. Uncertain. Quantif. **12**(3) (2022)
33. Chen, S., Feng, L., Rickels, M.R., Peleckis, A., Sokolsky, O., Lee, I.: A data-driven behavior modeling and analysis framework for diabetic patients on insulin pumps. In: 2015 International Conference on Healthcare Informatics, pp. 213–222. IEEE (2015)
34. in Children Network (DirecNet) Study Group, D.R.: Evaluation of factors affecting cgms calibration. Diab. Technol. Ther. **8**(3), 318–325 (2006)
35. in Children Network (DirecNet) Study Group, D.R.: The accuracy of the guardian® rt continuous glucose monitor in children with type 1 diabetes. Diab. Technol. Ther. **10**(4), 266–272 (2008)
36. in Children Network (DirecNet) Study Group, D.R., et al.: The accuracy of the cgms in children with type 1 diabetes - results of the diabetes research in children network (direcnet) accuracy study. Diab. Technol. Ther. **5**(5), 781 (2003)
37. Clarke, W.L., Kovatchev, B.: Continuous glucose sensors: continuing questions about clinical accuracy (2007)
38. Cobelli, C., Dalla Man, C., Sparacino, G., Magni, L., De Nicolao, G., Kovatchev, B.P.: Diabetes: models, signals, and control. IEEE Rev. Biomed. Eng. **2**, 54–96 (2009)
39. Cobelli, C., Renard, E., Kovatchev, B.: Artificial pancreas: past, present, future. Diabetes **60**(11), 2672–2682 (2011)
40. Daneshkhah, A., Bedford, T.: Sensitivity analysis of a reliability system using gaussian processes. Advances in Mathematical Modeling for Reliability. IOS Press, Amsterdam, The Netherlands p. 46 (2008)
41. Daneshkhah, A., Bedford, T.: Probabilistic sensitivity analysis of system availability using gaussian processes. Reliab. Eng. Syst. Saf. **112**, 82–93 (2013)
42. Daneshkhah, A., Hosseinian-Far, A., Chatrabgoun, O.: Sustainable maintenance strategy under uncertainty in the lifetime distribution of deteriorating assets. Strategic Engineering for Cloud Computing and Big Data Analytics, pp. 29–50 (2017)

43. De Falco, I., Cioppa, A.D., Marcelli, A., Scafuri, U., Stellaccio, L., Tarantino, E.: Prediction of personalized blood glucose levels in type 1 diabetic patients using a neuroevolution approach. In: Proceedings of the Genetic and Evolutionary Computation Conference Companion, pp. 1708–1716 (2021)
44. De Paula, M., Avila, L.O., Martinez, E.C.: Controlling blood glucose variability under uncertainty using reinforcement learning and gaussian processes. Appl. Soft Comput. **35**, 310–332 (2015)
45. Efanova, I.B., Zaitsev, S.V., Zhivotovsky, B., Köhler, M., Efendić, S., Orrenius, S., Berggren, P.O.: Glucose and tolbutamide induce apoptosis in pancreatic β-cells: a process dependent on intracellular ca2+ concentration. J. Biol. Chem. **273**(50), 33501–33507 (1998)
46. El Idrissi, T., Idri, A., Bakkoury, Z.: Data mining techniques in diabetes self-management: a systematic map. In: Trends and Advances in Information Systems and Technologies: Volume 2 6, pp. 1142–1152. Springer (2018)
47. El-Khatib, F.H., Russell, S.J., Magyar, K.L., Sinha, M., McKeon, K., Nathan, D.M., Damiano, E.R.: Autonomous and continuous adaptation of a bihormonal bionic pancreas in adults and adolescents with type 1 diabetes. J. Clin. Endocrinol. Metab. **99**(5), 1701–1711 (2014)
48. El-Khatib, F.H., Russell, S.J., Nathan, D.M., Sutherlin, R.G., Damiano, E.R.: A bihormonal closed-loop artificial pancreas for type 1 diabetes. Sci. Transl. Med. **2**(27), 27ra27–27ra27 (2010)
49. Ewings, S.M., Sahu, S.K., Valletta, J.J., Byrne, C.D., Chipperfield, A.J.: A Bayesian network for modelling blood glucose concentration and exercise in type 1 diabetes. Stat. Methods Med. Res. **24**(3), 342–372 (2015)
50. Farhy, L.S., McCall, A.L.: Pancreatic network control of glucagon secretion and counterregulation. Methods Enzymol. **467**, 547–581 (2009)
51. Fehlmann, M., Carpentier, J.L., Van Obberghen, E., Freychet, P., Thamm, P., Saunders, D., Brandenburg, D., Orci, L.: Internalized insulin receptors are recycled to the cell surface in rat hepatocytes. Proc. Natl. Acad. Sci. **79**(19), 5921–5925 (1982)
52. Ferrannini, E., Gastaldelli, A., Miyazaki, Y., Matsuda, M., Pettiti, M., Natali, A., Mari, A., DeFronzo, R.: Predominant role of reduced beta-cell sensitivity to glucose over insulin resistance in impaired glucose tolerance. Diabetologia **46**, 1211–1219 (2003)
53. Flannick, J., Johansson, S., Njølstad, P.R.: Common and rare forms of diabetes mellitus: towards a continuum of diabetes subtypes. Nat. Rev. Endocrinol. **12**(7), 394–406 (2016)
54. Gao, C., Chen, F., Yu, P.: Dynamics of a new delayed glucose–insulin model with obesity. Int. J. Bifurcat. Chaos 2450070 (2024)
55. Garg, S., Zisser, H., Schwartz, S., Bailey, T., Kaplan, R., Ellis, S., Jovanovic, L.: Improvement in glycemic excursions with a transcutaneous, real-time continuous glucose sensor: a randomized controlled trial. Diab. Care **29**(1), 44–50 (2006)
56. Garg, S.K., Smith, J., Beatson, C., Lopez-Baca, B., Voelmle, M., Gottlieb, P.A.: Comparison of accuracy and safety of the seven and the navigator continuous glucose monitoring systems. Diab. Technol. Ther. **11**(2), 65–72 (2009)
57. Georga, E.I., Fotiadis, D.I., Tigas, S.K.: Personalized predictive modeling in type 1 diabetes. Academic Press (2017)
58. Grodsky, G.M.: Further studies on the dynamic aspects of insulin release in vitro with evidence for a two-compartmental storage system. Acta Diabet. Lat. **6**, 554–559 (1969)
59. Grodsky, G.M., et al.: A threshold distribution hypothesis for packet storage of insulin and its mathematical modeling. J. Clin. Investig. **51**(8), 2047–2059 (1972)
60. Group, U.P.D.S.U., et al.: Intensive blood-glucose control with sulphonylureas or insulin compared with conventional treatment and risk of complications in patients with type 2 diabetes (ukpds 33). The Lancet **352**(9131), 837–853 (1998)
61. Hadj-Abo, A., Enge, S., Rose, J., Kunte, H., Fleischhauer, M.: Individual differences in impulsivity and need for cognition as potential risk or resilience factors of diabetes self-management and glycemic control. PloS One **15**(1), e0227995 (2020)
62. Hashiguchi, Y., Sakakida, M., Nishida, K., Uemura, T., Kajiwara, K.I., Shichiri, M.: Development of a miniaturized glucose monitoring system by combining a needle-type glucose sensor

with microdialysis sampling method: long-term subcutaneous tissue glucose monitoring in ambulatory diabetic patients. Diab. Care **17**(5), 387–396 (1994)
63. Hellman, B., Gylfe, E., Bergsten, P., Grapengiesser, E., Lund, P., Berts, A., Tengholm, A., Pipeleers, D., Ling, Z.: Glucose induces oscillatory ca 2+ signalling and insulin release in human pancreatic beta cells. Diabetologia **37**, S11–S20 (1994)
64. Henquin, J.C.: Regulation of insulin secretion: a matter of phase control and amplitude modulation. Diabetologia **52**(5), 739–751 (2009)
65. Henquin, J.C., Ishiyama, N., Nenquin, M., Ravier, M.A., Jonas, J.C.: Signals and pools underlying biphasic insulin secretion. Diabetes **51**(suppl_1), S60–S67 (2002)
66. Hernandez, R.D., Lyles, D.J., Rubin, D.B., Voden, T.B., Wirkus, S.A., et al.: A model of beta-cell mass, insulin, glucose, and receptor dynamics with applications to diabetes (2001)
67. Holly, J., Amiel, S., Sandhu, R., Rees, L., Wass, J.: The role of growth hormone in diabetes mellitus. J. Endocrinol. **118**(3), 353–364 (1988)
68. Hoorens, A., Van de Casteele, M., Klöppel, G., Pipeleers, D., et al.: Glucose promotes survival of rat pancreatic beta cells by activating synthesis of proteins which suppress a constitutive apoptotic program. J. Clin. Investig. **98**(7), 1568–1574 (1996)
69. Hovorka, R., Canonico, V., Chassin, L.J., Haueter, U., Massi-Benedetti, M., Federici, M.O., Pieber, T.R., Schaller, H.C., Schaupp, L., Vering, T., et al.: Nonlinear model predictive control of glucose concentration in subjects with type 1 diabetes. Physiol. Meas. **25**(4), 905 (2004)
70. Huang, Z., Huang, L., Waters, M.J., Chen, C.: Insulin and growth hormone balance: implications for obesity. Trends Endocrinol. Metab. **31**(9), 642–654 (2020)
71. Hueckstеadt, T., Olefsky, J.M., Brandenberg, D., Heidenreich, K.: Recycling of photoaffinity-labeled insulin receptors in rat adipocytes. dissociation of insulin-receptor complexes is not required for receptor recycling. J. Biol. Chem. **261**(19), 8655–8659 (1986)
72. Hugl, S.R., White, M.F., Rhodes, C.J.: Insulin-like growth factor i (igf-i)-stimulated pancreatic β-cell growth is glucose-dependent: synergistic activation of insulin receptor substrate-mediated signal transduction pathways by glucose and igf-i in ins-1 cells. J. Biol. Chem. **273**(28), 17771–17779 (1998)
73. IDF: United Arab Emirates diabetes report 2000 - 2045, diabetes report 2000 - 2045. (2022). https://diabetesatlas.org/data/en/country/208/ae.html. Accessed: 13 June 2024
74. Jacobs, P.G., Herrero, P., Facchinetti, A., Vehi, J., Kovatchev, B., Breton, M., Cinar, A., Nikita, K., Doyle, F., Bondia, J., et al.: Artificial intelligence and machine learning for improving glycemic control in diabetes: best practices, pitfalls and opportunities. IEEE Rev. Biomed, Eng (2023)
75. Jain, A., Roy, P.: A mathematical model for assessing how obesity-related factors aggravate diabetes. Acta Applicandae Mathematicae **191**(1), 8 (2024)
76. Kim, S.H., Park, M.J.: Effects of growth hormone on glucose metabolism and insulin resistance in human. Ann. Pediatr. Endocrinol. Metab. **22**(3), 145 (2017)
77. Klein, H.H., Freidenberg, G., Matthaei, S., Olefsky, J.: Insulin receptor kinase following internalization in isolated rat adipocytes. J. Biol. Chem. **262**(22), 10557–10564 (1987)
78. Kovatchev, B.: Automated closed-loop control of diabetes: the artificial pancreas. Bioelectron. Med. **4**(1), 14 (2018)
79. Kovatchev, B., Anderson, S., Heinemann, L., Clarke, W.: Comparison of the numerical and clinical accuracy of four continuous glucose monitors. Diab. Care **31**(6), 1160–1164 (2008)
80. Kovatchev, B., Clarke, W.: Peculiarities of the continuous glucose monitoring data stream and their impact on developing closed-loop control technology (2008)
81. Kushner, T., Bortz, D., Maahs, D.M., Sankaranarayanan, S.: A data-driven approach to artificial pancreas verification and synthesis. In: 2018 ACM/IEEE 9th International Conference on Cyber-Physical Systems (ICCPS), pp. 242–252. IEEE (2018)
82. LeBlanc, H., Chauvet, D., Lombrail, P., Robert, J.J.: Glycemic control with closed-loop intraperitoneal insulin in type i diabetes. Diab. Care **9**(2), 124–128 (1986)
83. Lee, J., Kim, J.H., et al.: Clinical utilities of continuous glucose monitoring and insulin pumps in pediatric patients with type 1 diabetes. Ewha Med. J. **44**(3), 55–62 (2021)

84. Leelarathna, L., Choudhary, P., Wilmot, E.G., Lumb, A., Street, T., Kar, P., Ng, S.M.: Hybrid closed-loop therapy: where are we in 2021? Diab. Obes. Metab. **23**(3), 655–660 (2021)
85. Lei, M., Chen, D., Ling, P., Wang, C., Yang, D., Deng, H., Yang, X., Xu, W., Yan, J.: Effect of artificial pancreas system use on glycaemic control among pregnant women with type 1 diabetes mellitus: A meta-analysis of randomized controlled trials. Diab. Obes. Metab. **26**(2), 673–681 (2024)
86. Lewis, D.: How it started, how it is going: the future of artificial pancreas systems (automated insulin delivery systems). J. Diab. Sci. Technol. **15**(6), 1258–1261 (2021)
87. Li, J., Johnson, J.D.: Mathematical models of subcutaneous injection of insulin analogues: a mini-review. Discrete Contin. Dyn. Syst. Ser. B **12**(2), 401 (2009)
88. Lu, M., Flanagan, J.U., Langley, R.J., Hay, M.P., Perry, J.K.: Targeting growth hormone function: strategies and therapeutic applications. Signal Transduct. Target. Ther. **4**(1), 3 (2019)
89. Mari, A., Tura, A., Grespan, E., Bizzotto, R.: Mathematical modeling for the physiological and clinical investigation of glucose homeostasis and diabetes. Front. Physiol. **11**, 575789 (2020)
90. Marshall, S., Green, A., Olefsky, J.M.: Evidence for recycling of insulin receptors in isolated rat adipocytes. J. Biol. Chem. **256**(22), 11464–11470 (1981)
91. McClain, D., Maegawa, H., Lee, J., Dull, T., Ulrich, A., Olefsky, J.: A mutant insulin receptor with defective tyrosine kinase displays no biologic activity and does not undergo endocytosis. J. Biol. Chem. **262**(30), 14663–14671 (1987)
92. Messer, L.H., Berget, C., Vigers, T., Pyle, L., Geno, C., Wadwa, R.P., Driscoll, K.A., Forlenza, G.P.: Real world hybrid closed-loop discontinuation: predictors and perceptions of youth discontinuing the 670g system in the first 6 months. Pediatr. Diab. **21**(2), 319–327 (2020)
93. Ming, W., Guo, X., Zhang, G., Liu, Y., Wang, Y., Zhang, H., Liang, H., Yang, Y.: Recent advances in the precision control strategy of artificial pancreas. Med. Biol. Eng. Comput. 1–24 (2024)
94. MohammadRidha, T., Moog, C., Delaleau, E., Fliess, M., Join, C.: A variable reference trajectory for model-free glycemia regulation. SIAM Conference on Control and Its Applications, Paris (2015)
95. Morgan, D.O., Roth, R.A.: Acute insulin action requires insulin receptor kinase activity: introduction of an inhibitory monoclonal antibody into mammalian cells blocks the rapid effects of insulin. Proc. Natl. Acad. Sci. **84**(1), 41–45 (1987)
96. Nauck, M.A., Wefers, J., Meier, J.J.: Treatment of type 2 diabetes: challenges, hopes, and anticipated successes. Lancet Diab. Endocrinol. **9**(8), 525–544 (2021)
97. Ni Ki, C., Hosseinian-Far, A., Daneshkhah, A., Salari, N.: Topic modelling in precision medicine with its applications in personalized diabetes management. Expert Syst. **39**(4), e12774 (2022)
98. Nwokolo, M., Hovorka, R.: The artificial pancreas and type 1 diabetes. J. Clin. Endocrinol. Metab. **108**(7), 1614–1623 (2023)
99. Oakley, J.E., O'Hagan, A.: Probabilistic sensitivity analysis of complex models: a Bayesian approach. J. R. Stat. Soc. Ser. B: Stat. Methodol. **66**(3), 751–769 (2004)
100. Olarescu, N.C., Gunawardane, K., Hansen, T.K., Møller, N., Jørgensen, J.O.L.: Normal physiology of growth hormone in adults (2015)
101. Ortmann, L., Shi, D., Dassau, E., Doyle, F.J., Leonhardt, S., Misgeld, B.J.: Gaussian process-based model predictive control of blood glucose for patients with type 1 diabetes mellitus. In: 2017 11th Asian Control Conference (ASCC), pp. 1092–1097. IEEE (2017)
102. Palumbo, P., Ditlevsen, S., Bertuzzi, A., De Gaetano, A.: Mathematical modeling of the glucose-insulin system: a review. Math. Biosci. **244**(2), 69–81 (2013)
103. Paoletti, N., Liu, K.S., Smolka, S.A., Lin, S.: Data-driven robust control for type 1 diabetes under meal and exercise uncertainties. In: Computational Methods in Systems Biology: 15th International Conference, CMSB 2017, Darmstadt, Germany, Sept. 27–29, 2017, Proceedings 15, pp. 214–232. Springer (2017)
104. Pedersen, M.G.: Contributions of mathematical modeling of beta cells to the understanding of beta-cell oscillations and insulin secretion. J. Diab. Sci. Technol. **3**(1), 12–20 (2009)

105. Pfeiffer, E.F., Thum, C., Clemens, A.: The artificial beta cell-a continuous control of blood sugar by external regulation of insulin infusion (glucose controlled insulin infusion system). Hormone Metab. Res. **6**(05), 339–342 (1974)
106. Pickup, J., Keen, H.: Continuous subcutaneous insulin infusion at 25 years: evidence base for the expanding use of insulin pump therapy in type 1 diabetes. Diab. Care **25**(3), 593–598 (2002)
107. Porksen, N., Munn, S., Steers, J., Vore, S., Veldhuis, J., Butler, P.: Pulsatile insulin secretion accounts for 70% of total insulin secretion during fasting. Am. J. Physiol.-Endocrinol. Metab. **269**(3), E478–E488 (1995)
108. Pørksen, N., Nyholm, B., Veldhuis, J.D., Butler, P.C., Schmitz, O.: In humans at least 75% of insulin secretion arises from punctuated insulin secretory bursts. Am. J. Physiol.-Endocrinol. Metab. **273**(5), E908–E914 (1997)
109. Quesada, I., Tudurí, E., Ripoll, C., Nadal, A.: Physiology of the pancreatic α-cell and glucagon secretion: role in glucose homeostasis and diabetes. J. Endocrinol. **199**(1), 5–19 (2008)
110. Ritzel, R., Schulte, M., Pørksen, N., Nauck, M.S., Holst, J.J., Juhl, C., März, W., Schmitz, O., Schmiegel, W.H., Nauck, M.A.: Glucagon-like peptide 1 increases secretory burst mass of pulsatile insulin secretion in patients with type 2 diabetes and impaired glucose tolerance. Diabetes **50**(4), 776–784 (2001)
111. Rorsman, P., Ashcroft, F.M.: Pancreatic β-cell electrical activity and insulin secretion: of mice and men. Physiol. Rev. **98**(1), 117–214 (2018)
112. Rorsman, P., Eliasson, L., Renstrom, E., Gromada, J., Barg, S., Gopel, S.: The cell physiology of biphasic insulin secretion. Physiology **15**(2), 72–77 (2000)
113. Russell, S.J., El-Khatib, F.H., Sinha, M., Magyar, K.L., McKeon, K., Goergen, L.G., Balliro, C., Hillard, M.A., Nathan, D.M., Damiano, E.R.: Outpatient glycemic control with a bionic pancreas in type 1 diabetes. N. Engl. J. Med. **371**(4), 313–325 (2014)
114. Saha, T., Del Caño, R., Mahato, K., De la Paz, E., Chen, C., Ding, S., Yin, L., Wang, J.: Wearable electrochemical glucose sensors in diabetes management: a comprehensive review. Chem. Rev. **123**(12), 7854–7889 (2023)
115. Samuel, P., Khan, N., Klein, G., Skobkarev, S., Mammon, B., Fournier, M., Hawke, K., Weissinger, A., Elliott, T.: Open-source artificial pancreas systems are safe and effective when supported in-clinic: outcomes in 248 consecutive type 1 diabetes clients. Can. J. Diab. **48**(1), 59–65 (2024)
116. Sanchez, E.N.: The artificial pancreas: current situation and future directions (2019)
117. Shao, J., Liu, Z., Li, S., Wu, B., Nie, Z., Li, Y., Zhou, K.: Continuous glucose monitoring time series data analysis: a time series analysis package for continuous glucose monitoring data. J. Comput. Biol. **30**(1), 112–116 (2023)
118. Shichiri, M., Sakakida, M., Nishida, K., Shimoda, S.: Enhanced, simplified glucose sensors: long-term clinical application of wearable artificial endocrine pancreas. Artif. Organs **22**(1), 32–42 (1998)
119. Shichiri, M., Yamasaki, Y., Kawamori, R., Hakui, N., Abe, H.: Wearable artificial endocrine pancreas with needle-type glucose sensor. The Lancet **320**(8308), 1129–1131 (1982)
120. Silber, H.E., Nyberg, J., Hooker, A.C., Karlsson, M.O.: Optimization of the intravenous glucose tolerance test in t2dm patients using optimal experimental design. J. Pharmacokinet. Pharmacodyn. **36**, 281–295 (2009)
121. Sonne, O., Simpson, I.A.: Internalization of insulin and its receptor in the isolated rat adipose cell: Time-course and insulin concentration dependency. Biochimica et Biophysica Acta (BBA)-Molecular Cell Research **804**(4), 404–413 (1984)
122. Storch, M., Rössle, M., Kerp, L.: Pulsatile insulin secretion into the portal vein in liver cirrhosis. Deutsche Medizinische Wochenschrift (1946) **118**(5), 134–138 (1993)
123. Stožer, A., Dolenšek, J., Rupnik, M.S.: Glucose-stimulated calcium dynamics in islets of Langerhans in acute mouse pancreas tissue slices. PloS One **8**(1), e54638 (2013)
124. Sturis, J., Polonsky, K.S., Mosekilde, E., Van Cauter, E.: Computer model for mechanisms underlying ultradian oscillations of insulin and glucose. Am. J. Physiol.-Endocrinol. Metab. **260**(5), E801–E809 (1991)

125. Swenne, I.: The role of glucose in the in vitro regulation of cell cycle kinetics and proliferation of fetal pancreatic b-cells. Diabetes **31**(9), 754–760 (1982)
126. Tamborlane, W., Beck, R., Bode, B., Buckingham, B., Chase, H., Clemons, R., Fiallo-Scharer, R., Fox, L., Gilliam, L., Hirsch, I., et al.: Juvenile diabetes research foundation continuous glucose monitoring study group continuous glucose monitoring and intensive treatment of type 1 diabetes. N. Engl. J. Med. **359**(14), 1464–1476 (2008)
127. Tersey, S.A., Nishiki, Y., Templin, A.T., Cabrera, S.M., Stull, N.D., Colvin, S.C., Evans-Molina, C., Rickus, J.L., Maier, B., Mirmira, R.G.: Islet β-cell endoplasmic reticulum stress precedes the onset of type 1 diabetes in the nonobese diabetic mouse model. Diabetes **61**(4), 818–827 (2012)
128. Toffanin, C., Zisser, H., Doyle, F.J., III., Dassau, E.: Dynamic insulin on board: incorporation of circadian insulin sensitivity variation. J. Diab. Sci. Technol. **7**(4), 928–940 (2013)
129. Toffolo, G., Bergman, R.N., Finegood, D.T., Bowden, C.R., Cobelli, C.: Quantitative estimation of beta cell sensitivity to glucose in the intact organism: a minimal model of insulin kinetics in the dog. Diabetes **29**(12), 979–990 (1980)
130. Topp, B., Promislow, K., Devries, G., Miura, R.M., T FINEGOOD, D.: A model of β-cell mass, insulin, and glucose kinetics: pathways to diabetes. J. Theoret. Biol. **206**(4), 605–619 (2000)
131. Trevitt, S., Simpson, S., Wood, A.: Artificial pancreas device systems for the closed-loop control of type 1 diabetes: what systems are in development? J. Diab. Sci. Technol. **10**(3), 714–723 (2016)
132. Tsoukas, M.A., Majdpour, D., Yale, J.F., El Fathi, A., Garfield, N., Rutkowski, J., Rene, J., Legault, L., Haidar, A.: A fully artificial pancreas versus a hybrid artificial pancreas for type 1 diabetes: a single-centre, open-label, randomised controlled, crossover, non-inferiority trial. Lancet Digit. Health **3**(11), e723–e732 (2021)
133. Valletta, J.: Dynamic modelling of the effect of physical activity on capillary blood glucose concentration in people with type 1 diabetes. Ph.D. thesis, PhD Thesis, University of Southampton (2011)
134. Wang, W., Wang, S., Wang, X., Liu, D., Geng, Y., Wu, T.: A glucose-insulin mixture model and application to short-term hypoglycemia prediction in the night time. IEEE Trans. Biomed. Eng. **68**(3), 834–845 (2020)
135. Wang, X., Zhang, Y., Song, X.Y.: Mathematical model for diabetes mellitus with impulsive injections of glucose-insulin. Chin. Q. J. Math. **32**(2), 118 (2017)
136. Ware, J., Boughton, C.K., Allen, J.M., Wilinska, M.E., Tauschmann, M., Denvir, L., Thankamony, A., Campbell, F.M., Wadwa, R.P., Buckingham, B.A., et al.: Cambridge hybrid closed-loop algorithm in children and adolescents with type 1 diabetes: a multicentre 6-month randomised controlled trial. Lancet Digit. Health **4**(4), e245–e255 (2022)
137. Weinzimer, S.A., Steil, G.M., Swan, K.L., Dziura, J., Kurtz, N., Tamborlane, W.V.: Fully automated closed-loop insulin delivery versus semiautomated hybrid control in pediatric patients with type 1 diabetes using an artificial pancreas. Diab. Care **31**(5), 934–939 (2008)
138. Weir, G.C., Bonner-Weir, S.: Five stages of evolving beta-cell dysfunction during progression to diabetes. Diabetes **53**(suppl_3), S16–S21 (2004)
139. Xu, N.Y., Nguyen, K.T., DuBord, A.Y., Pickup, J., Sherr, J.L., Teymourian, H., Cengiz, E., Ginsberg, B.H., Cobelli, C., Ahn, D., et al.: Diabetes technology meeting 2021. J. Diab. Sci. Technol. **16**(4), 1016–1056 (2022)
140. Zhang, Y., Holt, T.A., Khovanova, N.: A data driven nonlinear stochastic model for blood glucose dynamics. Comput. Methods Programs Biomed. **125**, 18–25 (2016)
141. Zhao, Y., Jing, W., Li, L., Zhao, S., Yamasaki, M.: Dynamical modeling the effect of glucagon-like peptide on glucose-insulin regulatory system based on mice experimental observation. Math. Biosci. **366**, 109090 (2023)

Mathematical Modeling in Diabetes Care and Innovation

Maghnia Hamou-Maamar

Abstract This chapter reviews mathematical models utilized for assessing physiological parameters, primarily insulin sensitivity over the short term. Subsequently, recent advancements in models designed to manage diabetes and monitor long-term care progress are discussed. The review also addresses challenges and limitations inherent in mathematical modeling within this domain. Furthermore, it explores new techniques and innovations, highlighting the potential of mathematical approaches in advancing technologies such as the artificial pancreas. This evolution is driven by innovations in insulin pumps and closed-loop systems, promising enhanced management of diabetes.

1 Introduction

Diabetes is a chronic metabolic disorder characterized by elevated blood glucose levels due to insufficient insulin production or impaired insulin function. It manifests in two primary forms: insulin-dependent diabetes mellitus (type 1) and non-insulin-dependent diabetes mellitus (type 2). Poorly managed diabetes can lead to various acute and chronic complications, such as harm to the eyes, kidneys, nerves, and blood vessels, significantly contributing to global mortality rates.

Type 1 diabetes mellitus (T1DM) typically begins in childhood or adolescence and results from an autoimmune destruction of β-cells in the pancreas, leading to little or no insulin production. In contrast, type 2 diabetes mellitus (T2DM), the most common form, usually develops in adults and is characterized by insulin resistance and a relative insulin deficiency. This progressive loss of adequate β-cell insulin secretion frequently occurs against a backdrop of insulin resistance.

T2DM is considered one of the top ten life-threatening diseases worldwide, representing approximately 90% of diabetes mellitus cases. T1DM, which accounts for about 10% of cases, has also seen a dramatic rise across countries of all income lev-

M. Hamou-Maamar (✉)
Faculty of Exact Sciences and Computer Science, University of Mostaganem, Mostaganem, Algeria
e-mail: maghnia.hamoumaamar@gmail.com

A. Boutayeb and M. E. N. Lamlili (eds.), *Computational Mathematics and Modelling for Diabetes*, Industrial and Applied Mathematics,
https://doi.org/10.1007/978-981-96-1925-2_4

els. According to the International Diabetes Federation 2021 Atlas, the prevalence of diabetes among people aged 20 to 79 years is estimated at around 537 million, which is approximately 10.5% of the global population in this age group. This number is projected to rise to 643 million by 2030 and further to 783 million by 2045. Alarmingly, nearly half of all diabetes cases are undiagnosed, amounting to approximately 240 million people worldwide, predominantly in low- and middle-income countries. In regions such as Africa, Southeast Asia, and the Western Pacific, more than half of those with diabetes remain undiagnosed. Additionally, over 1.2 million children and adolescents are affected by T1DM, with more than half being under the age of 15 [36].

In the same year, the expenditure for diabetes care varied significantly across different regions. The North American and Caribbean region had the highest expenditure at $414.5 billion, followed by the Western Pacific at $241.3 billion, and Europe at $189.3 billion. South and Central America spent $65.3 billion, while the Middle East and North Africa spent $32.6 billion. Africa and Southeast Asia had the lowest expenditures at $12.6 billion and $10.1 billion, respectively (see Fig. 1) [69].

Improving diabetes care involves implementing various strategies and adhering to established standards. Key components include screening and diagnosis, which play a significant role in categorizing diabetes and tailoring effective management programs for each individual. Regular assessments of blood pressure, lipid profile, kidney function, and eye health aid in customizing personalized care as part of a comprehensive medical evaluation. Another essential standard is the assessment of comorbidities by evaluating conditions concurrent with diabetes, such as hypertension or obesity, ensuring holistic management [36].

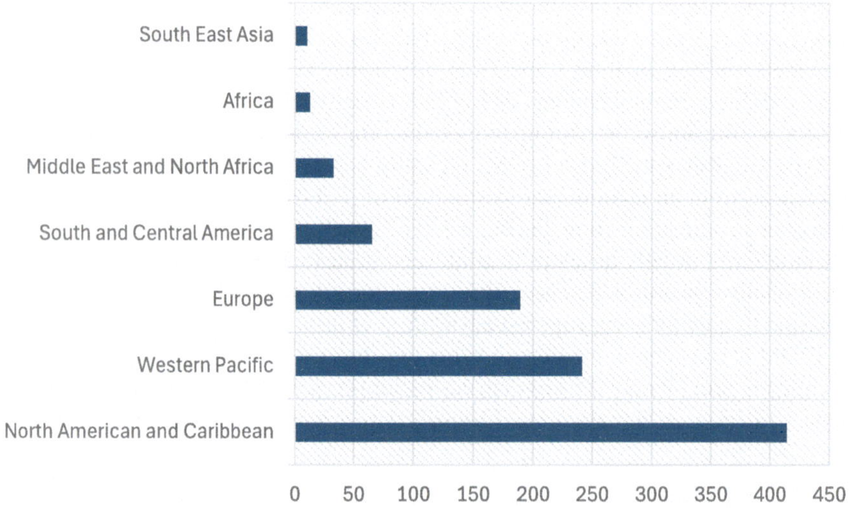

Fig. 1 Expenditure in billion U.S. dollars in 2021 across different regions

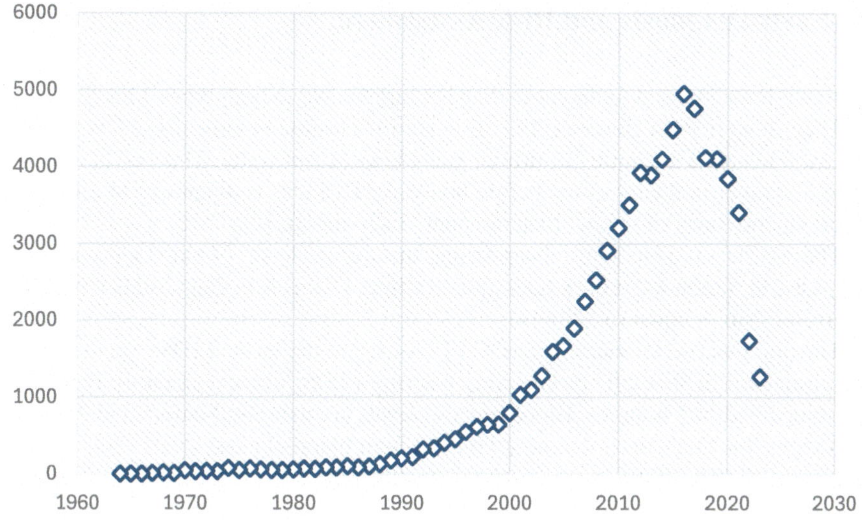

Fig. 2 Number of publications on diabetes mathematical models between 1964 and 2023. This graph shows the yearly distribution of publications related to diabetes mathematical models based on PubMed search results. It highlights the trend in research activity over the years, indicating a growing recognition of the importance of mathematical models in diabetes research

Mathematical modeling in the field of diabetes plays an increasingly important role in understanding different aspects of the disease (see Fig. 2). It becomes an interesting tool, and the beauty of mathematics appears when it can describe a complex system such as glucose-insulin regulation with simple equations. Through the richness of theory, we can develop a rigorous analysis. The domain of diabetology extensively uses mathematical approaches, ranging from theoretical models to predictive ones, and includes advanced approaches that study the effects of drugs on individuals with diabetes.

This chapter provides a structured review of mathematical models applied in diabetes research and management. It begins with an overview of foundational models used to assess short-term physiological parameters, particularly insulin sensitivity, essential for understanding immediate responses to interventions like diet and medication. Subsequently, recent advancements in models tailored for long-term diabetes management are explored, focusing on their ability to predict disease progression, personalized treatment outcomes, and complications over extended periods. The chapter also addresses challenges in mathematical modeling within this domain, including data variability and model validation against clinical data. Furthermore, it discusses emerging technologies such as the artificial pancreas, highlighting how innovations in insulin delivery systems and closed-loop control algorithms promise to revolutionize diabetes care by automating glucose regulation and enhancing patient quality of life.

2 Early Diagnosis and Risk Assessment

Different tests are used to diagnose diabetes, applicable to both types 1 and 2.

The fasting plasma glucose (FPG) test measures fasting blood glucose levels after at least 8 hours of not eating or drinking anything except water. This test is usually done first thing in the morning, before breakfast. Diabetes is diagnosed at a fasting blood glucose level of greater than or equal to 126 mg/dL [2].

The A1C test is a blood test that provides the average levels of blood glucose over the past 2 to 3 months. It does not require fasting. Diabetes is diagnosed at an A1C of greater than or equal to 6.5% [2].

The oral glucose tolerance test (OGTT) is used to diagnose T2DM, prediabetes, and gestational diabetes. It involves administering a 75 g glucose dose orally, typically consumed within 5 minutes. Blood glucose levels are measured before and 2 hours after ingestion. Diabetes is diagnosed if the 2-hour blood glucose level is 200 mg/dL or higher [2]. The test assesses how effectively the body clears glucose from the bloodstream.

The random (also called casual) plasma glucose test checks blood glucose at any time of the day when severe diabetes symptoms are present. Diabetes is diagnosed at a blood glucose level of greater than or equal to 200 mg/dL [2].

Table 1 summarizes the test results for each of the above tests.

The tests described above can confirm if a person has diabetes, but they cannot identify the type: T1DM, T2DM, or monogenic diabetes (a rare form of diabetes caused by a mutation in a single gene, unlike T1DM and T2DM which result from multiple gene mutations). Treatment depends on the type of diabetes, so knowing which type is important. For this, other tests like genetic testing can diagnose most forms of monogenic diabetes, caused by a mutation in a gene. Monogenic diabetes is more common in babies developing diabetes before 6 months old, some teens, and some young adults. Another method is detecting autoantibodies to see if the diabetes is type 1. Autoantibodies are proteins in the immune system that mistakenly attack healthy tissues and cells [2].

Various mathematical models have been developed to describe glucose-insulin dynamics, contributing to different diagnostic methods. These models quantify phys-

Table 1 Diabetes diagnosis tests [2]

	Non-diabetic	Prediabetes	Diabetes
A1C	Below 5.7%	5.7% to 6.4%	6.5% or above
Fasting plasma glucose	99 mg/dL or below	100 to 125 mg/dL	126 mg/dL or above
Oral glucose tolerance test	139 mg/dL or below	140 to 199 mg/dL	200 mg/dL or above
Random plasma glucose test	N/A	N/A	200 mg/dL or above

iological parameters such as insulin sensitivity, glucose effectiveness, insulin secretion, and β-cell function, crucial for understanding biomarkers predicting the progression to both T1DM and T2DM. β-cell function is critical in the pathogenesis of diabetes [12, 17, 18, 41, 75]. A historically important model in diabetes research is the Bergman minimal model, introduced in 1989 [12]. Its widespread use in numerous clinical studies is due to its ability to accurately estimate parameters related to insulin secretion during the intravenous glucose tolerance test (IVGTT). This model helps evaluate first-phase insulin release and provides precise insulin response assessments. The model is given by

$$\frac{dG}{dt} = -(S_G + X)G, \tag{1}$$

$$\frac{dX}{dt} = p_2 I(t) - p_3 X. \tag{2}$$

Here, G represents plasma glucose, X is the "remote insulin" compartment, which can be identified with the interstitial fluid where insulin exerts its action in glucose disposal, and $I(t)$ is the plasma insulin. The parameter p_2 is the fractional rate of insulin appearance in interstitial fluid; p_3 represents the fractional clearance of insulin from the interstitial compartment; and S_G is the glucose effectiveness which accounts for glucose's ability to self-normalize independent of insulin. The total glucose effectiveness, E, is defined as the rate of change of plasma glucose concentration:

$$E = -\frac{\partial \dot{G}}{\partial G} = S_G + X,$$

and insulin sensitivity (in steady state) is defined as the quantitative influence of insulin to enhance glucose disappearance:

$$S_I = \frac{\partial I_{steadystate}}{\partial E_{steadystate}} = \frac{p_2}{p_3}.$$

In addition, the model uses first-phase insulin release to measure β-cell function as follows: $\text{AIR}_{\text{glucose}}$ is the average insulin response above basal concentration over the first 2–10 min. A useful characteristic is the following hyperbolic relationship:

$$S_I \times \text{AIR} = DI.$$

DI is called the disposition index and is influenced by both the environment and genetics [11]. The disposition index has been used to identify at-risk individuals; those with low DI are at risk for diabetes.

The minimal model suffers from some problems such as it may underestimate S_I when AIR is large and therefore be unreliable in comparing S_I between groups with very different characteristic levels of AIR [31]. A variety of ameliorated versions of the minimal model have been introduced [5, 34, 40, 64, 67, 70, 74], such as in [67], in order to address this limitation of the original version and to adapt the model to simulate very ill patients [74]. Another extended version of the model, incorporating the effect of insulin on lipolysis and glucose disposal, was introduced by Periwal et al. in 2008 [64]. For a comprehensive review of modified versions up until 2014, refer to Cobelli et al. [16]. For recent studies, see Alkhateeb et al. [5], Nath et al. [58], and Bartlette et al. [9].

Another simple method used to estimate insulin secretion is the Homeostatic Model Assessment (HOMA), introduced by Matthews and colleagues in 1985 [50]. HOMA detects insulin resistance even before the onset of diabetes, enabling early intervention. It computes insulin sensitivity (HOMA-IR or HOMA%S) and β-cell function (HOMA-b or HOMA%B) using fasting glucose and fasting insulin:

$$\text{HOMA-IR} = \frac{\text{Glucose}_{\text{fasting}} \times \text{Insulin}_{\text{fasting}}}{22.5},$$

$$\text{HOMA-b} = \frac{20 \times \text{Insulin}_{\text{fasting}}}{\text{Glucose}_{\text{fasting}} - 3.5}\%.$$

Another accurate method for assessing insulin sensitivity is the Quantitative Insulin Sensitivity Check Index (QUICKI) [42]:

$$\text{QUICKI} = \frac{1}{\log(\text{Insulin}_{\text{fasting}}) + \log(\text{Glucose}_{\text{fasting}})}.$$

More recent mathematical models for estimating these parameters include the SPINA method [23]. SPINA incorporates fundamental physiological motifs such as saturation kinetics, non-competitive inhibition, and pharmacokinetics. It provides biomarkers of carbohydrate homeostasis, specifically the secretory capacity of β-cells (SPINA-GBeta) and insulin receptor gain (SPINA-GR):

$$\text{SPINA-GBeta} = \frac{I_{\text{steady state}}(D_\beta + G_{\text{steady state}})}{G_3[G_{\text{steady state}}]},$$

$$\text{SPINA-GR} = \frac{G_1 P_{\text{steady state}}(D_R + I_{\text{steady state}})}{G_E I_{\text{steady state}} G_{\text{steady state}}} - \frac{D_R}{G_E I_{\text{steady state}}} - \frac{1}{G_E},$$

where the parameters and variables are given in Table 2.

In summary, Table 3 summarizes methods used to estimate insulin secretion, and Table 4 summarizes methods used to estimate (β)-cell function using mathematical models. Our understanding of insulin sensitivity is crucial in evaluating β-cell

Table 2 Parameters and their descriptions of SPINA model [23]

Parameter	Explanation
α_G	Dilution factor (1/VD) for glucose
β_G	Clearance exponent for glucose
α_I	Dilution factor (1/VD) for insulin
β_I	Clearance exponent for insulin
G_1	$\frac{\alpha_G}{\beta_G}$
D_β	EC50 of glucose at beta cells
G_3	$\frac{\alpha_I}{\beta_I}$
D_R	EC50 of insulin at its receptor
G_E	Effector gain (Calibration factor)
P	Constitutive endogenous glucose production

Table 3 Summary of methods used to estimate insulin sensitivity using mathematical models

Parameter	Description	Reference
HOMA2-%S	Nonlinear model that utilizes either C-peptide or insulin with paired glucose measurement to provide an estimate of insulin sensitivity	[11]
OGIS (oral glucose-insulin sensitivity)	A simple method for calculating insulin sensitivity using glucose and insulin data that is in good agreement with clamp-based measurements	[54]
QUICKI	Quantitative insulin sensitivity check index, based on logarithmic transformations of fasting insulin and glucose levels	[42]
SPINA-GR	Estimates insulin receptor gain using a nonlinear model incorporating steady-state parameters	[23]

function, as it reveals how β-cells adjust insulin secretion relative to tissue sensitivity. This relationship, captured by the disposition index, underscores the interplay between insulin sensitivity and β-cell response, crucial for understanding glycemic regulation and diabetes pathophysiology.

Table 4 Summary of methods used to estimate β-cell function using mathematical models

Parameter	Description	Reference
HOMA2-%B	Based on fasting samples, a nonlinear model that utilizes either C-peptide or insulin with paired glucose measurement to provide an estimate of the β-cell's function. This version incorporates an estimate of proinsulin secretion and thus allows use of either total or specific insulin assays. Further, it accounts for renal glucose losses, thus allowing its use in hyperglycemic subjects	[50]
Glucose sensitivity	The slope of the curve relating the rate of insulin secretion to standardized glucose concentrations during the test. It is calculated from the glucose and C-peptide data using a model developed by Mari and colleagues	[54]
Rate sensitivity	An index of early insulin release that represents the dependence of the rate of insulin secretion on the rate of change of glucose concentration. It is calculated from the glucose and C-peptide data using a model developed by Mari and colleagues	[54]
Potentiation	A time-varying factor expressing a potentiation effect upon insulin secretion. It accounts for the physiological processes that can acutely modify insulin secretion. It is calculated from the glucose and C-peptide data using a model developed by Mari and colleagues	[54]
Φ_d	"Dynamic sensitivity," a measure of the stimulatory effect of the rate of increase in glucose on the secretion of stored insulin, calculated using a model developed by Cobelli and colleagues	[15]
Φ_s	"Static sensitivity" is a measure of the effect of glucose on β-cell secretion calculated using a model developed by Cobelli and colleagues	[15]

3 Disease Progression Modeling

Modeling diabetes, whether T1DM or T2DM, involves capturing the complex physiological processes that regulate blood glucose levels. However, there are key differences and challenges unique to each type of diabetes. We begin by reviewing the mathematical models used for T2DM and then proceed to discuss those used for T1DM.

As defined previously, T2DM is caused by a combination of two primary factors: defective insulin secretion by pancreatic β-cells and the inability of insulin-sensitive tissues to respond appropriately to insulin. Therefore, any model describing the progression of this type should consider these two key factors. Various mathematical models have been introduced to describe the progression of diabetes in a patient. The reader is invited to consult the review papers in [3, 14, 77].

Starting with the prominent approach introduced by Topp et al. in 2001 [72], this model significantly contributed to our understanding of the physiological basis of the development of T2DM. It includes glucose (G), insulin (I), and pancreatic β-cell mass dynamics (β). This model is considered a fast-slow model, meaning perturbations in glucose and insulin relax to quasi-equilibrium on a timescale of hours, which is much faster than changes in mass, occurring over months or years. We recall the system of Topp et al. given by

The original Topp model is

$$\frac{dI}{dt} = \sigma \frac{G^2}{\alpha + G^2} \beta - kI, \tag{3}$$

$$\frac{dG}{dt} = R - (E + S_I I)G, \tag{4}$$

$$\frac{d\beta}{dt} = (d_0 + r_1 G - r_2 G^2)\beta. \tag{5}$$

The parameters of the model are presented in Table 5.

Several extensions have stemmed from this model [30, 32, 65, 76]. Ribbing (2010) [65] introduced a semi-mechanistic model, building upon the Topp model, with the aim of enhancing the understanding of various dynamics related to fasting plasma glucose, fasting insulin, insulin sensitivity, and β-cell mass. Their work targeted a comprehensive description of these dynamics across a heterogeneous population, including individuals with insulin resistance, and patients at various stages of T2DM, both with and without antidiabetic treatment.

Table 5 Parameters of the Topp model [72]

Parameter	Description
σ	Glucose-induced insulin secretion rate
α	Half-saturation constant for glucose
β	Insulin sensitivity parameter
k	Insulin clearance rate
R	Glucose appearance rate
E	Glucose effectiveness
S_I	Insulin sensitivity
d_0	Basal degradation rate of β-cells
r_1	Glucose-induced increase in β-cell replication
r_2	Glucose-induced inhibition of β-cell replication

A recent extension was introduced by Goel in [30]. Two modified versions were presented to incorporate insulin resistance and hypersecretion of insulin as causes of diabetogenic events. The first one is the Topp-IR model, where the β-cell mass is allowed to depend on I, and not on G alone. Consequently, a significant implication emerges; a novel pathway to diabetes unfolds within the model, contingent solely on changes in insulin resistance, rendering Topp-IR a mathematical model of insulin resistance-dependent diabetogenesis, a concept not initially encompassed by Topp's original model. So, the third equation (5) becomes

$$\frac{d\beta}{dt} = (d_0 + r_1 G - r_2 G^2 - d_I I)\beta.$$

The second approach is called the Topp-HS model, where the hypersecretion is modeled by assuming that $S_I(\sigma)$ represents how insulin sensitivity changes in response to secretory capacity. The second Topp equation (3) is modified to be

$$\frac{dG}{dt} = R(E + S_I(\sigma)I)G,$$

where $S_I(\sigma) = -8.33 \times 10^{-3}\sigma + 1.079$, while the other parameters are as in the Topp-IR model above.

A more recent model, introduced by Yang et al. in 2023 [76], is based on the Topp model and addresses the progression of T2DM with underlying obesity as an epigenetic-related diabetogenic factor, denoted by X, which is a bounded variable with a real value. As the progression of obesity leads to a worse impact on the Glucose-Insulin-β-cell mass regulatory system, X represents the severity of the obesity-related factor and is quantified as a variable with an upper bound of 1. The model is formulated as follows:

$$\frac{dG}{dt} = G_{\text{in}} + p_1(X) - f_2(G) - C(I)GI,$$

$$\frac{dI}{dt} = f_1(G)p_2(X)\beta - kI,$$

$$\frac{d\beta}{dt} = (f_3(I) + p_3(X))\beta,$$

where all variables in the system are on the time scale of days. The functions are defined as

$$f_2(G) = g_1 G, \quad C(I) = r_0 + \frac{r_1}{r_2 + e^{r_3 I}}, \quad f_1(G) = \frac{s_1 G^2}{G^2 + s_2}, \quad f_3(I) = \frac{m_1 I}{I^2 + (m_2)^2} - m_3,$$

and all parameters in the model are positive.

The function $p_1(X)$ represents the increased hepatic glucose production caused by the pathological factor; $p_2(X)$ symbolizes the impact of the factor on the insulin secretion rate; and $p_3(X)$ describes the abnormal response of β-cells to a hostile environment that develops on a slow time scale. For non-diabetic subjects, $p_1(X) = 0$, $p_2(X) = 1$, and $p_3(X) = 0$ when $X = 0$, so the model aligns with the undisturbed glucose-insulin regulatory system in the absence of diabetogenic factors. The expressions of these functions are

$$p_1(X) = h_1 X^{\alpha},$$

$$p_2(X) = 1 + h_2 X,$$

$$p_3(X) = q_1 X(q_2 - X),$$

$$\frac{dX}{dt} = n_1 X(n_2 - X).$$

Numerical simulations of the Yang model over a long-term period (42 years) provided valuable insights into disease progression in obesity-related diabetes. The model showed that severe obesity-induced hyperinsulinemia within 7 years, preceding diabetes onset by almost a decade. Insulin levels then decreased after 19 years, followed by β-cell failure. In contrast, mild obesity mitigated disease severity and slowed progression rates, highlighting the benefits of weight loss interventions. Sensitivity tests on key parameters underscored the model's robustness and its implications for optimizing treatment strategies. Additionally, controlling diabetogenic factors demonstrated therapeutic efficacy, aligning with clinical observations.

Another recent and advanced model was introduced by De Gaetano and colleagues in 2024 [19], resulting from extensive research efforts by him and his team from the early 2000s to the present [20, 21]. The model, consisting of only six equations, explicitly considers the contributions of physical exercise $Y(t)$ and food intake $F(t)$ toward variations in insulin sensitivity. It stands out for its simplicity while incorporating many features relevant to the long-term development of T2DM, focusing on the dynamics of the interaction between glycemia, insulinemia, and β-cell mass. By considering representative daily average glycemia and insulinemia, the model is given by the system:

$$\frac{dG(t)}{dt} = k^{G}_{\max} e^{-L(t)I(t)} G(t) - S(t)I(t)G(t) - R(G(t)), \tag{6}$$

$$\frac{dB(t)}{dt} = k_B e^{-\lambda_B A(t)} (A(t) - A_0) - k_{OBG} G(t) B(t), \tag{7}$$

$$\frac{dI(t)}{dt} = -k_{OI} I(t) + \frac{B(t)}{B_{ON}} k^{I}_{\max} \frac{G(t)^{\gamma}}{G^{\gamma}_{50} + G(t)^{\gamma}}, \tag{8}$$

$$\frac{dS(t)}{dt} = k_S - k_{OS}S(t) + k_{SY}Y(t) - k_{OSF}F(t)S(t) - k_{OSA}(A(t) - A_0)S(t), \tag{9}$$

$$\frac{dL(t)}{dt} = k_L - k_{OL}L(t) + k_{LY}Y(t) - k_{OLF}F(t)L(t) - k_{OLA}(A(t) - A_0)L(t), \tag{10}$$

$$\frac{dA(t)}{dt} = 1. \tag{11}$$

The model also incorporates the effect of urinary glucose elimination at high glycemia, represented by

$$R(G(t)) = s_{GR}\left(\sqrt{4\rho_{GR}G(t) + (G(t) - G_R - \rho_{GR})^2} + (G(t) - G_R - \rho_{GR})\right). \tag{12}$$

The meaning of the state variables is summarized in Table 6. The glucose dynamics are described by Eq. (6), where hepatic glucose production, represented by the term $k_{\max}^{G}e^{-L(t)I(t)}G(t)$, depends on hepatic insulin sensitivity L, as well as plasma glucose and insulin concentrations, modeled as a decreasing exponential function. In the same equation, glucose utilization due to the insulin effect on the periphery is represented by $S(t)I(t)G(t)$, where $S(t)$ is the peripheral insulin sensitivity, which varies over time and can be affected by aging. The glucose dynamics also consider renal elimination, represented by the function R, which can also incorporate treatments. The β-cell mass dynamics are represented by Eq. (7), where the production rate of β-cells is assumed to decrease with age, following a slow exponential decay with rate $\lambda_B A$. The dynamics also incorporate the toxicity of high levels of glucose on β-cells, represented by the term $-k_{OBG}G(t)B(t)$, leading to their dysfunction and death. This likely occurs because chronic exposure to high glucose levels causes β-cells to produce excessive amounts of insulin, gradually exhausting and damaging them. Consequently, β-cells lose their ability to produce insulin in response to

Table 6 Parameters of the De Gaetano model [19]

Variable	Units	Meaning
t	Year	Time after A_0
G	mM	Glycemia
B	Mc	β-cell mass in millions of β-cells
I	pM	Insulinemia
S	/year/pM	Systemic or peripheral insulin sensitivity
L	mM/pM	Liver or central insulin sensitivity
A	Year	Age
R	mM/year	Renal glucose elimination

glucose, leading to hyperglycemia and ultimately diabetes. The model distinguishes between hepatic insulin sensitivity and peripheral insulin sensitivity, enhancing its ability to detect critical stages in the development of T2DM.

After discussing the most recent mathematical approaches for the progression of T2DM, we now turn to those dedicated to T1DM. In T1DM, the pancreas produces little or no insulin. This condition is primarily characterized by the autoimmune destruction of pancreatic β-cells. The immune system mistakenly identifies β-cell proteins as foreign and destroys them, leading to insulin deficiency. Glucose metabolism involves various cell types, including β-cells, T-cells, macrophages, and dendritic cells. T-cells, originating from the bone marrow, play a crucial role in distinguishing between the body's cells and invading pathogens. Macrophages, a type of white blood cell, engulf and digest pathogens that lack specific proteins found on healthy cells. Dendritic cells process antigen material and present it to T-cells for immune detection. The autoimmune response depletes the pool of β-cells, causing insulin deficiency and resulting in severe complications such as hypoglycemia.

This destruction leads to an absolute deficiency of insulin, making patients with T1DM dependent on insulin for life. Despite rigorous disease management, these patients face an estimated 10–17.7 years decrease in lifespan [8, 35, 52, 61]. The risk for T1DM is attributed to a combination of genetic variants, environmental factors, and interactions between genes and the environment. Genetic susceptibility plays a significant role, with familial transmission evident in about 10% of T1DM patients. Environmental risk factors also contribute to T1DM susceptibility [45, 56, 62]. The consumption of a Western-type diet is associated with an increased risk of developing T1DM. These dietary components can trigger immune responses that may contribute to the autoimmune attack on β-cells [24, 45]. Various mathematical models have been developed to enhance our understanding of T1DM, drawing insights from experimental studies. One of the earliest models attempting to describe disease dynamics was proposed by Freiesleben De Blasio et al. in 1999 [29]. This spatial model incorporates parameters to predict the system's response to triggering events while minimizing damage to remaining β-cells, formulated as a system of ordinary differential equations (ODEs):

$$\dot{M} = a + (k + b)M_A - cM - gMA, \tag{13}$$

$$\dot{M}_A = gMA - kM_A, \tag{14}$$

$$\dot{A} = lM_A + pT + qM_AT - mA, \tag{15}$$

$$\dot{T} = sM_AT - tT, \tag{16}$$

where A denotes free β-cell antigenic proteins, M represents macrophages, and M_A denotes antigen-expressing macrophages. The model distinguishes between activated and non-activated macrophages, with cytokine production associated only with activated macrophages containing antigenic peptides. The rate of β-cell death, $\dot{B}$, is assumed to be proportional to activated macrophage levels, $\dot{B} = -l_B M_A$, where l_B is

Table 7 Parameters and their descriptions [29]

Parameter	Description
a	Recruitment rate of macrophages to volume
b	IL-1, TNF-α-induced recruitment rate of macrophages
c	Efflux rate of macrophages from volume
g	Uptake rate of β-cell antigens in macrophages
k	Decay rate of activated macrophages
l	Effective rate of IL-1, TNF-α-induced β-cell damage
m	Clearance rate of free β-cell proteins
p	Effective rate of IFN-γ-induced β-cell damage
q	Effective rate of β-cell damage caused by effector-macrophages
s	Proliferation rate of T-cells
t	Death rate of activated T-cells

IL-1, IFN-γ, and TNF-α are cytokines involved in inflammatory responses. IL-1 and TNF-α are induced by macrophages, while IFN-γ is induced by T-cells [73]

constant. The release of proteins due to β-cell death is modeled by $\dot{A} = -n\dot{B} = lM_A$, where $l = nl_B$.The set of parameters is described in Table 7.

In addition to this foundational model, Amrani et al. (2000) identified that T1DM progression hinges on specific T-cell populations [6], while Khadra et al. (2009) demonstrated bistability in T1DM dynamics, showing competition between high-affinity and low-affinity T-cells [43]. Jaberi-Douraki et al. (2015) introduced a model of T-cell avidity, elucidating mechanisms where effector T-cells evade Treg control at high avidities, thereby influencing autoimmune responses [38].

4 Enhancing Diabetes Management with Mathematical Models

Diabetes treatment involves a combination of strategies to manage blood sugar levels effectively. Medications play a crucial role in controlling diabetes. For individuals with T1DM, insulin is essential and is typically administered via injections or insulin pumps [26]. In contrast, those with T2DM often start with oral medications to improve insulin sensitivity, enhance insulin release, or reduce glucose production by the liver. Over time, some individuals with T2DM may require insulin to maintain optimal blood sugar control [26, 28, 37].

Regular monitoring of blood sugar levels is fundamental to adjusting treatment and maintaining optimal control. Methods include the A1C test, which measures average blood sugar levels over the past 2 to 3 months, the random blood sugar test for checking levels at any time, the fasting blood sugar test after an overnight fast, and the glucose tolerance test to evaluate responses after consuming a sugary liquid [25, 26, 63].

Lifestyle modifications are also critical components of diabetes management. These include dietary adjustments, such as managing portion sizes and controlling carbohydrate intake, along with adopting a healthy eating plan. Regular physical activity helps regulate blood sugar levels, and maintaining a healthy weight is essential for overall diabetes management [1, 25, 26, 63].

In T2DM, particularly with concurrent obesity, managing both hyperglycemia and obesity becomes paramount. Effective glucose-lowering medications include GLP-1 receptor agonists, dual GIP and GLP-1 receptor agonists like tirzepatide, SGLT-2 inhibitors, metformin, and amylin mimetics, which also promote weight loss. Conversely, insulin secretagogues, thiazolidinediones, and insulin tend to be associated with weight gain [25, 26, 63]. Preference is given to medications that offer both glucose-lowering and weight loss effects in individuals with T2DM and overweight or obesity. Continuous monitoring of weight management goals is recommended, and adjustments to treatment may include metabolic surgery or structured lifestyle management programs if goals are not achieved with initial therapies [1].

The era of mathematical modeling provides powerful tools for quantifying crucial aspects of glucose metabolism, encompassing physiological parameters like insulin sensitivity, glucose effectiveness, insulin secretion, and β-cell function. Understanding these biomarkers is pivotal for predicting the progression of both T1DM and T2DM [3, 13, 14].

Early models by Derouich and Boutayeb (2002) investigated glucose and insulin dynamics during physical exercise, laying foundational insights into how exercise impacts glucose utilization, insulin sensitivity, and overall metabolic responses [22]. Romeres et al. (2021) and Ntaganda et al. (2021) extended this work by examining exercise effects in healthy individuals and those with T1DM, contributing further to our understanding of glucose homeostasis during physical activity [59, 66]. Additionally, Landersdorfer et al. (2008) explored the influence of diet on glucose metabolism, considering factors such as meal composition, timing, and nutrient absorption [47].

Meta-models, such as those developed by Scharbarg et al. (2022), focus on optimizing insulin therapy during aerobic activity for managing T1DM, providing tailored dosing strategies for physical exercise scenarios [68]. Pharmacokinetic-Pharmacodynamic models by Lee et al. (2004) and Sun et al. (2011) deepen our understanding of insulin dynamics, including absorption, distribution, and its effects on blood glucose levels [49, 71].

Recent advancements include models linking carbohydrate and lipid intake to predict dietary effects on biomarkers and health outcomes, including T2DM risk, as demonstrated by Hassell-Sweatman et al. (2020) [33]. Nelida Elizabeth López-Palau's (2020) model for T2DM integrates gastric emptying, incretin effects, and metabolic rates, simulating glucose tests to provide insights for personalized treatment strategies [53]. Lazebnik et al. (2023) introduce mathematical models for the BCG vaccine, exploring its potential in preventing autoimmune diseases like T1DM [48]. Furthermore, Ferdous et al. (2023) highlight the impact of lifestyle factors on the evolution of T2DM prevalence, emphasizing the role of lifestyle modifications in diabetes prevention and management [27].

5 Limitations and Challenges

Although mathematical models have extensive applications, they still face several limitations and challenges. We categorize these limitations into three classes:

5.1 Data Quality and Availability

Challenges related to the quality and availability of data needed for accurate modeling:

- For diagnostic models, improving parameter identifiability requires more frequent and extended sampling compared to the standard 2-hour oral glucose or meal tolerance test. However, minimizing sample size or test duration may have practical advantages, but it increases the risk of type II errors or the need for larger sample sizes [39].
- Accounting for insulin sensitivity when interpreting model-derived β-cell parameters is not always performed, but it is advisable and can reveal differences between groups [31].
- Many of the data sources used in model development are older datasets in diabetes models. For instance, the SPINA Carb model, which supports fast in vivo estimation of insulin sensitivity and β-cell function, utilized records from 2472 subjects selected from the NHANES cohort of 2007–2008, with an additional 100 subjects from a different cohort included in the clamp study [23]. The model of DeGaetano is based on data obtained from the NHANES 2003–2004 survey and data obtained from the NHANES 2005–2006 survey [19].
- It's noteworthy that most data inputted into models are derived from European populations, with only a limited number of models developed using non-European population data. However, in the era of real-world evidence, the increasing availability of registry data from clinical practice settings offers an opportunity for model validation. Incorporating modern epidemiological data diabetes mellitus into disease progression equations for simulation represents a significant technological advancement. This approach may help mitigate the impact of limitations on model simulation. Nevertheless, additional validation studies are essential to assess findings and generalize results across diverse ethnic populations [51].
- Long-term modeling, particularly in disease progression sections, faces challenges due to the lack of data characterizing such evolution over extended periods. The absence of long-term data for model validation remains a problem, especially for longitudinal records of single patients [19].

5.2 *Complexity of Models*

Challenges related to the complexity of models:

- Mathematical models are increasingly used to inform public health policy, but a major dilemma faced by readers is how to evaluate the quality of models. All models require simplifying assumptions, and there are trade-offs between creating models that are more "realistic" versus those that are grounded in more well-characterized data on the behavior of disease processes [10].
- Complex models are not necessarily more accurate or reliable simply because they can more easily fit real-world data than simpler models; complex models can suffer parameter estimation problems that can be difficult to detect and often cannot be fixed by "calibrating" models to external data. Conversely, complexity can be important to include when uncertain factors are central to a disease process or research question.

5.3 *Emerging Trends*

In addition to the aforementioned limitations, mathematical models for diabetes face ongoing challenges in our modern era. Many mechanisms underlying diabetes need to be targeted by mathematical modeling in the future. We refer to

- Future research in T1DM aims to explore several key areas using mathematical modeling:
  Understanding these future research areas could pave the way for novel treatments and interventions in T1DM. For instance, targeting the melatonergic pathway in pancreatic β-cells may optimize mitochondrial function and mitigate oxidative stress-induced damage. Additionally, exploring the effects of melatonin and related compounds on immune responses and inflammation could offer new therapeutic avenues. Investigating the role of gut microbiome-derived factors and environmental influences like glyphosate-based herbicides may provide insights into disease prevention and management strategies.
  These research directions underscore the potential of mathematical modeling to integrate diverse aspects of T1DM pathophysiology and guide innovative approaches for treatment and prevention [7].
- Another domain where mathematical models can be implicated:
  The role of insulin in the brain, particularly in the hippocampal region, has been demonstrated to be critical for functional and structural changes in the brain for cognitive processes. Insulin plays a trophic role in the brain and serves as a metabolic homeostasis regulator, promoting neuroplasticity and high energy regulation. Understanding the molecular mechanisms of insulin on brain plasticity is critical for identifying the mechanisms that regulate neural plasticity in health and metabolic disease, such as diabetes-induced cognitive decline, as well as in neurodegenerative diseases, particularly Alzheimer's disease [4].

6 Innovation in Diabetic Technologies

The development of diabetes technology is driven by interdisciplinary collaboration among physicians, chemists, engineers, and mathematicians, leading to significant advancements in patient care. A notable achievement is the wearable artificial pancreas, now approaching outpatient usability [60]. Progress in mathematical models and control algorithms has played a crucial role in developing markers for diabetes control and sophisticated monitoring techniques. Advances in closed-loop systems, fueled by mathematical modeling, along with innovations in continuous glucose monitoring (CGM) and their integration with predictive models, have enabled proactive diabetes management [46].

An artificial pancreas, also known as a physiological closed-loop system, represents an automated insulin delivery system that continuously monitors glucose levels and adjusts insulin delivery in real time to maintain optimal blood glucose levels. This system comprises three main components: a CGM for real-time glucose measurements, an insulin pump for delivering insulin, and a control algorithm that calculates the required insulin dose based on CGM data. By integrating these components, the closed-loop system mimics the glucose-regulating function of a healthy pancreas, thereby reducing patient burden and improving glycemic control [44].

Existing simulators for T1DM utilize physiological models of the glucose-insulin system to simulate glycemic scenarios akin to real-life T1DM patients. Integrating these physiological models with closed-loop control methodologies enables the creation of simulation environments that faithfully replicate the glucose-insulin dynamics observed in T1DM patients. Such simulators have proven instrumental in the development and testing of various treatment strategies for T1DM [57].

Recent similators for T1DM have contributed insights into the nonlinearity of insulin action in the hypoglycemic range and the kinetics of glucagon during hypoglycemic responses. These findings have led to enhancements in the UVA/PADOVA T1DM Simulator [44, 75]. The updated model incorporates nonlinear insulin-dependent utilization as glucose levels decrease, thereby improving the simulation of hypoglycemic events. Additionally, it integrates a single-compartment model for glucagon kinetics, where glucagon secretion is influenced by plasma insulin and glucose dynamics, with a built-in delay that affects endogenous glucose production. This extension allows the simulator to accurately model scenarios ranging from single-meal to multiple-meal days, making it a valuable tool for in silico testing of cutting-edge diabetes therapies (see Fig. 3). The mathematical representation of these dynamics using nonlinear differential equations is detailed in [75].

7 Conclusion

In summary, the collaborative efforts of researchers, clinicians, and engineers have propelled diabetic technology forward. Mathematical models, predictive algo-

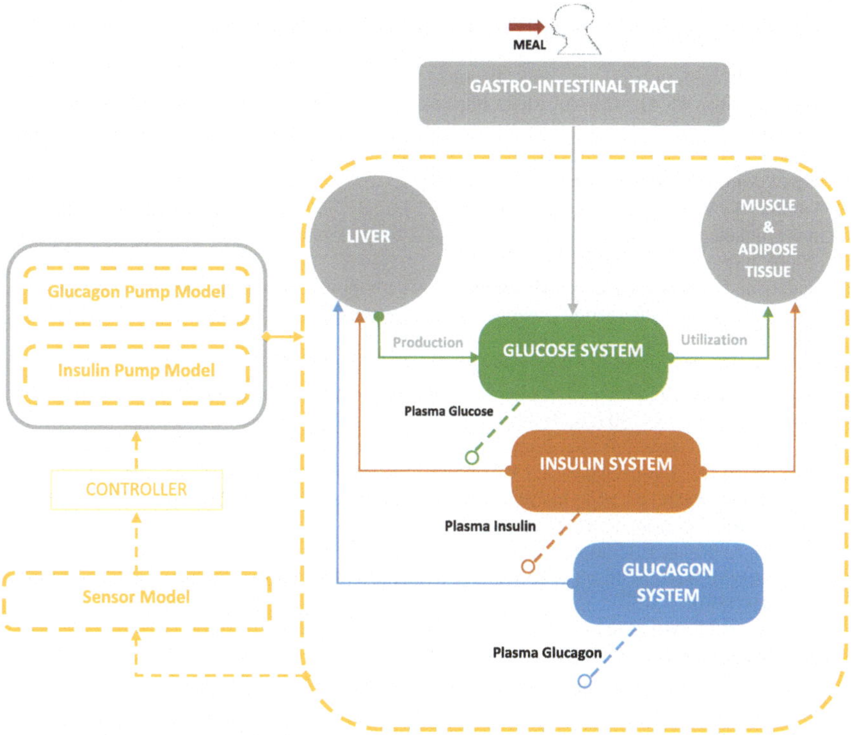

Fig. 3 Principal components of computer simulation environment: a model of the glucose-insulin system; a model of sensor error; the controller to be tested; and a model of insulin pump and s.c. insulin kinetics [75]

rithms, and continuous glucose monitoring have revolutionized diabetes management, improving patient outcomes and quality of life. These models have allowed for development of a wearable artificial pancreas and other sophisticated closed-loop systems. As regulatory guidelines continue to evolve, we anticipate further breakthroughs in this critical field, paving the way for enhanced therapies and better management strategies for diabetes.

References

1. American Diabetes Association Professional Practice Committee.: 8. Obesity and weight management for the prevention and treatment of type 2 diabetes: Standards of care in diabetes–2024. Diabet. Care **47**(Suppl 1), S145–S157 (2024). https://doi.org/10.2337/dc24-S008
2. Committee, American Diabetes Association Professional Practice.: Standards of medical care in diabetes–2022. Diabet. Care **45**(Suppl 1), 1–16 (2022)
3. Ajmera, I., Swat, M., Laibe, C., Le Novère, N., Chelliah, V.: The impact of mathematical modeling on the understanding of diabetes and related complications. CPT Pharmacomet.

Syst. Pharmacol. **2**(7), e54 (2013). https://doi.org/10.1038/psp.2013.30
4. Al Haj Ahmad, R.M., Ababneh, N.A., Al-Domi, H.A.: Brain insulin resistance as a mechanistic mediator links peripheral metabolic disorders with declining cognition. Diabet. Metab. Syndr. Clin. Res. Rev. **16**(4), 102468 (2022). https://doi.org/10.1016/j.dsx.2022.102468
5. Alkhateeb, H., El Fathi, A., Ghanbari, M., Haidar, A.: Modeling glucose dynamics during moderate exercise in individuals with type 1 diabetes. PLoS One **16**(3), e0248280 (2021). https://doi.org/10.1371/journal.pone.0248280
6. Amrani, A., Verdaguer, J., Serra, P., Tafuro, S., Tan, R., Santamaria, P.: Progression of autoimmune diabetes driven by avidity maturation of a T-cell population. Nature **406**(6797), 739–742 (2000). https://doi.org/10.1038/35021081
7. Anderson, G.: Type I diabetes pathoetiology and pathophysiology: Roles of the gut microbiome, Pancreatic cellular interactions, and the Bystander activation of memory CD8+ T Cells. Int. J. Mol. Sci. **24**(4), 3300 (2023). https://doi.org/10.3390/ijms24043300
8. Arffman, M., Hakkarainen, P., Keskimäki, I., Oksanen, T., Sund, R.: Long-term and recent trends in survival and life expectancy for people with type 1 diabetes in Finland. Diabetes Res. Clin. Pract. **198**, 110580 (2023). https://doi.org/10.1016/j.diabres.2023.110580
9. Bartlette, K., Carreau, A.-M., Xie, D., Garcia-Reyes, Y., Rahat, H., Pyle, L., Nadeau, K.J., Cree-Green, M., Diniz Behn, C.: Oral minimal model-based estimates of insulin sensitivity in obese youth depend on oral glucose tolerance test protocol duration. Metabol. Open **9**, 100078 (2021). https://doi.org/10.1016/j.metop.2021.100078
10. Basu, S., Andrews, J.: Complexity in mathematical models of public health policies: A guide for consumers of models. PLoS Med. **10**(10), e1001540 (2013). https://doi.org/10.1371/journal.pmed.1001540
11. Bergman, R.N.: Minimal model: Perspective from 2005. Horm. Res. **64**(Suppl 3), 8–15 (2005). https://doi.org/10.1159/000089312
12. Bergman, R.N.: Toward physiological understanding of glucose tolerance: Minimal-model approach. Diabetes **38**(12), 1512–1527 (1989). https://doi.org/10.2337/diab.38.12.1512
13. Boutayeb, A., Lamlili, M.E.N., Boutayeb, W.: A review of compartmental mathematical models used in diabetology. In: Boutayeb, A. (eds.) Disease Prevention and Health Promotion in Developing Countries. Springer, Cham (2020). https://doi.org/10.1007/978-3-030-34702-4_14
14. Boutayeb, A., Chetouani, A.: A critical review of mathematical models and data used in diabetology. Biomed. Eng. Online **5**, 43 (2006). https://doi.org/10.1186/1475-925X-5-43
15. Breda, E., Cavaghan, M.K., Toffolo, G., Polonsky, K.S., Cobelli, C.: Oral Glucose Tolerance Test Minimal Model Indexes of Beta-Cell Function and Insulin Sensitivity. Diabetes **50**(1), 150–158 (2001). https://doi.org/10.2337/diabetes.50.1.150
16. Cobelli, C., Dalla Man, C., Toffolo, G., Basu, R., Vella, A., Rizza, R.: The Oral Minimal Model Method. Diabetes **63**(4), 1203–1213 (2014). https://doi.org/10.2337/db13-1198
17. Cobelli, C., Toffolo, G.M., Dalla Man, C., Campioni, M., Denti, P., Caumo, A., Butler, P., Rizza, R.: Assessment of Beta-Cell Function in Humans, Simultaneously With Insulin Sensitivity and Hepatic Extraction, From Intravenous and Oral Glucose Tests. Am. J. Physiol. Endocrinol. Metab. **293**(1), E1–E15 (2007). https://doi.org/10.1152/ajpendo.00421.2006
18. Dalla Man, C., Rizza, R.A., Cobelli, C.: Meal Simulation Model of the Glucose-Insulin System. IEEE Trans. Biomed. Eng. **54**(10), 1740–1749 (2007). https://doi.org/10.1109/TBME.2007.893506
19. De Gaetano, A., Nagy, I., Kiss, D., Romanovski, V.G., Hardy, T.A.: A simplified longitudinal model for the development of type 2 diabetes mellitus. J. Theor. Biol. **587**, 111822 (2024). https://doi.org/10.1016/j.jtbi.2024.111822
20. De Gaetano, A., Hardy, T.A.: A novel fast-slow model of diabetes progression: Insights into mechanisms of response to the interventions in the diabetes prevention program. PLoS One **14**(10), e0222833 (2019). https://doi.org/10.1371/journal.pone.0222833
21. De Gaetano, A., Hardy, T.A., Beck, B., Abu-Raddad, E., Palumbo, P., Bue-Valleskey, J., Pørksen, N.: Mathematical models of diabetes progression. Am. J. Physiol.-Endocrinol. Metab. **295**(6), E1462–E1479 (2008). https://doi.org/10.1152/ajpendo.00654.2007

22. Derouich, M., Boutayeb, A.: The effect of physical exercise on the dynamics of glucose and insulin. J Biomech **35**(7), 911–917 (2002). https://doi.org/10.1016/s0021-9290(02)00055-6
23. Dietrich, J.W., Dasgupta, R., Anoop, S., et al.: SPINA carb: A simple mathematical model supporting fast In-Vivo estimation of insulin sensitivity and beta cell function. Sci. Rep. **12**, 17659 (2022). https://doi.org/10.1038/s41598-022-22531-3
24. Du, C., Whiddett, R.O., Buckle, I., Chen, C., Forbes, J.M., Fotheringham, A.K.: Advanced glycation end products and inflammation in type 1 diabetes development. Cells **11**(21), 3503 (2022). https://doi.org/10.3390/cells11213503
25. ElSayed, N.A., Aleppo, G., Aroda, V.R., Bannuru, R.R., Brown, F.M., Bruemmer, D., Collins, B.S., Hilliard, M.E., Isaacs, D., Johnson, E.L., Kahan, S., Khunti, K., Leon, J., Lyons, S.K., Perry, M.L., Prahalad, P., Pratley, R.E., Seley, J.J., Stanton, R.C., Gabbay, R.A.: American Diabetes Association. Obesity and Weight Management for the Prevention and Treatment of Type 2 Diabetes: Standards of Care in Diabetes-2023. Diabet. Care **46**(Suppl 1), S128–S139 (2023). https://doi.org/10.2337/dc23-S008
26. ElSayed, N.A., Aleppo, G., Aroda, V.R., Bannuru, R.R., Brown, F.M., Bruemmer, D., Collins, B.S., Hilliard, M.E., Isaacs, D., Johnson, E.L., Kahan, S., Khunti, K., Leon, J., Lyons, S.K., Perry, M.L., Prahalad, P., Pratley, R.E., Seley, J.J., Stanton, R.C., Gabbay, R.A.; on behalf of the American Diabetes Association. Pharmacologic Approaches to Glycemic Treatment: Standards of Care in Diabetes-2023. Diabetes Care, **46**(Suppl 1), S140-S157 (2023). https://doi.org/10.2337/dc23-S009
27. Ferdous, A. An Ordinary Differential Equation Model for Assessing the Impact of Lifestyle Intervention on Type 2 Diabetes Epidemic. Healthcare Analytics 2023;4:100271 (2023). https://doi.org/10.1016/j.health.2023.100271
28. Freckmann, G., Buck, S., Waldenmaier, D., Kulzer, B., Schnell, O., Gelchsheimer, U., Ziegler, R., Heinemann, L.: Insulin Pump Therapy for Patients With Type 2 Diabetes Mellitus: Evidence, Current Barriers, and New Technologies. J. Diabetes Sci. Technol. **15**(4), 901–915 (2021). https://doi.org/10.1177/1932296820928100
29. Freiesleben De Blasio, B., Bak, P., Pociot, F., Karlsen, A.E., Nerup, J. Onset of Type 1 Diabetes: A dynamical instability. Diabetes **48**(9), 1677-1685 (1999). https://doi.org/10.2337/diabetes.48.9.1677
30. Goel, P.: Insulin resistance or hypersecretion? The βIG Picture Revisited. J. Theor. Biol. **384**, 131–139 (2015). https://doi.org/10.1016/j.jtbi.2015.07.033
31. Ha, J., Muniyappa, R., Sherman, A.S., Quon, M.J.: When MINMOD artifactually interprets strong insulin secretion as weak insulin action. Front. Physiol. **12**, 601894 (2021). https://doi.org/10.3389/fphys.2021.601894
32. Hamou-Maamar, M., Belhamiti, O.: Leptin Effect's on Glucose and Insulin Kinetics: A Mathematical Model. Commun. Nonlinear Sci. Numer. Simul. **113**, 106591 (2022). https://doi.org/10.1016/j.cnsns.2022.106591
33. Hassell Sweatman, C.Z.W. Mathematical Model of Diabetes and Lipid Metabolism Linked to Diet, Leptin Sensitivity, Insulin Sensitivity and VLDLTG Clearance Predicts Paths to Health and Type II Diabetes. J Theor Biol **486**, 110037 (2020). https://doi.org/10.1016/j.jtbi.2019.110037
34. Huard, B., Kirkham, G.: Mathematical Modelling of Glucose Dynamics. Curr. Opinion Endocrine Metabol. Res. **25**, 100379 (2022). https://doi.org/10.1016/j.coemr.2022.100379
35. Huo, L., Harding, J.L., Peeters, A., et al.: Life Expectancy of Type 1 Diabetic Patients During 1997–2010: A National Australian Registry-Based Cohort Study. Diabetologia **59**(6), 1177–1185 (2016). https://doi.org/10.1007/s00125-015-3857-4
36. International Diabetes Federation.: IDF Diabetes Atlas, 10th edn. International Diabetes Federation, Brussels (2021). https://diabetesatlas.org
37. InformedHealth.org [Internet]. Cologne, Germany: Institute for Quality and Efficiency in Health Care (IQWiG); 2006-. Type 2 diabetes: Learn More–What medications help in type 2 diabetes? [Updated 2023 Dec 18]. Available from: https://www.ncbi.nlm.nih.gov/books/NBK279506/

38. Jaberi-Douraki, M., Pietropaolo, M., Khadra, A.: Continuum Model of T-Cell Avidity: Understanding Autoreactive and Regulatory T-Cell Responses in Type 1 Diabetes. J. Theor. Biol. **383**, 93–105 (2015). https://doi.org/10.1016/j.jtbi.2015.07.032
39. Kahn, S.E., Chen, Y.-C., Esser, N., Taylor, A.J., van Raalte, D.H., Zraika, S., Verchere, C.B.: The β-Cell in Diabetes: Integrating Biomarkers With Functional Measures. Endocrine Rev. **42**(5), 528–583 (2021). https://doi.org/10.1210/endrev/bnab021
40. Kartono, A., Mufid, H. M., Wahyudi, S. T. Modified Glucose Absorption Equation in an Oral Minimal Model for Type 2 Diabetes Mellitus. J. Phys.: Confer. Ser. **1491**(1), 012023 (2020). https://dx.doi.org/10.1088/1742-6596/1491/1/012023
41. Kartono, A., Nurullaeli, Syafutra, H., Wahyudi, S. T., Sumaryada, T. A Mathematical Model of the Intravenous Glucose Tolerance Test Illustrating an n-Order Decay Rate of Plasma Insulin in Healthy and Type 2 Diabetic Subjects. IOP Conference Series: Materials Science and Engineering **532**(1), 012016 (2019). https://doi.org/10.1088/1757-899X/532/1/012016
42. Katz, A., Nambi, S.S., Mather, K., Baron, A.D., Follmann, D.A., Sullivan, G., Quon, M.J.: Quantitative Insulin Sensitivity Check Index: A Simple, Accurate Method for Assessing Insulin Sensitivity in Humans. J. Clin. Endocrinol. Metab. **85**(7), 2402–2410 (2000). https://doi.org/10.1210/jcem.85.7.6661
43. Khadra, A., Santamaria, P., Edelstein-Keshet, L.: The Role of Low Avidity T Cells in the Protection Against Type 1 Diabetes: A Modeling Investigation. Journal of Theoretical Biology **256**(1), 126–141 (2009). https://doi.org/10.1016/j.jtbi.2008.09.019
44. Khodaei, M.J., Candelino, N., Mehrvarz, A., Jalili, N.: Physiological Closed-Loop Control (PCLC) Systems: Review of a Modern Frontier in Automation. IEEE Access **8**, 23965–24005 (2020). https://doi.org/10.1109/ACCESS.2020.2966928
45. Kieleväinen, V., Turtinen, M., Luopajärvi, K., Härkönen, T., Ilonen, J., Knip, M., & Finnish Pediatric Diabetes Register. Increased HLA Class II Risk is Associated With a More Aggressive Presentation of Clinical Type 1 Diabetes. Acta Paediatrica (Oslo, Norway: 1992) **112**(3), 522-528 (2023). https://doi.org/10.1111/apa.16621
46. Kovatchev, B.P.: Diabetes Technology: Markers, Monitoring, Assessment, and Control of Blood Glucose Fluctuations in Diabetes. Scientifica **2012**, 283821 (2012). https://doi.org/10.6064/2012/283821
47. Landersdorfer, C.B., Jusko, W.J.: Pharmacokinetic/Pharmacodynamic Modelling in Diabetes Mellitus. Clin Pharmacokinet **47**, 417–448 (2008). https://doi.org/10.2165/00003088-200847070-00001
48. Lazebnik, T., Bunimovich-Mendrazitsky, S., Kiselyov, A.: Mathematical Model for BCG-Based Treatment of Type 1 Diabetes. Physica A: Statistical Mechanics and its Applications **622**, 128891 (2023). https://doi.org/10.1016/j.physa.2023.128891
49. Lee, S.H., Kwon, K.I.: Pharmacokinetic-Pharmacodynamic Modeling for the Relationship Between Glucose-Lowering Effect and Plasma Concentration of Metformin in Volunteers. Arch. Pharm. Res. **27**(7), 806–810 (2004). https://doi.org/10.1007/BF02980152
50. Levy, J.C., Matthews, D.R., Hermans, M.P.: Correct Homeostasis Model Assessment (HOMA) Evaluation Uses the Computer Program. Diabetes Care **21**(12), 2191–2192 (1998). https://doi.org/10.2337/diacare.21.12.2191
51. Li, J., Bao, Y., Chen, X., et al.: Decision Models in Type 2 Diabetes Mellitus: A Systematic Review. Acta Diabetol. **58**(11), 1451–1469 (2021). https://doi.org/10.1007/s00592-021-01742-6
52. Livingstone, S.J., Levin, D., Looker, H.C., Lindsay, R.S., Wild, S.H., Joss, N., Leese, G., Leslie, P., McCrimmon, R.J., Metcalfe, W., McKnight, J.A., Morris, A.D., Pearson, D.W., Petrie, J.R., Philip, S., Sattar, N.A., Traynor, J.P., Colhoun, H.M., Scottish Diabetes Research Network Epidemiology Group, & Scottish Renal Registry. Estimated Life Expectancy in a Scottish Cohort With Type 1 Diabetes, 2008–2010. JAMA **313**(1), 37–44 (2015). https://doi.org/10.1001/jama.2014.16425
53. López-Palau, N.E., Olais-Govea, J.M.: Mathematical Model of Blood Glucose Dynamics by Emulating the Pathophysiology of Glucose Metabolism in Type 2 Diabetes Mellitus. Sci Rep **10**, 12697 (2020). https://doi.org/10.1038/s41598-020-69629-0

54. Mari, A., Schmitz, O., Gastaldelli, A., Oestergaard, T., Nyholm, B., Ferrannini, E.: Meal and Oral Glucose Tests for Assessment of Beta-Cell Function: Modeling Analysis in Normal Subjects. Am. J. Physiol. Endocrinol. Metab. **283**(6), E1159–E1166 (2002). https://doi.org/10.1152/ajpendo.00093.2002
55. Mari, A., Pacini, G., Murphy, E., Ludvik, B., Nolan, J.J.: A Model-Based Method for Assessing Insulin Sensitivity from the Oral Glucose Tolerance Test. Diabetes Care **24**(3), 539–548 (2001). https://doi.org/10.2337/diacare.24.3.539. Erratum. In: Diabetes Care 37(7), 2063 (2014)
56. Morran, M.P., Vonberg, A., Khadra, A., Pietropaolo, M.: Immunogenetics of Type 1 Diabetes Mellitus. Molecular Aspects of Medicine **42**, 42–60 (2015). https://doi.org/10.1016/j.mam.2014.12.004
57. Mujahid, O., Contreras, I., Beneyto, A., et al.: Generative Deep Learning for the Development of a Type 1 Diabetes Simulator. Communications Medicine **4**(1), 51 (2024). https://doi.org/10.1038/s43856-024-00476-0
58. Nath, A., Biradar, S., Balan, A., Dey, R., Padhi, R.: Physiological Models and Control for Type 1 Diabetes Mellitus: A Brief Review. IFAC-PapersOnLine **51**(1), 289–294 (2018). https://doi.org/10.1016/j.ifacol.2018.05.077
59. Ntaganda, J.M., Minani, F., Banzi, W., et al.: Analysis of Physical Activity Effects on Plasma Glucose-Insulin System Dynamics: A Mathematical Model. Transactions of the Institute of Measurement and Control **43**(14), 3272–3281 (2021). https://doi.org/10.1177/01423312211022867
60. Nuzzo, M.G., Schettino, M.: Advanced Technology (Continuous Glucose Monitoring and Advanced Hybrid Closed-Loop Systems) in Diabetes from the Perspective of Gender Differences. Diabetology **4**(4), 519–526 (2023). https://doi.org/10.3390/diabetology4040045
61. Ou, H.-T., Yang, C.-Y., Wang, J.-D., Hwang, J.-S., Wu, J.-S.: Life Expectancy and Lifetime Health Care Expenditures for Type 1 Diabetes: A Nationwide Longitudinal Cohort of Incident Cases Followed for 14 Years. Value in Health **19**(8), 976–984 (2016). https://doi.org/10.1016/j.jval.2016.05.017
62. Parkkola, A., Härkönen, T., Ryhänen, S. J., Ilonen, J., Knip, M.; Finnish Pediatric Diabetes Register. Extended Family History of Type 1 Diabetes and Phenotype and Genotype of Newly Diagnosed Children. Diabetes Care **36**(2), 348–354 (2013). https://doi.org/10.2337/dc12-0445
63. Pereira, M.J., Eriksson, J.W.: Emerging Role of SGLT-2 Inhibitors for the Treatment of Obesity. Drugs **79**(3), 219–230 (2019). https://doi.org/10.1007/s40265-019-1057-0
64. Periwal, V., Chow, C.C., Bergman, R.N., Ricks, M., Vega, G.L., Sumner, A.E.: Evaluation of Quantitative Models of the Effect of Insulin on Lipolysis and Glucose Disposal. Am J Physiol Regul Integr Comp Physiol **295**(4), R1089–R1096 (2008). https://doi.org/10.1152/ajpregu.90426.2008
65. Ribbing, J., Hamrén, B., Svensson, M.K., Karlsson, M.O.: A Model for Glucose, Insulin, and Beta-Cell Dynamics in Subjects with Insulin Resistance and Patients with Type 2 Diabetes. J Clin Pharmacol **50**(8), 861–872 (2010). https://doi.org/10.1177/0091270009349711
66. Romeres, D., Schiavon, M., Basu, A., Cobelli, C., Basu, R., Dalla Man, C.: Exercise Effect on Insulin-Dependent and Insulin-Independent Glucose Utilization in Healthy Individuals and Individuals with Type 1 Diabetes: A Modeling Study. Am J Physiol Endocrinol Metab **321**(1), E122–E129 (2021). https://doi.org/10.1152/ajpendo.00084.2021
67. Roy, A., Parker, R.S.: Dynamic Modeling of Free Fatty Acid, Glucose, and Insulin: An Extended "Minimal Model" Diabetes Technology & Therapeutics **8**(6), 617-626 (2006). https://doi.org/10.1089/dia.2006.8.617
68. Scharbarg, E., Greck, J., Le Carpentier, E., et al.: A Metamodel-Based Flexible Insulin Therapy for Type 1 Diabetes Patients Subjected to Aerobic Physical Activity. Sci Rep **12**, 8017 (2022). https://doi.org/10.1038/s41598-022-11772-x
69. Health Care Expenditure Due to Diabetes Worldwide in 2021, by Region. Statista. Available at: https://www.statista.com/statistics/241831/health-care-costs-due-to-diabetes-worldwide-by-region/
70. Subramanian, V., Bagger, J.I., Harihar, V., Holst, J.J., Knop, F.K., Vilsbøll, T.: An Extended Minimal Model of OGTT: Estimation of α- and β-Cell Dysfunction, Insulin Resistance, and

the Incretin Effect. Am J Physiol Endocrinol Metab **326**(2), E182–E205 (2024). https://doi.org/10.1152/ajpendo.00278.2023
71. Sun, L., Kwok, E., Gopaluni, B., Vahidi, O.: Pharmacokinetic-Pharmacodynamic Modeling of Metformin for the Treatment of Type II Diabetes Mellitus. Open Biomed Eng J **5**, 1–7 (2011). https://doi.org/10.2174/1874120701105010001
72. Topp, B., Promislow, K., Devries, G., Miura, R.M., Finegood, D.T.: A Model of β-Cell Mass, Insulin, and Glucose Kinetics: Pathways to Diabetes. J Theor Biol **206**(4), 605–619 (2000). https://doi.org/10.1006/jtbi.2000.2150
73. Turner, M. D., Nedjai, B., Hurst, T., Pennington, D. J. Cytokines and Chemokines: At the Crossroads of Cell Signalling and Inflammatory Disease. Biochimica et Biophysica Acta (BBA) - Molecular Cell Research **1843**(11), 2563-2582 (2014). https://doi.org/10.1016/j.bbamcr.2014.05.014
74. Van Herpe, T., Pluymers, B., Espinoza, M., Van den Berghe, G., De Moor, B.A., Model, Minimal, for Glycemia Control in Critically Ill Patients. In,: International Conference of the IEEE Engineering in Medicine and Biology Society. New York, NY, USA **2006**, 5432–5435 (2006). https://doi.org/10.1109/IEMBS.2006.260613
75. Visentin, R., Campos-Náñez, E., Schiavon, M., et al.: The UVA/Padova Type 1 Diabetes Simulator Goes From Single Meal to Single Day. J Diabetes Sci Technol **12**(2), 273–281 (2018)
76. Yang, B., Li, J., Haller, M.J., Schatz, D.A., Rong, L.: Modeling the Progression of Type 2 Diabetes with Underlying Obesity. PLoS Comput. Biol. **19**(2), e1010914 (2023). https://doi.org/10.1371/journal.pcbi.1010914
77. Yang, B., Li, J., Haller, M.J., Schatz, D.A., Rong, L. The Progression of Secondary Diabetes: A Review of Modeling Studies. Frontiers in Endocrinology (Lausanne) **13**, 1070979 (2022). https://doi.org/10.3389/fendo.2022.1070979

An Overview of Data Mining and Machine Learning Models for Diabetes

Abdesslam Boutayeb and Yasser Lamlili E. N.

Abstract This chapter is presenting an overview of data mining and machine learning models devoted to different aspects of diabetes during the last two decades. Thirty-six machine learning models have been used by 31 authors for different aspects of diabetes, including prediction, diagnosis, classification and complications of diabetes. Table 1 summarizes the different algorithms used with at least one reference for readers who need to know how these algorithms are defined and implemented. Table 8 lists 30 authors with the names of the algorithms implemented, the data for training and testing these algorithms and the algorithm giving the best accuracy.

1 Introduction

In parallel with the huge increase of datasets in all domains and the crucial need for decision-makers to select the efficient strategy among a vast basket of possibilities, the use of data mining, artificial intelligence and machine learning increased significantly and regularly during the last decades. These new concepts are based on exploratory, diagnostic, predictive, prescriptive, quantitative, qualitative and descriptive or inferential statistical analysis in general.

Based on methods of data analysis, data mining uses the correlation between variables and clustering between individuals or groups of individuals to discover and highlight unknown links and/or patterns. Using mapping and different methods of data analysis such as principal component analysis (PCA), linear regression (LR), linear discriminant analysis (LDA), clustering, classification as well as complex algorithms, these methods can help decision-makers to adopt the cost-efficient strategy based on segregated data, especially in the case of huge datasets.

A. Boutayeb (✉)
Emirates Aviation University, Dubai, UAE
e-mail: x.boutayeb@gmail.com

Y. Lamlili E. N.
Mohammed First University, Oujda, Morocco

A. Boutayeb and M. E. N. Lamlili (eds.), *Computational Mathematics and Modelling for Diabetes*, Industrial and Applied Mathematics,
https://doi.org/10.1007/978-981-96-1925-2_5

Artificial intelligence (AI) is a technique trying to mimic human knowledge and behaviour by non-human machines coded or programmed to solve real-life problems going through the successive stages of learning, planning, testing, deciding and proposing different solutions. The speech recognition, language processing and academic research are well-known domains where AI has proved its power.

The use of machine learning algorithms to predict the onset of diabetes and its complications as well as the main risk factors associated with this silent epidemic saw a significant development during the last decade. A large number of studies are regularly published using different datasets dealing with diabetes [1–7, 9–11, 14–31, 33–40].

This chapter is an overview of machine learning approaches devoted to different aspects of diabetes, including prediction, diagnosis, classification and detection of the main risk factors. An example of such approaches is given in Chap. 2 of this book where the reader may find how machine learning algorithms are applied to predict diabetes risk factors and early complications of diabetes [31].

2 General Presentation of Machine Learning Approaches

A multitude of machine learning algorithms are being used by different authors aiming to predict the onset of diabetes and its complications and/or to characterize the main risk factors of type 2 diabetes. The main purpose of this chapter is not a systematic review but rather a brief summary of recent methods used for diabetes in its different aspects [39].

All machine learning algorithms need the availability of a database with indication on the population or sample considered and the features/attributes measured. Then, according to the type of available data and features (numeric, binomial, qualitative), authors may go through different stages (data cleaning, missing data, correlated data...) before opting for one algorithm or comparing between many algorithms in order to choose the best of them in terms of performance metrics.

2.1 Machine Learning Algorithms

The main purpose of Machine Learning (ML) is to transform the information acquired from a data sample into general concepts through two steps: (1) analysing available data to discover (hidden) dependencies and (2) using the discovered dependencies to generate corresponding outputs for new inputs. Generally, machine learning is divided into supervised learning, unsupervised learning and semi-supervised learning [27].

In order to select the best model based on its accuracy, authors usually split data into a first subset (70%) for training and a second subset (30%) for testing. Consequently, the results given by any machine learning model are highly dependent

Table 1 Features (attributes) used in PIDD

Feature	Description	Range
Age	Years	[21, 81]
Preg	Number of times pregnant	[0, 17]
Gluc	Plasma glucose at 2 hours in an oral GTTT	[0, 199]
BP	Diastolic blood pressure (mmHg)	[0, 122]
Skin	Triceps skin fold thickness (mm)	[0, 99]
Insulin	2-hour serum insulin (mh/ml)	[0, 846]
BMI	Body mass index (weight in kg)/(height in m^2)	[0, 67.1]
DPF	Diabetes pedigree function	[0.078, 242]
Outcome	Binary value indicating non-diabetic/diabetic	0 or 1

Source Dua et al. [20]

on the data used. Unfortunately, diabetes data are limited in availability and most existing data-driven diabetes detection models are trained using the Pima Indians Diabetes Database (PIDD) providing 768 records and the 9 attributes indicated in Table 1 [20]. For example, the PIDD was used by 75% of the authors cited by We et al. in a comprehensive review on diabetes detection based on machine learning and deep learning approaches published in 2024 [39].

A (non-exhaustive) summary of machine learning algorithms used for diabetes is given in Table 2. With each algorithm, at least one reference is also indicated for readers who are willing to know how the algorithm is mathematically defined and how it is usually implemented.

2.2 *Classification Performance Metrics*

Machine learning algorithms are usually associated with some performance metrics that allow comparing between different algorithms and eventually choosing the best adapted model. The metrics Accuracy, Precision, Specificity, Recall, F-Measure, and MCC are defined as follows:

$$\text{Accuracy} = \frac{\text{TN} + \text{TP}}{\text{TN} + \text{TP} + \text{FN} + \text{FP}} \tag{1}$$

$$\text{Recall} = \frac{\text{TP}}{\text{TP} + \text{FN}} \tag{2}$$

$$\text{Specificity} = \frac{\text{TN}}{\text{TN} + \text{FP}} \tag{3}$$

Table 2 Summary of machine learning and deep learning algorithms used for diabetes

Machine Learning (ML) algorithms	Reference for explanation	Machine Learning (ML) algorithms	Reference for explanation
K-Nearest Neighbour (K-NN)	[24, 26]	Radial Basis Function Network (RBF)	[34]
Decision Tree (DT)	[24, 26, 37]	Linear Discriminant Analysis (LDA)	[2]
Random Forest (RF)	[24, 37, 40]	Artificial Immune System (AIS	[14]
Logistic Regression (LR)	[24, 32, 40]	Artificial Neural Network (ANN)	[14, 30]
Naïve Bayes (NB)	[26, 30, 32]	Probabilistic Neural Networks (PNNs),	[14]
Support Vector Machine (SVM)	[24, 37, 40]	Time Delay Networks (TDNs)	[14]
AdaBoost (AB)	[27, 37, 40]	Feedforward Networks (FFNs)	[14]
XGBoost (XGB)	[26, 27]	Cascade-Forward Networks (CFNs),	[14]
Gradient Boost (GB)	[37]	Distributed Time Delay Networks (DTDNs)	[14]
Gaussian naïve Bayes (GNB)	[37]	Learning Vector Quantization (LVQ)	[14]
Multilayer Perception (MLP)	[34, 37]	Gaussian Mixture Model (GMM)	[32]
Random Tree (RT)	[35]	Elaboration Likelihood Model (ELM)	[32]
IBk decision tree	[35]	Ensemble Majority Voting (EMV)	[3]
J48	[24, 37, 40]	Neural Network Pattern Recognition (NPR)	[7]
Genetic Algorithm (GA)	[30]	Iterative Decision Tree (ID3)	[40]
Permutation Importance (PI)	[30]	Recursive Feature Elimination (RFE)	[29]
LASSO	[30]	Gradient Boosting Machine (GBM)	[31]
AdaboostM1	[6]	Principal Component Analysis (PCA)	[13]

$$\text{Precision} = \frac{\text{TP}}{\text{TP} + \text{FP}} \tag{4}$$

$$\text{F- Measure} = 2\text{x}\left(\frac{\text{Precision x Recall}}{\text{Precision } + \text{ Recall}}\right) \tag{5}$$

$$\text{MCC} = \frac{((\text{TN} \times \text{TP})) - (FP \times FN))}{((TP + FP) \times (TP + FN) \times (TN + FP) \times (TN + FN))1/2}. \tag{6}$$

ROC AUC: Graphical representation of the relationship between TP and FP is Receiver Operating Characteristics Curve (ROC). As the area under the ROC curve (ROC AUC) value approaches 1, the diagnostic value increases.

Here
'True Positive (TP)' is the correctly predicted positive class value,
'False Negative (FN)' is the incorrectly predicted negative class value,
'False Positive (FP)' is the incorrectly predicted positive class value, and
'True Negative (TN)' represents the correctly predicted negative class value.

Most of papers dealing with machine learning models give details on how these metrics are defined.

Other criteria include Mean Absolute Error (MAE) and Root Mean Square Error (RMSE):

$$\text{MAE} = \frac{1}{n}\sum_{j=1}^{n}\left|y_j - \hat{y}_j\right|$$

$$\text{RMSE} = \sqrt{\frac{1}{n}\sum_{j=1}^{n}\left(y_j - \hat{y}_j\right)^2}.$$

3 An Overview of Studies Devoted to Data Mining and Diabetes

During the last two decades, diverse data and methods of analysis have been used to predict the incidence and prevalence of diabetes [24, 34].

3.1 Data Mining and Diabetes

The Arab populations have many similarities and dissimilarities. They share culture, language and religion but they are also subject to economic, political and social differences. The comparison between Arab countries in terms of health, education and economic indicators shows that an inequitable pattern is seen in indicators like maternal mortality ratio, infant mortality rate, life expectancy at birth, human devel-

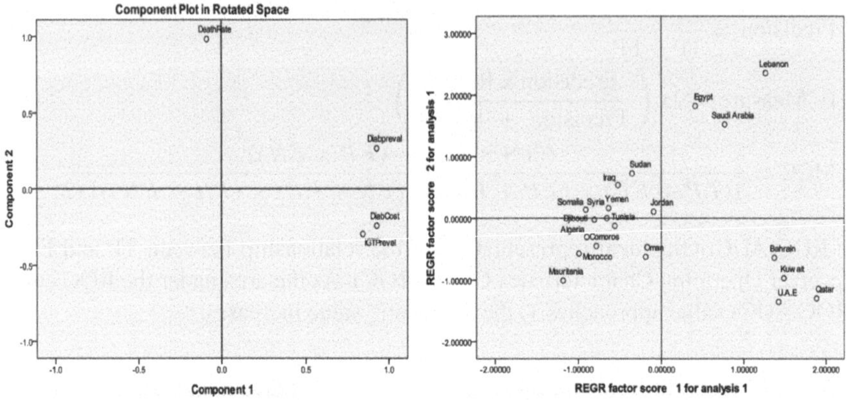

Fig. 1 Principal component analysis projections of variables and countries on the 1st plan *Source* Boutayeb et al. [13]

opment index, children nutritional status, early marriage, adolescent fertility, different levels of education, labour force participation of women and others [12].

In 2012, Boutayeb et al. dedicated a paper to the rise of diabetes prevalence in the Arab region. They used Principal Component Analysis (PCA) to illustrate similarities and differences between Arab countries according to four variables: (1) the prevalence of diabetes, (2) impaired glucose tolerance (IGT), (3) diabetes-related deaths and (4) diabetes-related expenditure per person. A linear regression was also used to study the correlation between human development index and diabetes prevalence [13]. The results given by the PCA are given in Fig. 1.

The authors also used a linear regression to investigate the correlation between the Human Development Index (HDI) and diabetes comparative prevalence. They found that a high correlation ($R = 0.81$) exist between the variable prevalence of diabetes (Diabpreval) and the Human Development Index (HDI), indicating the extreme opposition between the group of six countries with the highest HDI and highest prevalence of diabetes (Kuwait, Lebanon, Qatar (0.831, 20.2%), Saudi Arabia, Bahrain and UAE) and another group of six countries with low HDI and low diabetes prevalence (Somalia, Mauritania, Djibouti, Comoros, Sudan and Yemen). The remaining nine countries had a combined HDI and diabetes prevalence between those of the lowest and highest groups (Libya, Jordan, Oman, Tunisia, Algeria, Syria, Egypt, Morocco and Iraq).

The authors concluded that the majority of Arab countries are engaged in a multidimensional transition (demographic, economic, epidemiological and geographic). This four-dimensional transition engendered many direct and indirect factors explaining the high prevalence of diabetes. The shift from rural-to-urban dominance leads to sedentary life and less physical exercise. The socioeconomic development allowed for higher income and more consumption especially of fast food and Western diet. Finally, with epidemiological and demographic transitions, ageing with a different history of early childhood conditions fostered the development of obesity and impaired glucose tolerance.

In 2020, Almutairi et al. developed different stochastic models in order to study the trends in incidence rates of diabetes over 15 years during the period 1999 – 2013, and to predict the incidence and prevalence of diabetes in Saudi Arabia for the future level up to 2025. They proposed the following five models: Multiple Linear Regression (MLR), Adaptive Neuro-Fuzzy Interference System (ANFIS), Artificial Neural Networks (ANN), Support Vector Regression (SVR) and Bayesian Linear Regression (BLR) and implemented these algorithms using data provided by the Saudi Ministry of Health and obtained via a health interview survey concerning adults aged 15 years and over. Using the five variables (age, sex, obesity, inactivity and smoking), they computed and compared the results given by each model. Indicating that smoking, obesity and physical inactivity were the most significant contributing factors in diabetes, they stressed 'the urgency need for seeking more strategies and preventive programmes to overcome this disease and reduce its complications, or at least delay them' [8].

In 2021, Boutayeb et al. used Data Mining for the prediction of risk factors of type 2 diabetes mellitus in Gulf countries. They performed a case-control epidemiological study on a dataset of 4061 patients. The population of this study is characterized by cardiac patients living in the four Gulf countries (UAE, Oman, Kuwait, Bahrain) followed in their corresponding hospitals. Principal Component Analysis (PCA) was carried out on cleaned data and then different risk factors of type 2 diabetes were detected and analysed by data mining technics (Decision Tree) (Fig. 2). The principal component analysis method allowed to divide the population into three distinctive groups, namely, non-diabetic persons, pre-diabetics and type 2 diabetic persons [15].

The authors also used data mining technics (Decision Tree) to detect the main risk factors of Type 2 diabetes in the four Gulf countries based on the dataset of 4061 patients. As illustrated in Fig. 3, the major risk factors are hypertension and

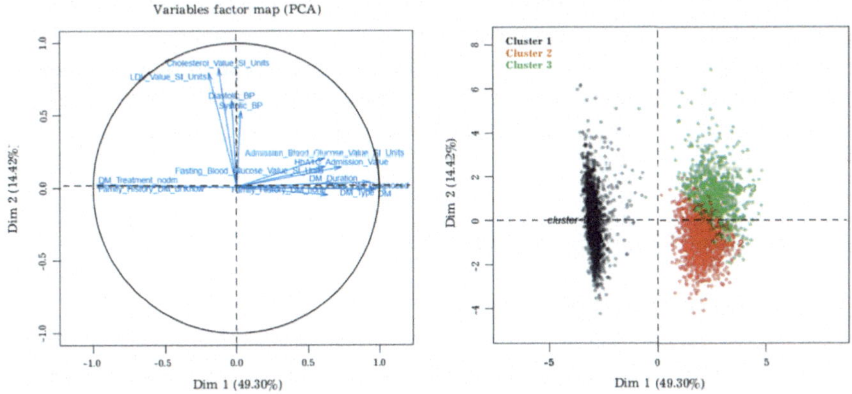

Fig. 2 Principal component analysis projections of 13 variables and individuals on the 1st plan *Source* Boutayeb et al. [15]

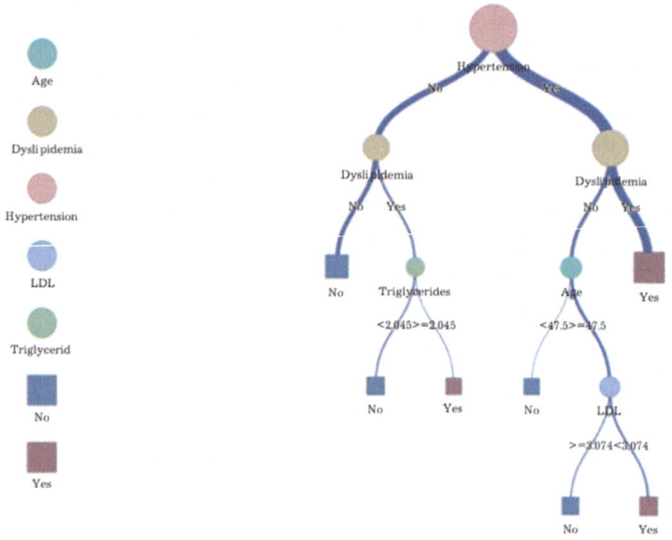

Fig. 3 Decision tree: Cart algorithm *Source* Boutayeb et al. [15]

dyslipidaemia followed by less important ones as: high rate of triglyceride, advanced age and low rate of cholesterol (LDL). In fact, a patient with hypertension and dyslipidaemia has a high risk of developing diabetes. Also, a hypertensive middle-aged person may develop diabetes if his/her rate of cholesterol (LDL) is less than 3.07. However, a patient has reduced chances of developing diabetes if he/she doesn't have any associated problems with hypertension or dyslipidaemia [15].

In 2024, Khriss et al. assessed the influence of lifestyle factors on the risk of type 2 diabetes in adults over the age of 50. Using data from the 'GOLF COAST', they concentrated on the effect of three lifestyle attributes (marital status, education and employment status) on diabetes in adults aged 50 years or over. Besides the descriptive statistics, they conducted an analytic analysis by combining chi-squared (χ^2) test and gradient boosting machines. Their results showed significant relationship between the likelihood of having type 2 diabetes and each of the three factors considered (marital status, education and work) [30].

3.2 *Machine Learning and Diabetes*

3.2.1 Algorithms Predicting Diabetes

In 2014, Ali et al. focussed on the prediction of diabetes types and proposed a boosting ensemble technique that internally uses random committee classifier. They used a local dataset of 100 patients suffering from type 1 (56) and type 2 (44) diabetes with their personal and clinical information given by 18 attributes (gender, diabetes type, fasting blood sugar, HbA1c, hypoglycaemia, systolic blood pressure, diastolic blood pressure, weight, height, total cholesterol, triglyceride, low-density lipoprotein, high-density lipoprotein, occupational therapy, physical therapy, vital symptoms, sentiments and activities). They recalled a number of ensemble learning techniques which have been proposed in literature for diabetes mellitus and implemented a novel ensemble method (AdaboostM1 with random committee) with a prediction accuracy of 81% [7].

The use of machine learning approaches for the prediction of diabetes was also considered in 2018 by Patil et al. [37]. The PIMA Indian data repository was used by the authors to analyse and compare the performance of classification algorithms by implementing eight machine learning models (LR, K-NN, SVM, GB, DT, MLP, RF and GNB). The highest accuracy was achieved by the LR model, with 79.54% and an RMSE of 46.52%; the lowest accuracy was given by MLP, with 64.07% and an RMSE of 59.94% (Table 3).

In 2018, El Jerjawi and Abu-Naser considered diabetes prediction using artificial neural network. Stressing that the most important issue is the prediction to be very accurate and to use a reliable method for that, they used Artificial Neural Networks (ANN) to predict whether a person is diabetic or not. The criterion was to minimize the error function in neural network training using a neural network model. They found that, after training the ANN model, the average error function of the neural network was equal to 0.01 and the accuracy of the prediction of whether a person is diabetics or not was 87.3% [22].

In 2018, Azrar et al. used the PID dataset and implemented three machine learning algorithms (DT, NB, K-NN) for diabetes prediction. Indicating that the tool used for testing and validation was Rapid Miner while all algorithms worked with 70:30 ratio for training and testing, they found that the best accuracy (75.65%) was given by DT which outperformed NB(71.74%) and K-NN (65.19%) [10].

In 2018, Zou et al. analysed two datasets with the purpose of using machine learning techniques to predict diabetes: (1 Luzhou) a dataset obtained from hospital

Table 3 Prediction performance for different algorithms

Algorithm	MAE (%)	RSME (%)	Accuracy score (%)
Logistic Regression (LR)	21.65	46.52	**79.54**
K-Nearest Neighbours (K-NN)	25.11	50.11	74.89
Linear SVM	32.03	56.60	67.97
Gradient Boosting (GB)	20.78	45.58	79.22
Decision Tree (DT)	26.84	51.81	73.16
Multilayer perception (MLP)	35.93	59.94	64.07
Random Forest (RF)	23.81	48.80	76.19
Gaussian Naïve Bayes (GNB)	23.81	48.80	76.00

Source Patilet al. [37]

physical examination data in Luzhou, China and (2 PIDd) Pima Indians diabetics data (PIDD). From the first dataset, they used a healthy people physical examination data of 164431 instances as the training set. The data considered included 14 attributes, namely, age, height, weight, physique index, fasting glucose, waistline, pulse rate, breathe, Left Systolic Pressure (LSP), Right Systolic Pressure (RSP), Left Diastolic Pressure (LDP), Right Diastolic Pressure (RDP), Low-Density Lipoprotein (LDL) and High-Density Lipoprotein (HDL). They implemented three machine learning classifiers (RF, J48, NN) and found that with PIDD, NN was the best of the three algorithms when using all features (76.76%) or only blood glucose (71.98%) while with Luzhou data, the best accuracy was given by RF (80.84%) when using all features and J48 (76.10%) when using blood glucose (Table 4) [40].

In 2019, Lai et al. proposed predictive models for diabetes mellitus using two machine learning algorithms (LR, GBM). They explored a Canadian database of 13309 patients aged between 18 and 90 years with the following 8 attributes: age, sex, Blood Pressure (BP), Fasting Blood Glucose (FBG), Body Mass Index (BMI), high-density lipoprotein, triglycerides and low-density lipoprotein. They used the Area under the Receiver Operating Characteristic curve (AROC) to evaluate the discriminatory capability of these models and found that the AROC for GBM and LR was 84.7% and 84.0%, respectively, while sensitivity was of 71.6% for GBM and 73.4% for LR. A comparison showed that the GBM and LR models performed better than the RF and DT models [33].

In 2019, Faruque et al. used machine learning techniques to predict diabetes mellitus. They compared the results yielded by four algorithms (SVM, NB, K-NN and C4.5 decision tree) and used three evaluation metrics (accuracy, recall and precision) in order to compare the performance of the classification models on predicting diabetes. The authors collected data of 200 patients from the diagnostic of Medical Centre Chittagong (MCC), Bangladesh. This comparative study revealed that the C 4.5 decision tree model achieved the best performance with an accuracy of 73% [23].

In 2019, Alam et al. used the dataset PIDD with 768 records and 9 attributes and proposed machine learning techniques to predict diabetes by investigating the effect of different attributes and the relationship between them. Implementing three algorithms (K-means clustering, RF, ANN) and comparing their performance revealed

Table 4 Predicting diabetes by using all features and blood glucose in the two datasets

Dataset	Classifier	Accuracy using all features (%)	Accuracy using blood glucose (%)
Luzhou	RF	**80.84**	75.97
	J48	78.53	**76.10**
	NN	78.41	75.72
PIDD	RF	76.04	67.28
	J48	72.75	68.95
	NN	**76.67**	**71.98**

Source Zou et al. [40]

that, with an accuracy of 75.7%, the ANN algorithm was better than RF (74.7%) and K-means clustering method (73.6%). The authors concluded that the ANN method may be useful to assist medical professionals with treatment decisions [5].

In 2020(b), Hasan et al. proposed a study on diabetes diagnosis and prediction, using a total of 768 female patients, including 268 diabetic patients (positive) and 500 non-diabetic patients (negative). Their analysis was based on the results given by six machine learning models (AB, K-NN, DT, XGB, NB and RF), involving the following eight different variables: age, BMI, glucose, insulin, pregnant, pressure, triceps and pedigree. The authors also implemented an ensemble method using multiple classifiers to enhance the performance and choose the Area Under the Curve (AUC) as an evaluation metric. The results obtained showed that AdaBoost and XGBoost were the best models used together achieving an AUC score of 0.95 [25].

Another study using machine learning models to predict diabetes in women was proposed in 2021 by Abdulhadi et al. Exploiting the PIDD dataset, the authors rescaled all variables using a standardization method and compared four machine learning models (LDA, LR, SVM and RF). This study revealed that the highest accuracy score was achieved by the RF model, with 82% [3].

In 2021, Bukhari et al. predicted female diabetes using Artificial Neural Network (ANN). They collected data from the National Institute of Diabetes and Digestive and Kidney Diseases and considered eight features to train the ANN model. Using the evaluation metrics of accuracy and mean squared error (MSE), the results of their predictive model indicated an accuracy of 93% [17].

In 2021, Ahmad et al. considered machine learning to investigate health-related features and their impact on the prediction of diabetes. They compared the role of HbA1c and FPG as input features, using the following five machine learning classifiers: Logistic Regression (LR), Support Vector Machines (SVM), Decision Tree (DT), Random Forest (RF) and Ensemble Majority Voting (EMV). The features were evaluated through accuracy, precision recall and F1-score. Stressing that their analysis agreed with the American Diabetes Association (ADA) and other health institutions worldwide in terms of the risk factors of diabetes and pre-diabetes, the authors concluded that important factors specific to the Saudi population can be identified by performing analysis of the disease using selected features [4].

In 2024, Khriss et al. published a chapter, using machine learning and deep learning approaches to predict diabetes risk factors and early complications. They used data from the 'GOLF COAST'. They investigated a sample of 3372 patients with a large set of 28 attributes including age, gender, marital status, work, education, BMI, HbA1C, hypertension and 20 others. Given the large number of patients and attributes, the authors used seven feature selection algorithms (LASSO, DT, LR, RF, RFE, PI, GA) and then executed six machine learning algorithms (ANN, K-NN, DT, LR, NB, SVM) to compute the values provided by each algorithm in terms of Accuracy, Precision, Recall and F1-Score. As indicated by Table 5 below, the results of this study showed that SVM was the best in terms of Accuracy (89.33%), Precision (92.24%) and F1-Score (90.24%) while the best Recall value was given by ANN with a value of 89.12% [31].

Table 5 Performance evaluation and comparison between the six ML algorithms

Algorithm	Accuracy (%)	Precision (%)	F1-Score (%)	Recall (%)
SVM	**89.33**	**92.24**	**90.24**	88.33
ANN	88.74	90.57	89.84	**89.12**
K-NN	74.67	78.77	76.73	74.80
LR	87.26	91.45	88.19	85.15
NB	86.07	90.08	87.12	84.35
DT	85.48	86.42	87.11	87.80

Source Khriss et al. [31]

3.2.2 Algorithms for Diagnosis and Detection of Diabetes

In 2014, Bozkurt et al. applied machine learning algorithms to the Pima Indian diabetes dataset (PIDD) for determining diabetes. They used the following eight classifiers: Probabilistic Neural Network (PNN), Learning Vector Quantization (LVQ), Feedforward Networks (FFN), Cascade-Forward Networks (CFN), Distributed Time Delay Networks (DTDN), Time Delay Networks (TDN), the Artificial Immune System (AIS), and the Gini decision tree algorithms. When they compared the performance of all these classifiers, they found that the best accuracy value (76.00%) and the best sensitivity value (63.33%) occurred with DTDN while the best specificity value (88.75%) was achieved with PNN [16].

In 2019, Alaoui et al. explored the dataset PIDD with 768 records and 9 attributes and proposed 4 machine learning algorithms (NB, DT, SVM, ANN) for diagnosing diabetes in women aged 21 years and above. Using the two data mining technologies Weka and Orange, they found that the algorithm SVM implemented in Weka as SMO showed the highest accuracy with 76.8%, sensitivity with 77.3% and precision with 76.9% [6].

In 2020(a), Hassan et al. proposed machine learning-based Web Application for Early Diagnosis of Diabetes. They used the Prima Indian Diabetes Dataset (PIDD) and compared the performance of six machine learning algorithms (NB, LR, K-NN, DT, RF and SVM). Their results showed that SVM was the best among the six mentioned algorithms, with an accuracy of 96.6%. Consequently, they adopted this model for the development of the intelligent Web application for the diabetes [26].

In 2020, Oleiwi et al. used machine learning algorithms for the early detection of diabetes. In this study, the authors collected data from the Diabetes Hospital of Sylhet, Bangladesh, through a direct questionnaire filled by patients. Using reports of diabetes-related symptoms of 520 instances with 16 attributes, the authors trained 3 classification models (MLP, RBF and RF) to determine the best classifier model for predicting diabetes. A comparison between the models proposed showed that, with an accuracy of 98.80%, the RBF was the best model, outperforming the other models% [35].

In 2022, Chang et al. proposed an e-diagnosis system based on Machine Learning (ML) algorithms for diagnosing diabetes mellitus (type 2 diabetes). They used the

Table 6 A comparison between the three machine learning algorithms

Algorithm	Accuracy (%)	Precision (%)	F1-Score (%)	Sensitivity (%)	Specificity (%)	AUC (%)
RF	**79.57**	**89.40**	**85.17**	81.33	**75.00**	**86.24**
J48	74.78	70.86	78.68	**88.43**	59.63	78.55
NB	78.67	81.88	84.24	86.75	63.29	84.63

Source Chang et al. [18]

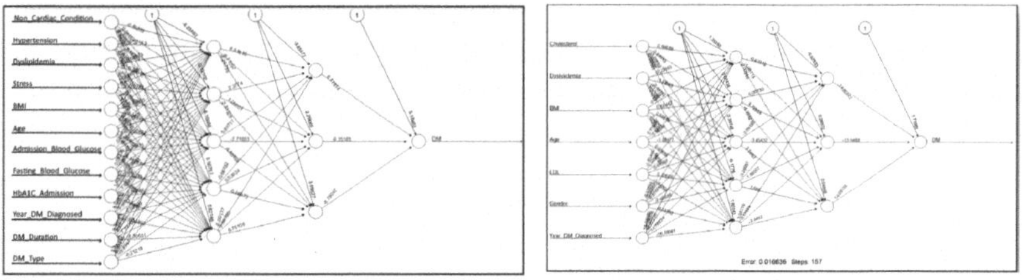

(a) (b)

Fig. 4 **a** Multilayer perceptron/Olden **b** Multilayer perceptron/cart algorithm *Source* Boutayeb et al. [14]

Pima Indians diabetes dataset (PIDD) to test and train three ML models (NB, RF and J48DT). Analysing the performance of each algorithm in terms of accuracy, precision, sensitivity, specificity, F1-Score and AUC, they concluded that the RF classifier outperformed both the NB and J48DT with accuracy metric (79.57%), precision (89.40%), specificity (75.00%), F1-Score (85.17%) and AUC (86.24%), while the J48 had the best sensitivity (88.43%) of the three models (Table 6) [18].

In 2023, Boutayeb et al. considered machine learning methods in the detection of type 2 diabetes mellitus (T2DM) risk factors. The authors compared between the performance of three supervised data mining algorithms (DT, BN, NN) according to the accuracy rate and the model sensitivity. The comparative study was applied on the 'Gulf Coast' cleaned database of 3327 individuals and 33 attributes out of 4061 patients and 317 attributes provided by the initial database. They found that the best accuracy was given by neural networks (100%).

In order to ensure a good interpretation, variables that have no effect on the results were eliminated using Olden and Decision tree methods which lead to the two decision trees given in Fig. 4. Variables that were maintained by both methods were 'Dyslipidaemia', 'Hypertension', 'Age', 'BMI', 'Cholesterol' and 'LDL' [14].

The authors concluded that Multilayer Perceptron gives the best prediction of T2DM risk factors in terms of sensitivity and accuracy.

3.2.3 Algorithms for the Classification of Diabetes

In 2016, Tarig M. Ahmed developed a predicted model using data mining. Recalling that patients with diabetes need a regular control of their blood glucose by keeping the HbA1c less than 7%, the aim of the proposed model was to classify the diabetic patients into two classes, namely, under control (HbAlc < 7%) and out of control (HbA1c > 7%). Consequently, the author performed a classification based on HbAlc measurement and selected three algorithms (NB, Logistic and J48). The model was built after using logistic algorithm and 10061 medical records. According to the author, nearly 75% accuracy was obtained as a best result after implementing the model. The author suggested additional measurements such as the nutrition system and the exercise for the patients in the dataset in order to improve the model's accuracy [38].

In 2021, Pamuk and Kaya dealt with the classification of type 2 diabetes using machine learning techniques. The aim of their study was to use machine learning techniques to diagnose type 2 diabetes using medical laboratory data. They selected the following machine learning techniques to use: J48, Random Forest (RF), Random Tree (RT) and Ibk algorithms in the WEKA programme. The data investigated concerned 400 patients and the following 6 laboratory tests: age, gender, glucose, HbA1C, HGB and urine. The authors indicated that the four algorithms used were successfully trained and the highest accuracy value was 96.97% given by the Random Forest algorithm, with recall and F-measure values of 98.47% and 96.24%, respectively [36].

In 2022, Edeh et al. proposed '*A Classification Algorithm-Based Hybrid Diabetes Prediction Model*'. Their aim was to develop a model that can accurately predict the likelihood of developing diabetes in patients with the greatest amount of precision. They selected four machine learning classification algorithms, namely, supervised learning algorithms Random Forest (RF), Support Vector Machine (SVM), Naïve Bayes (NB) and Decision Tree (DT), and unsupervised learning algorithm (k-means) to identify diabetes in its early stages. The experiments were performed on two databases, one extracted from the Frankfurt Hospital in Germany and the other from the database PIMA Indian Diabetes (PIDD) provided by the UCI machine learning repository [21].

According to the authors, the results obtained from the database extracted from Frankfurt Hospital, Germany, showed that the RF algorithm outperformed with the highest accuracy of 97.6%, and the results obtained from the Pima Indian database showed that the SVM algorithm outperformed with the highest accuracy of 83.1% compared to other algorithms. The validity of these results was confirmed by the process of separating the dataset into two parts: a training set and a test set [21].

In a study published in 2023, Almutairi et al. investigated 'the ability of different classification methods to classify diabetes prevalence rates and the predicted trends in the disease according to associated behavioural risk factors (smoking, obesity, and inactivity) in Saudi Arabia'. They developed classification models for diabetes prevalence, using the following four machine learning algorithms: (1) Linear Discriminant (LD), (2) Support Vector Machine (SVM), (3) K-Nearest Neighbour (K-NN) and (4)

Neural Network Pattern Recognition (NPR). More precisely, they used four kernel functions of SVM (linear SVM, Gaussian SVM, quadratic SVM and cubic SVM) and two types of K-NN algorithms (fine K-NN and weighted K-NN). They compared their results with previous published similar studies and concluded that the experimental results on the predictive performance analysis of the classification models showed that weighted K-NN performed well in the prediction of diabetes prevalence rate, with the highest average accuracy of 94.5% and less training time than the other classification methods, for both men and women datasets [9].

3.2.4 Algorithms Dealing with Complications of Diabetes

Machine learning approaches were proposed to predict diabetes complications. In 2018, Dagliati et al. proposed different classification models in order to predict the onset of retinopathy, neuropathy and nephropathy in patients with type 2 diabetes. They constructed four machine learning models (LR, NB, SVMs and RF) and used data collected by Istituto Clinico Scientifico Maugeri (ICSM), Hospital of Pavia, Italy, to train the proposed algorithms. On the basis of a total of 943 records including seven features (gender, age, BMI, time from diagnosis, hypertension, glycated haemoglobin (HbA1c) and smoking habit), they found that the highest accuracy score was achieved by LR with 77.7% [19].

In 2015, Abdar et al. used machine learning techniques to predict the likelihood of patients getting a heart disease. They compared the results given by five algorithms (C5.0, ANN, SVM, K-NN, LR) and found that C5.0 DT was the model with the greatest accuracy 93.02% while the accuracy given by K-NN, SVM and ANN was, respectively, 88.37%, 86.05% and 80.23% [2].

In 2018, Azrar et al. used the PID dataset and implemented three machine learning algorithms (DT, NB, K-NN) for diabetes prediction. Indicating that the tool used for testing and validation was Rapid Miner while all algorithms worked with a ratio 70:30 for training and testing, they found that the best accuracy (75.65%) was given by DT which outperformed NB (71.74%) and K-NN (65.19%) [10].

In 2019, Birjais et al. dealt with prediction and diagnosis of future diabetes risk using the PIDD Dataset and the algorithm K-NN imputation to predict and fill in the missing values in the PIDD dataset to reduce bias. They implemented three machine learning algorithms (LR, GB, NB) and found that gradient boosting achieved the best accuracy (86%) compared to logistic regression (79.2%) and naive Bayes (77%) [11].

Predicting diabetes complications was also considered in 2020 by Kantawong et al. who proposed a machine learning model aiming to predict kidney disease, eye disease, hyperlipidaemia and coronary heart disease. For this purpose, the authors started with an initial dataset of 455 records which was reduced after selection and cleaning processes. An accuracy of 92.35% was reached by the iterative decision tree (ID3) [29].

In 2021, Jian et al. conducted a study on the use of machine learning approaches to predict diabetes complications. They explored a dataset (884 records and 79 features) collected by the Rashid Center for Diabetes and Research (RCDR) located in Ajman,

UAE. Given the huge number of attributes, the authors went through the following steps: (1) handling missing values and unbalanced data problems; (2) selecting the top five and ten features for each complication; (3) fixing the final number of records to train and build the binary classifiers for each complication as follows: metabolic syndrome (428), dyslipidemia (836), neuropathy (223), nephropathy (233), diabetic foot (240), hypertension (586), obesity (498) and retinopathy (228); (4) employing repeated stratified k-fold cross-validation (with $k = 10$ and a total of 10 repetitions) for a better estimation of the performance.

In order to classify the eight complications, the authors used the following machine learning models: Logistic Regression (LR), Support Vector Machine (SVM), Decision Tree (DT-CART), Random Forest (RF), AdaBoost and XGBoost. A summary of the results obtained is given in Table 7 [28].

In 2021, Ahmad et al. considered machine learning to investigate health-related features and their impact on the prediction of diabetes. They compared the role of HbA1c and FPG as input with 16 features measured on 3000 patients from 5 different Saudi hospitals across 3 regions, using the following 5 machine learning classifiers: Logistic Regression (LR), Support Vector Machines (SVM), Decision Tree (DT), Random Forest (RF) and Ensemble Majority Voting (EMV). The features were evaluated through accuracy, precision recall and F1-score. Their results showed that the best accuracy was given by SVM (82.10%) in the case of HbA1c while RF (88.27%) outperformed the authors' algorithms in the case of FPG. Stressing that their analysis agreed with the American Diabetes Association (ADA) and other health institutions worldwide in terms of the risk factors of diabetes and pre-diabetes, the authors concluded that important factors specific to the Saudi population can be identified by performing analysis of the disease using selected features [3].

In 2021, Abaker built a predictive model for detecting diabetes complications. Using 29 attributes associated with 644 electronic health records from Alsukari Hospital, the author implemented the following three machine learning algorithms: Logistic Regression (LR), Random Forest (RF) and K-Nearest Neighbour (K-NN). He found that LR outperformed the two other models by achieving an accuracy of

Table 7 Machine learning algorithm giving the best accuracy

Complication	Algorithm with the best accuracy	All attributes accuracy (%)	Top 10 attributes accuracy (%)	Top 5 attributes accuracy (%)
MetS	LR	77.1	73.5	75.6
Obesity	RF	80.0	80.3	78.0
Hypertension	RF	73.6	71.6	70.0
Dyslididemia	RF	76.3	71.3	67.7
Neuropathy	XGBoost	82.9	80.4	73.9
Nephropathy	AdaBoost	91.7	88.7	85.0
Retinopathy	XGBoost	87.2	80.4	77.1
Diabetic foot	XGBoost	97.8	93.8	91

Source Jian et al. [28]

81%, recall of 81% and F1-score of 75%. He highlighted that attributes such as infection years, swelling, diabetic ketoacidosis and diabetic septic foot were significant in predicting diabetes complications [1].

3.2.5 A Summary of Machine Learning Models in Diabetes

The machine learning models presented in this section are summarized in Table 8 with the main information (author, date of publication, data used, features, machine learning algorithms and the best accuracy achieved).

Table 8 A summary of machine learning methods with the algorithms used by each author

Author	Date	Population/ databases	Features/ attributes	Methods	Results/ highest accuracy
Abaker & Saeed	2021	644 records Alsukari Hospital	infection years, swelling, diabetic ketoacidosis, diabetic septic foot + 25	LR, RF, K-NN	LR: 81%
Abdar et al.	2015	270 records	Age, sex, FBS, RBP, SC +8 others	C5.0, ANN, SVM, K-NN, LR	C5.0: 93.02%
Abdulhadi & Al-Mousa	2021	(PIDD): 768 records	Age, Preg, Gluc, BP, Skin, Insulin, BMI, DPF, outcome	LDA, LR, SVM, RF	RF: 82%
Ahmad et al.	21	3000 records from 5 Saudi hospitals	Gender, height, weight, HbA1c, BP, HDL, LDL +9 other attributes	LR, SVM, DT, RF, EMV	SVM: 82.10% (HbA1c) RF: 88.3% FPG
Alam et al.	2019	(PIDD): 768 records	Age, Preg, Gluc, BP, Skin, Insulin, BMI, DPF, outcome	K-means clustering RF, ANN	ANN: 75.7%
Alaoui et al.	2019	(PIDD): 768 records	Age, Preg, Gluc, BP, Skin, Insulin, BMI, DPF, outcome	NB, DT, SVM, ANN	SVM: 76.8%
Ali et al.	2014	Local dataset 100 diabetics	Gender, diabetes type, FBG, HBA1c +14 other	AdaboostM1	AdaboostM1: 81%
Almutari et al.	2023	1272 men and women aged 25+	Age, gender, obesity, inactivity, smoking	LDA, NPR, SVM (linear, quadratic, cubic, Gaussian), K-NN (Fine, Weigh)	K-NN: 94.5%
Azrar et al	2018	(PIDD): 768 records	Age, Preg, Gluc, BP, Skin, Insulin, BMI, DPF, outcome	DT, NB, KNN	DT: 75.65%

(continued)

Table 8 (continued)

Author	Date	Population/ databases	Features/ attributes	Methods	Results/ highest accuracy
Birjais et al.	2919	(PIDD): 768 records	Age, Preg, Gluc, BP, Skin, Insulin, BMI, DPF, outcome	LR, GB, NB	GB: 86%
Boutayeb et al.	2021	Records Initial: 4061 Clean: 3327	Age, sex, Hypertension, BMI, Cholesterol, LDL, Dyslipidemia and others	PCA	
Boutayeb et al.	2023	Records Initial: 4061 Clean: 3327	Age, sex, Hypertension, BMI, Cholesterol, LDL, Dyslipidemia +26 others	DT, BN, NN	NN : 100 %
Bozkurt et al.	2014	(PIDD) 768 patients	Age, Preg, Gluc, BP, Skin, Insulin, BMI, DPF, outcome	AIS, ANN, PNN TDN, CFN, FFN DTDN, LQV	DTDN: 76%
Bukhari et al.	2021	(PIDD): 768 records	Age, Preg, Gluc, BP, Skin, Insulin, BMI, DPF, outcome	ANN	ANN: 93%
Chang	2022	(PIDD): 768 records	Age, Preg, Gluc, BP, Skin, Insulin, BMI, DPF, outcome	NB, RF, J48	RF: 79.57%
Dagliati et al.	2018	943 records	Age, gender, HbA1c BMI, hypertension, time from diagnosis, and smoking habit	LR, NB, SVM, RF	LR:77.7%
Edeh et al.	2022	Frankfurt		RF, SVM, NB DT	RF:97.6%
		Hospital (PIDD): 768 records	Age, Preg, Gluc, BP, Skin, Insulin, BMI, DPF, outcome		SVM: 83.1%
El Jerjawi & Abu-Naser	2018	(PIDD): 768 records	Age, Preg, Gluc, BP, Skin, Insulin, BMI, DPF, outcome	ANN	87.3%
Faruque et al.	2019	200 patients	Age, sex, weight, BP and other risk factors	SVM, C4.5, DT, NB, K-NN	C4.5 : 73%
Hassan et al. (a)	2020	PIDD: 768 records	Age, Preg, Gluc, BP, Skin, Insulin, BMI, DPF, outcome	LR, NB, K-NN, DT, RF, SVM	SVM: 96.6%
Hassan et al. (b)	2020	(PIDD): 768 records	Age, Preg, Gluc, BP, Skin, Insulin, BMI, DPF, outcome	AB, KNN, DT, XGB, NB, RF	AB & XGB together AUC: 0.95
Jian et al.	2012	RCDR in Ajman, UAE 884 records	79 features	RF, LR, SVM, DT, AdaBoost, XGBoost	XGBoost: 97.8 (diabetic foot)
Kantawong et al.	2020	455 records	Not mentioned in paper	ID3	ID3: 92.35%

(continued)

Table 8 (continued)

Author	Date	Population/ databases	Features/ attributes	Methods	Results/ highest accuracy
Khriss	2024	3372	Age, gender, work, marital status, education, hypertension, HbA1c, BMI and 20 others	ANN, K-NN, DT, LR, NB, SVM	SVM: 89.33%
Lai et al.	2019	13309	Age, sex, BP, FBG, BMI, triglycerides, high-density lipoprotein, low-density lipoprotein	GBM, LR, FR, DT	GBM: 84.7%
Oleiwi et al.	2020	520 persons	Age, sex, polydipsia, weakness, visual blurring itching, polyuria, obesity and 8 others	MLP, RBF, RF	RBF: 98.80%
Pamuk & Kaya	2021	400 patients	Age, gender, glucose, HbA1c, HGB, urine	J48, RF, RT, Ibk	RF: 96.97%
Patil & Tamane	2018	(PIDD): 768 records	Age, Preg, Gluc, BP, Skin, Insulin, BMI, DPF, outcome	LR, GB, K-NN, SVM, DT, MLP RF, GNB	LR: 79.54%
Tarig M	2016	10061 medical records	50 (demographic info, treatment plan and measurements related to control of diabetes)	NB, Logistic, J48 (Logistic combines LR and DT)	Logistic: 74.8%
Zou et al.	2018	Luzhou data 164431 PIDD 768 records	14 (age, height, weight, FG, HDL, LDL + 8 other Age, Preg, Gluc, BP, Skin, Insulin, BMI, DPF, outcome	RF, J48, NN	RF: 80.84% NN: 76.67%

4 Conclusion

In their comprehensive review published in 2024, We et al. found that, among all machine learning models cited in the review, three models (RF, LR, SVM) accounted for 53% with average accuracies of 81.9%, 79.2% and 81.4%, respectively.

In the present overview, 14 out of 30 authors used the Pima Indians Diabetes Database (PIDD) for training and testing their models. Among the 36 machine learning models used by one or more authors, the most implemented algorithms were RF, LR, SVM, DT, ANN and NB used 15, 12, 12, 11, 10 and 10 times, respectively.

Random Forest (RF) was the algorithm which achieved the greatest number of best accuracies (6 times) and with the best average accuracy of 87.55%.

References

1. Abaker, A.A., Saeed, K.A.: A comparative analysis of machine learning algorithms to build a predictive model for detecting diabetes complications. Informatica **45**, 117–125 (2021). https://doi.org/10.31449/inf.v45i1.3111
2. Abdar, M., Kalhori, N., Sutikno, T., Subroto, I.M.I., Arji, G.: Comparing performance of data mining algorithms in prediction heart diseases. International Journal of Electrical & Computer Engineering **5**(6), 1569–1576 (2015)
3. Abdulhadi, N., Al-Mousa, A.: Diabetes detection using machine learning classification methods. In: Proceedings of the 2021 International Conference on Information Technology (ICIT), pp. 350–354. Amman, Jordan (2021)
4. Ahmad, H.F., Mukhtar, H., Alaqail, H., Seliaman, M., Alhumam, A.: Investigating health-related features and their impact on the prediction of diabetes using machine learning. Applied Sciences **11**(3), 1173 (2021). https://doi.org/10.3390/app11031173
5. Alam, T.M., Iqbal, M.A., Ali, Y., Wahab, A., Ijaz, S., Baig, T.I., Hussain, A., Malik, M.A., Raza, M.M., Ibrar, S., Abbas, Z.: A model for early prediction of diabetes. Informatics in Medicine Unlocked **16**, 1–6 (2019). https://doi.org/10.1016/j.imu.2019.100204
6. Alaoui, S.S., Aksasse, B., Farhaoui, Y.: Data mining and machine learning approaches and technologies for diagnosing diabetes in women. In: International Conference on Big Data and Networks Technologies, Springer, Cham, pp. 59–72 (2019)
7. Ali, R., Siddiqi, M.H., Idris, M., Kang, B.H., Lee, S.: Prediction of diabetes mellitus based on boosting ensemble modelling. In: International Conference on Ubiquitous Computing and Ambient Intelligence, pp. 25–28. Springer International Publishing (2014)
8. Almutairi, E., Abbod, M., Itagaki, T.: Mathematical modelling of diabetes mellitus and associated risk factors in saudi arabia. International Journal of Simulation Science and Technology **21**, 1–7 (2020)
9. Almutairi, E.S., Abbod, M.F.: Machine learning methods for diabetes prevalence classification in saudi arabia. Modelling **4**, 37–55 (2023). https://doi.org/10.3390/modelling4010004
10. Azrar, A., Ali, Y., Awais, M., Zaheer, K.: Data mining models comparison for diabetes prediction. International Journal of Advanced Computer Science and Applications **9**(8), 320–323 (2018)
11. Birjais, R., Mourya, A.K., Chauhan, R., Kaur, H.: Prediction and diagnosis of future diabetes risk: A machine learning approach. SN Applied Sciences **1**(9), 1–8 (2019)
12. Boutayeb, A.: Inequalities within and between who-eastern mediterranean countries. In: A. Boutayeb, A. Maamri (eds.) Health Inequity: A Crucial Issue Worldwide, pp. 1–22. Cambridge Scholars Publishing (2023)
13. Boutayeb, A., Lamlili, E.N.M., Boutayeb, W., Maamri, A., Ziyyat, A., Ramdani, N.: The rise of diabetes prevalence in the arab region. Open Journal of Epidemiology **2**, 55–60 (2012)
14. Boutayeb, W., Badaoui, M., Al Ali, H., Boutayeb, A., Doukali, M.: Machine learning methods in the detection of type 2 diabetes mellitus risk factors. In: Lecture Notes in Networks and Systems, vol. 656, pp. 81–91 (2023)
15. Boutayeb, W., Badaoui, M., Al Ali, H., Boutayeb, A., Lamlili, E.N.M.: Use of data mining in the prediction of risk factors of type 2 diabetes mellitus in gulf countries. Mathematical Modelling and Computing **4**(8), 638–645 (2021)
16. Bozkurt, M.R., Yurtay, N., Yılmaz, Z., Sertkaya, C.: Comparison of different methods for determining diabetes. Turkish Journal of Electrical Engineering & Computer Sciences **22**, 1044–1055 (2014)

17. Bukhari, M.M., Alkhamees, B., Hussain, S., Gumaei, A., Assiri, A., Ullah, S.S.: An improved artificial neural network model for effective diabetes prediction. Complexity **2021**, 5525271 (2021). https://doi.org/10.1155/2021/5525271
18. Chang, V., Bailey, J., Xu, Q.A., Sun, Z.: Pima indians diabetes mellitus classification based on machine learning (ml) algorithms. Neural Computing and Applications **24**, 1–17 (2022). https://doi.org/10.1007/s00521-022-07049-z
19. Dagliati, A., Marini, S., Sacchi, L., Cogni, G., Teliti, M., Tibollo, V., De Cata, P., Chiovato, L., Bellazzi, R.: Machine learning methods to predict diabetes complications. Journal of Diabetes Science and Technology **12**, 295–302 (2018)
20. Dua, D., Graff, C.: Uci machine learning repository. http://archive.ics.uci.edu/ml (2017)
21. Edeh, M.O., Khalaf, O.I., Tavera, C.A., Tayeb, S., Ghouali, S., Abdulsahib, G.M., Richard-Nnabu, E., Louni, A.: A classification algorithm-based hybrid diabetes prediction model. Frontiers in Public Health **10**, 829519 (2022). https://doi.org/10.3389/fpubh.2022.829519
22. El Jerjawi, N.S., Abu-Naser, S.S.: Diabetes prediction using artificial neural network. International Journal of Advanced Science and Technology **121**, 55–64 (2018). https://doi.org/10.14257/ijast.2018.121.05
23. Faruque, M.F., Sarker, I.H.: Performance analysis of machine learning techniques to predict diabetes mellitus. In: Proceedings of the 2019 International Conference on Electrical, Computer and Communication Engineering (ECCE), pp. 1–4. Cox's Bazar, Bangladesh (2019)
24. Guariguata, L., Whiting, D.R., Hambleton, I., Beagley, J., Linnenkamp, U., Shaw, J.E.: Global estimates of diabetes prevalence for 2013 and projections for 2035. Diabetes Research and Clinical Practice **103**, 137–149 (2014)
25. Hasan, M.K., Alam, M., Das, D., Hossain, E., Hasan, M.: Diabetes prediction using ensembling of different machine learning classifiers. IEEE Access **8**, 76516–76531 (2020)
26. Hassan, F., Wardah, M., Yasir, M., Shahwani, H., Shah, S.A., Imran, M., Ashraf, M., Qasim, M., Akram, M., Rauf, Z.: Machine learning-based web application for early diagnosis of diabetes. Journal of Applied and Emerging Sciences **10**(2), 133–142 (2020). https://doi.org/10.3390/modelling4010004
27. He, B., Shu, K.I., Zhang, H.: Machine learning and data mining in diabetes diagnosis and treatment. In: IOP Conference Series: Materials Science and Engineering, vol. 490, pp. 42–49 (2019)
28. Jian, Y., Pasquier, M., Sagahyroon, A., Aloul, F.: A machine learning approach to predicting diabetes complications. Healthcare (Basel) **9**(12), 1712 (2021). https://doi.org/10.3390/healthcare9121712
29. Kantawong, K., Tongphet, S., Bhrommalee, P., Rachata, N., Pravesjit, S.: The methodology for diabetes complications prediction model. In: Proceedings of the 2020 Joint International Conference on Digital Arts, Media and Technology with ECTI Northern Section Conference on Electrical, Electronics, Computer and Telecommunications Engineering (ECTI DAMT & NCON), pp. 110–113. Pattaya, Thailand (2020)
30. Khriss, A., Badaoui, M., Kerkour Emiad, A., Boutayeb, W.: An assessment of the influence of lifestyle factors on the risk of type 2 diabetes in adults over the age of 50 years. In: Industrial and Computational Mathematics. vol pp. 327–352 (2024)
31. Khriss, A., Badaoui, M., Kerkour Emiad, A., Boutayeb, W.: Predictive modelling of diabetes risk factors and early complications using machine learning and deep learning. In: Industrial and Computational Mathematics. vol pp. 353–368 (2024)
32. Komi, M., Li, J., Zhai, Y., Zhang, X.: Application of data mining methods in diabetes prediction. In: 2017 2nd International Conference on Image, Vision and Computing (ICIVC), pp. 1006–1010. Chengdu, China (2017). https://doi.org/10.1109/ICIVC.2017.7984706
33. Lai, H., Huang, H., Keshavjee, K., Guergachi, A., Gao, X.: Predictive models for diabetes mellitus using machine learning techniques. BMC Endocrine Disorders **19**(101) (2019). https://doi.org/10.1186/s12902-019-0436-6
34. (NCD-RisC), N.R.F.C., Walton, J.: Worldwide trends in diabetes since 1980: A pooled analysis of 751 population-based studies with 4.4 million participants. Lancet **387**, 1513–1530 (2016)

35. Oleiwi, A., Shi, L., Tao, Y., Wei, L.: A comparative analysis and risk prediction of diabetes at early stage using machine learning approach. International Journal of Future Generation Communication and Networking **13**, 4151–4163 (2020)
36. Pamuk, Z., Kaya, C.: Classification of type 2 diabetes using machine learning techniques. European Journal of Science and Technology **28**, 1265–1268 (2021)
37. Patil, R., Tamane, S.: A comparative analysis on the evaluation of classification algorithms in the prediction of diabetes. International Journal of Electrical and Computer Engineering **8**, 3966 (2018)
38. Tarig, M.A.: Using data mining to develop model for classifying diabetic patient control level based on historical medical records. Journal of Theoretical and Applied Information Technology **87**(2), 316–350 (2016). https://www.semanticscholar.org/paper/USING-DATA-MINING-TO-DEVELOP-MODEL-FOR-CLASSIFYING-Ahmed/a28e022162cfa5129f9646e58d38191ec3bb3e8
39. Wee, B.F., Sivakumar, S., Lim, K.H., Wong, W.K., Juwono, F.H.: Diabetes detection based on machine learning and deep learning approaches. Multimedia Tools and Applications **83**, 24153–24185 (2024). https://doi.org/10.1007/s11042-023-16407-5
40. Zou, Q., Qu, K., Luo, Y., Yin, D., Ju, Y., Tang, H.: Predicting diabetes mellitus with machine learning techniques. Frontiers in Genetics **9**, 515 (2018)

Research Chapters

The Impact of Misinformation on Diabetes Management Using Mathematical Models

Woody Merat and Silvere Paul Nuiro

Abstract With the proliferation of social media platforms, the dissemination of misinformation occurs with disconcerting rapidity, significantly impacting diabetes management–a matter of escalating concern among researchers. In this study, one scrutinizes this impact through the application of both deterministic and stochastic models. These models consider the propagation of misinformation, and the efficacy of educational interventions in disease management, and incorporate the stochastic nature of phenomena associated with the progression of complications. The comprehensive methodological framework employed, integrating both quantitative and qualitative analyses, ensures the robustness and validity of the derived results. Our findings indicate that misinformation can precipitate suboptimal disease management, exacerbate complications, and diminish the quality of life for patients.

1 Introduction

According to the World Health Organization (WHO), diabetes mellitus is a chronic disease characterized by insufficient insulin production by the pancreas or ineffective use of insulin produced by the body. The incidence of this condition is increasing along with demographic aging, sedentary lifestyles, obesity, and even climate change [6, 22]. Almost a century after the discovery of insulin in 1921, diabetes accounted for 6.7 million fatalities in 2021, equating to 12.5 deaths per 1000 diabetic individuals. In addition, 44. 7% of adults with diabetes (240 million individuals) remain undiagnosed. As reported in the IDF Diabetes Atlas, 10th edition 2021, diabetes affects more than 537 million people worldwide, up from 463 million in 2019, marking an increase of 74 million within a mere two-year span. This figure is projected to increase to 578 million by 2030 and 700 million by 2045. In 2019,

W. Merat
Campus de Fouillole, French Universite des Antilles, Pointe-à -Pitre, France
e-mail: woody.merat@etu.univ-antilles.fr

S. P. Nuiro (✉)
Campus de Fouillole, Universite des Antilles, Pointe-à -Pitre, France
e-mail: paul.nuiro@univ-antilles.fr

A. Boutayeb and M. E. N. Lamlili (eds.), *Computational Mathematics and Modelling for Diabetes*, Industrial and Applied Mathematics,
https://doi.org/10.1007/978-981-96-1925-2_6

diabetes-related causes caused the deaths of more than four million people aged 20 to 79 years [47]. Two-thirds of the diabetic population reside in urban areas, and three-quarters are of working age. In particular, one in four adults in certain Caricom nations is diabetic [19]. These statistics underscore a significant global health crisis, indicating that diabetes is poised to become the seventh leading cause of mortality by 2030, underscoring the critical need for increased efforts to mitigate this growing public health challenge.

For decades, mathematical modeling of various real-world phenomena has significantly helped decision-makers worldwide anticipate social, economic, and health-related challenges. Specifically, modeling of diabetes and associated phenomena has elucidated numerous mechanisms, thus improving comprehension and explanation of this disease through extensive studies. Many of these studies are distinguished and leave a lasting impact. For example, the seminal work of Victor W. Bolie on insulin and glucose concentrations in 1961 [8]. Caspersen and colleagues have shown that the characteristic anatomical and physiological changes associated with the normal aging process are accelerated in individuals with diabetes, leading to a reduced life expectancy [14]. Holt et al. have shown that depressive symptoms affect up to a third of people with diabetes, not only reducing the quality of life but also compounding the challenges in self-management of diabetes [23]. M. Derouich and A. Boutayeb introduced the impact of physical activity through parameters in a mathematical model, facilitating the comparison of glycemic behavior among normal individuals, noninsulin-dependent diabetics, and insulin-dependent diabetics, both with and without physical exertion [16]. In the same year, A. Boutayeb and M. Derouich proposed models to monitor the number of patients developing complications according to the age structure and demonstrated the role of education and control parameters in modulating the number of diabetics with complications [10]. Two years later, A. Boutayeb and colleagues illustrated how effective and cost-effective strategies can be implemented by influencing diabetes incidence and/or controlling progression toward complication stages [11].

Numerous studies have shown that, along with cardiovascular diseases, cancers, and chronic respiratory diseases, diabetes mellitus is classified among the four main noncommunicable diseases (NCD) [46]. It is often accompanied by a multitude of complications, including nephropathy, retinopathy, neuropathy, and arteriosclerosis, among others [15, 17, 20, 28, 29, 32, 40]. Diabetes mellitus is increasingly becoming a global health problem. Consequently, mitigating the increase in the prevalence of diabetes and its associated complications, despite the various interventions proposed in the literature [2, 7, 9, 12, 18, 41, 43], remains two significant challenges that have not yet been fully addressed. Therefore, supporting people with diabetes and influencing their behaviors are critical components in the fight against diabetes, particularly its complications. The support system for diabetic people should function as a motivational agent, encouraging them to take responsibility and manage their diabetes by adopting a balanced lifestyle and a healthy diet. In addition, it should help them to find pertinent information about their condition and its treatment from more experienced diabetic individuals, thus enhancing their understanding of their disease [5].

Despite concerted efforts to increase public awareness about disease management, anticipated progress remains elusive. This lag can be attributed to the challenges individuals face in applying the knowledge disseminated or to the rampant spread of misinformation regarding disease management. As reported by the World Health Organization (WHO) and the Institute of International and Strategic Relations (IRIS) [26, 38], the propagation of false health information constitutes an increasing threat to societies, particularly as more and more people turn to search engines or social networks for health-related information. To mitigate this issue, extensive research has been conducted on modeling the dissemination of misinformation in the context of epidemic management. Recent studies have integrated two diffusion dynamics within a single model: the spread of disinformation and the transmission of the disease [35]. Another investigation has evaluated the role of awareness in the fight against diabetes and the efficacy of diabetes education in various regions [36]. However, as Bennett et al. [4], there is a paucity of literature on diabetes-related health among Caribbean-based populations. Consequently, it is imperative to examine the impact of diabetes education in this demographic. Furthermore, the issue of infectious diseases presents analogous challenges, as evidenced by a study that underscores the adverse effects of disseminating false information on vaccination intentions [42].

These references underscore the paramount importance of mathematical modeling in the study of diseases. Predominantly executed through deterministic models, this methodology is indispensable to comprehend and forecast the progression of a disease within a specified population. However, to more effectively address the uncertainty and stochastic fluctuations inherent in various parameters, it is occasionally imperative to employ stochastic models. These approaches provide a more precise description of the natural variability and intrinsic uncertainties in disease transmission, thereby improving our comprehensive understanding of these intricate phenomena.

2 The Formulation of the Deterministic Model

In contemporary society, the proliferation of misinformation is widespread, driven by indiscriminate sharing without prior verification. Faced with this increasing threat, platforms such as Facebook and Twitter are increasingly leveraging artificial intelligence to mitigate the dissemination of false information, meticulously selecting the appropriate methodologies and tools for this purpose [25]. Within the context of rumor propagation, as discussed in the literature [24, 33, 45], an uninformed individual, upon receiving a rumor from a disseminator, may become a conduit for its spread if they accept its veracity. Similarly, a diabetic patient exposed to inaccurate information regarding the management of their condition by someone experiencing complications can inadvertently propagate such misinformation, thus increasing the risk of complications if they integrate it into their own management strategy. In fact, people who subscribe to these erroneous beliefs are more susceptible to developing diabetes-related complications, which can lead to the emergence of new complications or the exacerbation of existing ones in certain diabetic patients. This issue persists despite the widely accepted understanding that diabetes is a manageable but not curable disease. Erroneous ideas continue to be propagated, such as the notion that insulin is an absolute cure for diabetes, the mandate for diabetics to completely eliminate sugar, the misconception that only children can be affected by type 1 diabetes and only the elderly by type 2, as well as the fallacious belief that maintaining a low weight would protect against diabetes or that significant weight loss could eradicate the disease.

To evaluate the ramifications of the dissemination of erroneous information on diabetes management, consider a diabetic population at a given time t. This population is divided into two cohorts: individuals without complications, indicated as $D(\cdot)$, and those with complications, indicated as $C(\cdot)$ so that the total population $N(\cdot) = D(\cdot) + C(\cdot)$. One defines β as the rate at which false information is disseminated, potentially leading to diabetes complications. Consequently, the product βDC quantifies the number of individuals developing complications as a result of adopting this misinformation. Additionally, one denotes the diabetes onset rate as $I(\cdot)$, the natural mortality rate as $\mu(\cdot)$, the complication development rate as $\lambda(\cdot)$, the complication cure rate as $\gamma(\cdot)$, the rate at which patients with complications become severely disabled as $\nu(\cdot)$, and the mortality rate due to complications as $\delta(\cdot)$. Drawing inspiration from the deterministic model of diabetes proposed by S. Bernard et al. [3], our model is depicted in the diagram in Fig. 1.

The scenario is consequently characterized by the ensuing system of differential equations:

$$\begin{cases} \dfrac{dD}{dt}(t) = I - (\lambda(t) + \mu(t))\, D(t) + \gamma(t)C(t) - \beta(t)D(t)C(t) & \text{(1a)} \\ \dfrac{dC}{dt}(t) = \lambda(t)D(t) - (\gamma(t) + \mu(t) + \nu(t) + \delta(t))\, C(t) + \beta(t)D(t)C(t), & \text{(1b)} \end{cases}$$

with initial conditions $D(0) = D_0 \in \mathbb{R}_+$ and $C(0) = C_0 \in \mathbb{R}_+$. Throughout, one assumes that

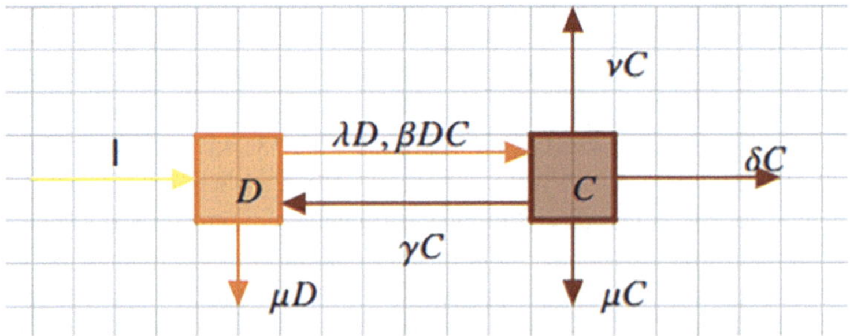

Fig. 1 This diagram represents the transfer of individual flows within the following system

$$(\mathbf{H}_1) : \lambda I > 0, \quad \delta + \nu > 0.$$

Following [3], one chooses to control the rate of healing of diabetes complications, represented by $\gamma(\cdot)$, with u denoting this control parameter. However, other parameters of the system could theoretically be under control.

The subsequent sections and subsections expound upon the spectrum of theoretical results and numerical simulations derived under the assumption of constant parameters.

2.1 *Initial Coefficient of Misinformation Propagation in Diabetes Management*

It is pertinent to recall that the parameter β is integral to a process in which misinformation about the disease, frequently disseminated by less experienced individuals, often supersedes empirical evidence and rational discourse. It is imperative that the insights of more experienced individuals are taken into account to educate those deficient in relevant knowledge about the disease. Consequently, the aim of this subsection is to determine whether there is a threshold at which the incidence of diabetic complications either decreases or increases. To this end, the initial coefficient of propagation of misinformation, denoted $\mathcal{T}_0$, regarding the disease shall be computed by considering the complication-free equilibrium $x_0 = \left(\mu^{-1} I, 0\right)^T$, with $\mu \neq 0$, of the model (1) and using the Next Generation Matrix method. one determine the matrices F and V^{-1} such that

$$F = \mu^{-1}\beta I \quad \text{and} \quad V^{-1} = (\gamma + \mu + \nu + \delta)^{-1}, \quad \text{with} \quad \mu > 0.$$

According to Van Den Driessche and Watmough (2002), $\mathcal{T}_0 = \rho\left(FV^{-1}\right)$, where $\rho\left(FV^{-1}\right)$ denotes the spectral radius of FV^{-1}, which leads to

$$\mathcal{T}_0 = \mu^{-1}(\gamma + \mu + \nu + \delta)^{-1}\beta I, \quad \text{with} \quad \mu > 0.$$

1. if $\beta I > \mu(\gamma + \mu + \nu + \delta)$, then the number of diabetics experiencing complications continues to increase.
2. if $\beta I < \mu(\gamma + \mu + \nu + \delta)$, then there will be a reduction in the number of diabetics suffering from complications.

3 Qualitative Study of the Deterministic Model

This section is devoted to the rigorous analysis of the existence and uniqueness of solutions, as well as the stability and optimal control issues pertinent to our model.

3.1 Existence of Solutions

The subsequent theorem guarantees the existence and uniqueness of a solution to the problem under consideration.

Theorem 1 *Let u be a fixed control. Then the Cauchy problem (1), with initial conditions $D_u(0) = D_0 \in \mathbb{R}_+$, $C_u(0) = C_0 \in \mathbb{R}_+$, and $t_m(u) \in \mathbb{R}_+ \cup \{+\infty\}$ has a unique maximal solution.*

Proof Consider the vector function $X(t) = \big(D(t), C(t)\big)$, such that its derivative with respect to time is given by $\frac{dX}{dt}(t) = \left(\frac{dD}{dt}(t), \frac{dC}{dt}(t)\right)$, with the initial condition $X(0) = \big(D(0), C(0)\big)$.

Observe that the system declined in (1) can be reformulated as the subsequent matrix equation system:

$$\begin{cases} \dfrac{dX}{dt}(t) = AX(t) + F(X(t)) \\ X(0) = X_0, \end{cases} \tag{2}$$

with $F(X(t)) = \begin{pmatrix} I - \beta D(t)C(t) \\ \beta D(t)C(t) \end{pmatrix}$ and $A = \begin{pmatrix} -(\lambda + \mu) & \gamma \\ \lambda & -(\gamma + \mu + \nu + \delta) \end{pmatrix}$.

Consider the function

$$\begin{aligned} W : I \times \mathbb{R}^2_+ &\longrightarrow \mathbb{R}^2_+, \\ (t, X(t)) &\longmapsto AX(t) + F(X(t)) \end{aligned}.$$

Let $X_1(t), X_2(t) \in \mathbb{R}^2_+$, $\| \cdot \|$ denote the Euclidean norm and $||| \cdot |||$ represent the matrix norm. The function W is locally Lipschitz continuous, as evidenced by the inequality

$$\| W(t, X_1(t)) - W(t, X_2(t)) \| \leq (|||A||| + c(t)) \cdot \| X_1(t) - X_2(t) \|$$

where

$$\| F(X_1(t)) - F(X_2(t)) \| \leq c(t) \cdot \| X_1(t) - X_2(t) \|$$

and

$$c(t) = \beta^2 \left(\| X_1(t) \| + \| X_2(t) \|\right).$$

Consequently, one establishes that the function W exhibits local Lipschitz continuity with respect to its second variable, thereby ensuring its continuity. The Cauchy-Lipschitz Theorem guarantees the stated result. □

3.2 Positivity of Solutions

Firstly, one will verify that the problem is mathematically well posed and that all variables D and C must remain positive at each instant, as they represent numbers of individuals.

Theorem 2 *For any fixed initial state $D_0 \in \mathbb{R}_+$, $C_0 \in \mathbb{R}_+$ fixed, and for any associated measurable control $u(\cdot)$, the system (1) is positively invariant in $\mathbb{R}^2_+$.*

Proof One postulate that $N(t) = D(t) + C(t) > 0$. Note that $D = D^+ - D^-$ and $C = C^+ - C^-$, where $D^+ = \max(D, 0) \geq 0$, $D^- = \max(-D, 0) \geq 0$, $C^+ = \max(C, 0) \geq 0$, and $C^- = \max(-C, 0) \geq 0$. By applying the operation of multiplication to equation (1a) with D^-, one derives the following expression:

$$\frac{D^- d(D^+ - D^-)}{dt} = ID^- + (\lambda + \mu)(D^-)^2 + \gamma C D^- + \beta C (D^-)^2$$

noting that $D^+ \cdot D^- = 0$. Hence,

$$\frac{D^- d(D^+ - D^-)}{dt} = \frac{D^- dD^+}{dt} - \frac{D^- dD^-}{dt} = -\frac{1}{2}\frac{d}{dt}[(D^-)^2]$$

from which it follows that

$$\frac{1}{2}\frac{d}{dt}[(D^-)^2] = -ID^- - (\lambda + \mu)(D^-)^2 - \gamma C D^- - \beta C (D^-)^2.$$

Analogously, by multiplying equation (1b) by C^-, one derives:

$$\frac{1}{2}\frac{d}{dt}[(C^-)^2] = -\lambda D C^- - (\gamma + \mu + \nu + \delta)(C^-)^2 + \beta D (C^-)^2.$$

Define $Z(t) = (D^-(t))^2 + (C^-(t))^2 \geq 0$. Summing the terms, one obtains the following:

$$\begin{aligned}\frac{dZ}{dt} &= 2\beta D Z + 2(\gamma + \lambda) D^- C^- - 2ID^- - 2(\lambda + \mu)(D^-)^2 \\ &\quad - 2(\gamma + \mu + \nu + \delta)(C^-)^2 - 2\beta N (D^-)^2 - 2(\gamma C^+ D^- + \lambda D^+ C^-).\end{aligned}$$

Given that $C^- D^- \leq \frac{1}{2}\left((C^-)^2 + (D^-)^2\right) = \frac{1}{2} Z$, it follows that

$$\frac{dZ}{dt} \leq (\gamma + \lambda + 2\beta D) Z$$

because all other terms are negative. By integrating over the interval $t \in [0, T]$ and applying Gronwall's inequality, one derives that

$$\forall t \in [0, T], \quad Z(t) \leq Z(0) \exp\left(\int_0^t (\gamma + \lambda + 2\beta D)\, ds \right).$$

Given that $Z(0) = 0$, it follows that $\forall t \in [0, T]$, $Z(t) = 0$, which implies $C^- \equiv 0$ and $D^- \equiv 0$. This result confirms that our system exhibits positive invariance within $\mathbb{R}^2_+$. □

The total population N adheres to the following double differential inequalities:

$$I - (\mu + u + \delta)N(t) \leq \frac{dN}{dt}(t) \leq I - \mu N(t)$$

which can be interpreted as a conservation law. Consequently, the model (1) will be analyzed within the compact region:

$$\Gamma_{d,1} = \left\{(D, C) \in \mathbb{R}^2_+ : 0 \leq D, C \leq \max\left\{N(0); \mu^{-1} I\right\}\right\}$$

If the mortality rate μ is zero, this condition will hold for a finite duration τ_m. Consequently, the model (1) will be examined within the compact region:

$$\Gamma_{d,2} = \left\{(D, C) \in \mathbb{R}^2_+ : 0 \leq D, C \leq I\tau_m + N(0)\right\}$$

Throughout this chapter, one defines:

$$\Lambda_d = \begin{cases} \max\left\{N(0); \mu^{-1} I\right\} & \text{if} \quad \mu > 0, \\ I\tau_m + N(0) & \text{if} \quad \mu = 0 \end{cases}$$

3.3 *The Pontryagin's Maximum Principle Corresponding to Our Problem*

Given the well-defined model, one presents a formulation of Pontryagin's maximum principle pertinent to our problem. Specifically, the objective is to maximize the defined objective function for a fixed discount rate $r > 0$.

$$\int_0^{+\infty} \exp(-rt) F\left(u(t), D(t), C(t)\right) dt \tag{3}$$

subject to the constraints of the system (1), where $F\left(u(t), D(t), C(t)\right)$ is a concave performance index and u is a control.

Theorem 3 *There exists an absolutely continuous application* $P(\cdot) = (P_D(\cdot), P_C(\cdot)) : [0, +\infty[\to \mathbb{R}^2$ *called the adjoint vector, with the limiting transversality conditions:*

$$\lim_{t\to+\infty} \exp(-rt)P_D(t)D(t) = 0 \quad and \quad \lim_{t\to+\infty} \exp(-rt)P_C(t)C(t) = 0,$$

and such that, for almost every $t \geq 0$, one has

$$\begin{cases} \dfrac{dP_D}{dt}(t) = & (r+\mu(t))P_D(t) - (\lambda(t)+\beta(t)C(t))\,(P_C - P_D)\,(t) \\ & \qquad -\dfrac{\partial F}{\partial D}(\gamma(t), D(t), C(t)) \\ \dfrac{dP_C}{dt}(t) = & (\gamma(t) - \beta(t)D(t))\,[P_C - P_D](t) \\ & +(r+\mu(t)+\nu(t)+\delta(t))P_C(t) - \dfrac{\partial F}{\partial C}(\gamma(t), D(t), C(t)). \end{cases} \tag{4}$$

Moreover, the optimal control $\bar{u}$, whose existence has been rigorously established in [3], adheres to the following maximization criterion:

$$H(u, D, C, P_D, P_C) = \max_{v\in L^\infty(\mathbb{R}_+)} H(v, D, C, P_D, P_C)$$

where H denotes the Hamiltonian derived from the state equations:

$$\begin{aligned} H(u, D, C, P_D, P_C) = F(u, D, C) + P_D[I - (\lambda+\mu)\,D(t) + \gamma C - \beta DC] \\ + P_C[\lambda D - (\gamma+\mu+\nu+\delta)\,C + \beta DC]. \end{aligned}$$

4 Study of Stability

In this section, our objective is to rigorously analyze the stability of the model and determine the conditions for the existence of equilibrium states. One defines $\sigma = \mu + \nu + \delta$.

Theorem 4 *Examine the ensuing system derived from the state and adjoint equations of our model:*

$$\begin{cases} \dfrac{dD}{dt}(t) & = I - (\lambda(t)+\mu(t))\,D(t) + \gamma(t)C(t) - \beta(t)D(t)C(t) \\ \dfrac{dC}{dt}(t) & = \lambda(t)D(t) - (\gamma(t)+\mu(t)+\nu(t)+\delta(t))\,C(t) + \beta(t)D(t)C(t), \\ \dfrac{dP_D}{dt}(t) & = (r+\mu(t))P_D(t) - (\lambda(t)+\beta(t)C(t))\,(P_C(t) - P_D(t)) \\ & \qquad -\dfrac{\partial F}{\partial D}(\gamma(t), D(t), C(t)), \\ \dfrac{dP_C}{dt}(t) & = (\gamma(t) - \beta(t)D(t))\,[P_C(t) - P_D(t)] \\ & \quad +(r+\mu(t)+\nu(t)+\delta(t))P_C(t) - \dfrac{\partial F}{\partial C}(\gamma(t), D(t), C(t)). \end{cases} \tag{5}$$

Subsequently, for any control u, and for any concave performance index F selected as previously declined, if

1. $\beta\mu > 0$, *there exists* $r^*_{\beta,M} \in \bar{\mathbb{R}}^*_+$ *such that for all* $r \in]0, r^*_{\beta,M}[$, *an equilibrium state* (D^*, C^*, P^*_D, P^*_C) *exists, satisfying* $D^* \neq (2\beta\mu)^{-1}(\beta I + \gamma\mu + \lambda\sigma + \mu\sigma)$ *for the system* (5).
2. $\beta\mu = 0$, *there exists* $r^*_{\beta,M} \in \bar{\mathbb{R}}^*_+$ *such that for all* $r \in]0, r^*_{\beta,M}[$, *an equilibrium state* (D^*, C^*, P^*_D, P^*_C) *exists for the system* (5).

Proof The equilibrium points (D^*, C^*, P^*_D, P^*_C) are characterized by the absence of temporal fluctuations in the variables D, C, P_D, P_C, implying that the subsequent equalities are satisfied:

$$\begin{cases} I - (\lambda + \mu)\, D^* + \gamma C^* - \beta D^* C^* = 0 & \text{(6a)} \\ \lambda D^* - (\gamma + \mu + \nu + \delta)\, C^* + \beta D^* C^* = 0 & \text{(6b)} \\ (r + \mu + \lambda) P^*_D - \lambda P^*_C - \beta C^*(P^*_C - P^*_D) - \dfrac{\partial F}{\partial D}(D^*, C^*) = 0 & \text{(6c)} \\ (\gamma - \beta D^*)(P^*_C - P^*_D) + (r + \mu + \nu + \delta) P^*_C - \dfrac{\partial F}{\partial C}(D^*, C^*) = 0 & \text{(6d)} \end{cases}$$

By synthesizing equations (6a) and (6b), one derives:

$$C^* = \sigma^{-1}(I - \mu D^*). \tag{7}$$

To determine the existence of an equilibrium state, one computes the determinant of the Jacobian matrix M_r of our system, as defined by

$$M_r = \begin{pmatrix} M_1 & M_2 \\ M_3 & M_4 \end{pmatrix},$$

with

$$M_1 = \begin{pmatrix} -(\lambda + \mu) - \beta C^* & \gamma - \beta D^* \\ \lambda + \beta C^* & -(\gamma + \sigma) + \beta D^* \end{pmatrix}, \quad M_2 = \begin{pmatrix} 0 & 0 \\ 0 & 0 \end{pmatrix},$$

$$M_3 = \begin{pmatrix} -\dfrac{\partial^2 F}{\partial D^2} & -\dfrac{\partial^2 F}{\partial D \partial C} - \beta(P^*_C - P^*_D) \\ -\beta(P^*_C - P^*_D) - \dfrac{\partial^2 F}{\partial C \partial D} & -\dfrac{\partial^2 F}{\partial C^2} \end{pmatrix} \quad \text{and}$$

$$M_4 = \begin{pmatrix} (r + \mu) + (\lambda + \beta C^*) & -(\lambda + \beta C^*) \\ -(\gamma + \beta D^*) & (\gamma - \beta D^*) + (r + \sigma) \end{pmatrix}.$$

It follows that

$$\det(M_r) = \left[\sigma(\mu + \lambda + \beta C^*) + \mu(\gamma - \beta D^*)\right] \cdot \left[(r + \sigma)(r + \mu + \lambda + \beta C^*) + (r + \mu)(\gamma - \beta D^*)\right]. \tag{8}$$

It is pertinent to observe that, by invoking relation (7), one obtains

$$\lim_{r\to 0^+} \det(M_r) = [(\beta I + \gamma\mu + \lambda\sigma + \mu\sigma) - 2\beta\mu D^*]^2 \geq 0.$$

Consequently, if $\beta\mu \neq 0$, then $\lim_{r\to 0^+} \det(M_r) > 0$ provided that

$$D^* \neq (2\beta\mu)^{-1}(\beta I + \gamma\mu + \lambda\sigma + \mu\sigma).$$

Conversely, if $\beta\mu = 0$, then $\lim_{r\to 0^+} \det(M_r) = (\beta I + \gamma\mu + \lambda\sigma + \mu\sigma)^2 > 0$, since $\lambda\sigma \neq 0$. □

Theorem 5 *When $\beta > 0$, one considers the half line (d) originating from the point $\left(\beta^{-1}(\sigma(\lambda+\mu)+\mu\gamma);\ 0\right)$, with the direction vector $(\mu;\ \sigma)$. one posit that*

$$D < \min\left\{\frac{\sigma^2}{2(\nu+\delta)\beta} + \frac{\gamma}{\beta};\ \Lambda_d\right\} \quad \textit{and} \quad \{(C(t), D(t))\,,\ t \in [0, t_m(u)]\} \cap (d) = \emptyset.$$

Consequently, there exists $r^*_\beta \in \bar{\mathbb{R}}^*_+$ such that for all $r \in]0, r^*_\beta[$, there is no limit cycle among the number of diabetics with complications, the number of diabetics without complications, and the controlled parameter chosen as specified previously.

Proof Initially, one examines $\det(J_r)$, the determinant of the Jacobian matrix

$$J_r = \begin{pmatrix} \frac{\partial \dot{D}}{\partial D} & \frac{\partial \dot{D}}{\partial C} & \frac{\partial \dot{D}}{\partial P_D} & \frac{\partial \dot{D}}{\partial P_C} \\ \frac{\partial \dot{C}}{\partial D} & \frac{\partial \dot{C}}{\partial C} & \frac{\partial \dot{C}}{\partial P_D} & \frac{\partial \dot{C}}{\partial P_C} \\ \frac{\partial \dot{P}_D}{\partial D} & \frac{\partial \dot{P}_D}{\partial C} & \frac{\partial \dot{P}_D}{\partial P_D} & \frac{\partial \dot{P}_D}{\partial P_C} \\ \frac{\partial \dot{P}_C}{\partial D} & \frac{\partial \dot{P}_C}{\partial C} & \frac{\partial \dot{P}_C}{\partial P_D} & \frac{\partial \dot{P}_C}{\partial P_C} \end{pmatrix}$$

pertaining to the system declined in equation (1), such that

$$\det(J_r) = \left[\sigma(\lambda + \mu + \beta C) + \mu(\gamma - \beta D)\right] \cdot \left[(r+\sigma)(r+\lambda+\mu+\beta C) + (r+\mu)(\gamma - \beta D)\right]. \quad (9)$$

one subsequently advances by means of case disjunction:

1. In the case where $\beta = 0$, it follows that $\det(J_r) > 0$, given that not all parameters are null, allowing us to define $r^*_{\beta,J} = \infty$.
2. Assuming $\beta > 0$, one observe that

$$\lim_{r\to 0^+} \det(J_r) = \left[\sigma(\lambda + \mu + \beta C) + \mu(\gamma - \beta D)\right]^2 \geq 0.$$

The limit reaches zero if the following condition holds.

$$\sigma C - \mu D - \beta^{-1}(\sigma(\lambda + \mu) + \mu\gamma) = 0. \tag{10}$$

Consequently, one infers the existence of $r^*_{\beta,J} > 0$ such that for all $0 < r < r^*_{\beta,J}$ and for all $(C, D) \notin (d)$, one has $\det(J_r) > 0$ by the continuity of $\det(J_r)$ with respect to the parameter r.

Conversely, the computation

$$K_r = \begin{vmatrix} \frac{\partial \dot{D}}{\partial D} & \frac{\partial \dot{D}}{\partial P_D} \\ \frac{\partial \dot{P}_D}{\partial D} & \frac{\partial \dot{P}_D}{\partial P_D} \end{vmatrix} + \begin{vmatrix} \frac{\partial \dot{C}}{\partial C} & \frac{\partial \dot{C}}{\partial P_C} \\ \frac{\partial \dot{P}_C}{\partial C} & \frac{\partial \dot{P}_C}{\partial P_C} \end{vmatrix} + 2 \begin{vmatrix} \frac{\partial \dot{D}}{\partial C} & \frac{\partial \dot{D}}{\partial P_C} \\ \frac{\partial \dot{P}_D}{\partial C} & \frac{\partial \dot{P}_D}{\partial P_C} \end{vmatrix}$$

leads to

$$K_r = -(\lambda + \gamma + \beta C - \beta D - r - 2\mu)^2 + 4^{-1}(r + 2\mu)^2 - \\ \left[\mu(r + \mu) + \sigma(r + \sigma) + 2(\nu + \delta)(\gamma - \beta D)\right]. \tag{11}$$

Analogously, one proceeds by case disjunction:

1. If $\beta = 0$, then $K_r < 0$, given that not all parameters are zero, and one subsequently sets $r^*_{\beta,K} = \infty$.
2. If $\beta > 0$, considering the polynomial function

$$\Phi(r) = 4^{-1}(r + 2\mu)^2 - \mu(r + \mu) - \sigma(r + \sigma) - 2(\nu + \delta)(\gamma - \beta D),$$

one observes that if $D < \dfrac{\sigma^2}{2(\nu + \delta)\beta} + \dfrac{\gamma}{\beta}$, then

$$\lim_{r \to 0^+} \Phi(r) = -\sigma^2 - 2(\nu + \delta)(\gamma - \beta D) < 0.$$

Thus, there exists $r^*_{\beta,K} > 0$ such that for all $0 < r < r^*_{\beta,K}$, one has $\Phi(r) < 0$, by continuity of Φ. Hence, $K_r < 0$.

Finally, setting $r^*_\beta = \min\left\{r^*_{\beta,J}, r^*_{\beta,K}\right\}$ leads to the announced result. □

The foregoing proof elucidates, as per Remark 1, that for $\beta = 0$, the outcome aligns with the results of Bernard et al. [3].

Remark 1 If $\beta = 0$ and for all $r > 0$, there is no limit cycle between the population of diabetics with complications, the population of diabetics without complications, and the selected control parameter.

Remark 2 For $\beta > 0$ and $r > r^*_\beta$, there might exist a limit cycle between the number of diabetics with complications, the number of diabetics without complications, and the chosen control parameter.

5 Characterization of the Equilibrium State

Our goal in this proposed control strategy is to maximize the number of people recovering from complications. One could choose to minimize the number of people progressing from the stage of diabetes without complications to the stage of diabetes with complications. For $u = \gamma$ and for a fixed discount rate $r > 0$, the cost function that is desired to maximize is

$$\int_0^{+\infty} \exp(-rt) F\,(u(t), D(t), C(t))\, dt, \tag{12}$$

using the same performance index F used by Séverine Bernard, Ténissia César, Silvere P. Nuiro, and Alain Piétrus [3], where $F(u, D, C) = \alpha \ln(u) + a \ln(C) + K(D)$, with the assumption

$(\mathbf{H}_2):\quad a > \alpha > 0$ and K is a differentiable concave and increasing function of D,

such that the optimality condition on H leads to

$$\bar{\gamma} = \frac{\alpha}{T \cdot C} \tag{13}$$

where $T = P_C - P_D > 0$ provided that α and C are positive.

The search for the equilibrium state associated with the system (1), leads us to solve the system (6), such that there is no variation of D, C, P_D, P_C with respect to t. Thus, equation (6a) implies that

$$D = \frac{IT + \alpha}{(\rho + \beta C)T} \tag{14}$$

with $(\rho + \beta C)T > 0$. Using equation (6b), one finds the following:

$$C = \frac{\lambda D}{\sigma - \beta D} - \frac{\alpha}{(\sigma - \beta D)T},$$

with $(\sigma - \beta D)T > 0$. This leads to

$$\beta \sigma T C^2 - (\beta I T - \sigma \rho T) C - \lambda (IT + \alpha) + \alpha \rho = 0. \tag{15}$$

One notices that the leading coefficient of the trinomial (15) in the variable C depends on β, and this trinomial has a degree less than or equal to 2. one will study, according to the values of β, the existence of solutions of equation (15), which is of first degree when $\beta = 0$, and of second degree when $\beta > 0$. So, initially, one will solve equation (15) when $\beta = 0$, and subsequently, when $\beta > 0$.

1. if $\beta = 0$, equations (6a), (6b), (6c), and (6d), respectively, yield

$$\begin{cases} D = \dfrac{IT + \alpha}{(\lambda + \mu)T} & (16a) \\ C = \dfrac{\lambda(IT + \alpha) - \alpha\rho}{\sigma\rho T} & (16b) \\ P_D = \dfrac{\lambda(a - \alpha)}{(r + \rho)(r + \sigma)} \times \dfrac{\sigma\rho T}{[\lambda(IT + \alpha) - \alpha\rho]} + \dfrac{K'(D)}{(r + \rho)} & (16c) \\ P_C = \dfrac{a - \alpha}{(r + \sigma)} \times \dfrac{\sigma\rho T}{[\lambda(IT + \alpha) - \alpha\rho]} & (16d) \end{cases}$$

Thus,

$$T = \left(\frac{a - \alpha}{r + \sigma}\right)\left(\frac{r + \rho - \lambda}{r + \rho}\right)\left(\frac{\sigma\rho T}{\lambda(IT + \alpha) - \alpha\rho}\right) + \frac{K'(D)}{(r + \rho)} \tag{17}$$

with $\lambda(IT + \alpha) - \alpha\rho > 0$.

2. If $\beta > 0$, the discriminant Δ of equation (15) is positive and is equal to

$$\Delta = [-(\beta IT - \sigma\rho T)]^2 - 4\beta\sigma T[-\lambda(IT + \alpha) + \alpha\rho]$$

or

$$\Delta = \left[4\lambda\beta\sigma I + (-\beta I + \rho\sigma)^2 T^2 + 4\alpha(\rho - \lambda)\beta\sigma T\right].$$

Consequently, equation (15) has the following two real roots:

$$C_1 = \frac{(\beta I - \sigma\rho)}{2\beta\sigma} - \frac{\sqrt{\left[4\lambda\beta\sigma I + (-\beta I + \rho\sigma)^2 T^2 + 4\alpha(\rho - \lambda)\beta\sigma T\right]}}{2\beta\sigma T} \leq 0$$

and

$$C_2 = C(T) = \frac{Y(T)}{2\beta\sigma} \tag{18}$$

with

$$Y(T) = (\beta IT - \sigma\rho T) + \sqrt{\left[4\lambda\beta\sigma I + (-\beta I + \rho\sigma)^2 T^2 + 4\alpha(\rho - \lambda)\beta\sigma T\right]} > 0,$$

provided that not all parameters are zero.

Subsequently, equations (14), (18), (6c), and (6d) respectively yield

$$
\begin{cases}
D = \dfrac{2\sigma(IT+\alpha)}{(2\sigma\rho T + Y(T))} & (19a)\\
C = \dfrac{Y(T)}{2\beta\sigma T} & (19b)\\
P_D = \dfrac{2\lambda\sigma\beta T}{(r+\sigma)(r+\rho)}\left(\dfrac{a-\alpha}{Y(T)} + \dfrac{IT+\alpha}{2\sigma\rho T + Y(T)}\right) & (19c)\\
\qquad + \dfrac{Y(T)}{2\sigma(r+\sigma)T} + \dfrac{K'(D)}{(r+\sigma)} & (19d)\\
P_C = \dfrac{2\sigma\beta T}{r+\sigma}\left(\dfrac{a-\alpha}{Y(T)} + \dfrac{IT+\alpha}{2\sigma\rho T + Y(T)}\right). & (19e)
\end{cases}
$$

Thus,

$$
T = \left(\frac{2\sigma\beta T}{r+\sigma}\right)\left(\frac{r+\sigma-\lambda}{r+\sigma}\right)\left(\frac{a-\alpha}{Y(T)} + \frac{IT+\alpha}{2\sigma\rho T + Y(T)}\right) - \frac{Y(T)}{2\sigma(r+\sigma)T} - \frac{K'(D)}{(r+\sigma)}. \quad (20)
$$

It is clear that the two previous cases require two distinct studies. Given the nature of equation (20), which is transcendental, the equilibrium state associated with the system (1) will be characterized in the case where β is zero. This choice leads us back to equation (17), which induces the resolution for T of the following equation:

$$
\lambda I(r+\lambda+\mu)(r+\sigma)T^2 + [\alpha\mu(r+\lambda+\mu)(r+\sigma) - (r+\mu)(a-\alpha)(\lambda+\mu)\sigma + \lambda I K'(D)(r+\sigma+)]T - \alpha\mu(r+\mu+\nu+\delta)K'(D) = 0. \quad (21)
$$

To simplify the notations, one assumes

$$
i(r) = \lambda I(r+\lambda+\mu)(r+\mu+\nu+\delta), \quad m(r) = \alpha\mu(r+\mu+\nu+\delta),
$$

$$
n_1(r) = -\alpha\mu(r+\lambda+\mu)(r+\mu+\nu+\delta) - (r+\mu)(a-\alpha)(\lambda+\mu)(\mu+\nu+\delta),
$$

$$
n_2(r) = \lambda I(r+\mu+\nu+\delta) \quad \text{and} \quad g(T) = K'(D).
$$

Thus, the equation (21) becomes

$$
i(r)T^2 + [n_1(r) + n_2(r)g(T)]T - m(r)g(T) = 0. \quad (22)
$$

If g is identically zero, then equation (22) implies that $T = -\dfrac{n_1(r)}{i(r)}$. In this case, assumptions ($\mathbf{H}_1$) and ($\mathbf{H}_2$) allow us to conclude that for all $r > 0$, one has: $T > 0$. For further details, see [1, 3]. Moreover, one assumes:

$$(\mathbf{H}_3):\ g \text{ is not identically zero and } K \text{ is twice differentiable on } \mathbb{R}_+^*.$$

Lemma 1 *The function* $g : T \mapsto K'\left(\dfrac{IT+\alpha}{(\lambda+\mu)T}\right)$ *is increasing on* $\mathbb{R}_+^*$ *and takes values in* $]l_1, l_2[$, *where*

$$l_1 = \lim_{T\longrightarrow 0^+} g(T) = \lim_{D\longrightarrow +\infty} K'(D) \geq 0 \tag{23}$$

and

$$l_2 = \lim_{T\longrightarrow +\infty} g(T) = K'\left(\frac{I}{\lambda+\mu}\right) > 0. \tag{24}$$

Proof Since K is twice differentiable in $\mathbb{R}_+^*$ and the function $T \mapsto \dfrac{IT+\alpha}{(\lambda+\mu)T}$ is differentiable in $\mathbb{R}_+^*$, then the function g is differentiable in $\mathbb{R}_+^*$.
Thus, one has

$$\forall T \in \mathbb{R}_+^*, \quad g'(T) = -\frac{\alpha}{(\lambda+\mu)T^2} K''\left(\frac{IT+\alpha}{(\lambda+\mu)T}\right) \geq 0,$$

as a consequence of the concavity of K. Consequently, g is increasing and one has: $\forall T \in \mathbb{R}_+^*, \quad l_1 \leq g(T) \leq l_2$. □

Remark 3 It can be noted that g is strictly increasing and bijective from $\mathbb{R}_+^*$ to $]l_1, l_2[$ provided that K' is strictly decreasing.

Lemma 2 *The function* $n : r \mapsto -\dfrac{n_1(r)}{n_2(r)}$ *increases strictly on* $]0, +\infty[$, *and is bijective from* $\mathbb{R}_+^*$ *to* $]n_0, n_\infty[$ *with*

$$n_0 = \lim_{r\to 0^+} n(r) = \frac{a\mu(\lambda+\mu)}{\lambda I} \text{ and } n_\infty = \lim_{r\to\infty} n(r) = \begin{cases} \infty, & \text{if } \mu > 0 \\ \dfrac{(a-\alpha)(\nu+\delta)}{I}, & \text{if } \mu = 0. \end{cases}$$

Lemma 3 *The function* $f : r \mapsto n(r) - 2n_2^{-2}(r)i(r)m(r)$, *defined on* $]0, +\infty[$, *admits the following limits:*

$$f_0 = \lim_{r\to 0^+} f(r) = \frac{\mu(\mu+\nu+\delta)(\lambda+\mu)(a-2\alpha)}{\lambda I},$$

$$f_\infty = \lim_{r \to \infty} f(r) = \begin{cases} -\infty, \ \textit{if} \ \ \mu > 0 \\ \frac{(a-\alpha)(\nu+\delta)}{I}, \ \textit{if} \ \ \mu = 0. \end{cases}$$

and one has

1. *if* $\mu = 0$*, f is positive and strictly increasing on* $]0, +\infty[$*,*
2. *otherwise, there exists* $r_f^{(0)} > 0$ *such that f is strictly*

 - *increasing on* $]0, r_f^{(0)}[$
 - *decreasing on* $]r_f^{(0)}, +\infty[$*.*

 Moreover,

 a. *if* $a > 2\alpha$*, there exists* $r_f^{(1)} > r_f^{(0)}$ *such that* $f(r_f^{(1)}) = 0$ *and*

  i. *for all* $r \in]0, r_f^{(1)}[$*, one has:* $f(r) > 0$*,*
  ii. *for all* $r \geq r_f^{(1)}$*, one has:* $f(r) \leq 0$

 b. *otherwise, there exists* $\lambda_f^{(1)}, \lambda_f^{(2)} > 0$ *such that*

  i. *for all* $\lambda \in [\lambda_f^{(1)}, \lambda_f^{(2)}]$*,* $f(r) \leq 0$ *for all* $r > 0$*,*
  ii. *for all* $\lambda \in]0, \lambda_f^{(1)}[\cup]\lambda_f^{(2)}, 1[$*, there exists* $0 < r_f^{(2)} < r_f^{(0)}$ *and* $r_f^{(3)} > r_f^{(0)}$ *such that* $f(r_f^{(2)}) = f(r_f^{(3)}) = 0$ *and*

  A. *for all* $r \in]r_f^{(2)}, r_f^{(3)}[$*, one has:* $f(r) > 0$*,*
  B. *for all* $r \in]0, r_f^{(2)}] \cup [r_f^{(3)}, +\infty[$*, one has:* $f(r) \leq 0$

 See Appendix 10 for the demonstration.

Remark 4 From Lemma 3, it follows that:

1. For $\mu = 0$, one considers $\mathcal{I}_0 =]0, +\infty[$. For all $r \in \mathcal{I}_0$, one has: $f(r) > 0$.
2. For all $\mu > 0$, one considers the following:

 - $\mathcal{I}_{\mu,\lambda}^{(1)} = [r_f^{(1)}, +\infty[$, for all $a > 2\alpha$ and for all $\lambda > 0$.
 - $\mathcal{I}_{\mu,\lambda}^{(2)} =]0, +\infty[$, for all $\lambda \in [\lambda_f^{(1)}, \lambda_f^{(2)}]$ and for all $a \leq 2\alpha$.
 - $\mathcal{I}_{\mu,\lambda}^{(3)} =]0, r_f^{(2)}] \cup [r_f^{(3)}, +\infty[$, for all $\lambda \in]0, \lambda_f^{(1)}[\cup]\lambda_f^{(2)}, 1[$ and for all $a \leq 2\alpha$.

Thus, one deduces that for all $k \in \{1, 2, 3\}$, $r \in \mathcal{I}_{\mu,\lambda}^{(k)}$, f is negative.

To summarize all situations:

$$\mathcal{I}_{\mu,\lambda,\alpha,a}^{f} = \begin{cases} \mathcal{I}_{\mu}, & \text{if} \quad \mu = 0 \quad \text{and} \quad a > \alpha \\ \mathcal{I}_{\mu,\lambda}^{(1)}, & \text{if} \quad \mu > 0, \quad a > 2\alpha \quad \text{and} \quad \lambda > 0 \\ \mathcal{I}_{\mu,\lambda}^{(2)}, & \text{if} \quad \mu > 0, \quad \alpha \leq 2a \quad \text{and} \quad \lambda \in [\lambda_f^{(1)}, \lambda_f^{(2)}] \\ \mathcal{I}_{\mu,\lambda}^{(3)}, & \text{if} \quad \mu > 0, \quad \alpha \leq 2a \quad \text{and} \quad \lambda \in]0, \lambda_f^{(1)}[\cup]\lambda_f^{(2)}, 1[. \end{cases}$$

5.1 Case Where g is Constant

In this subsection, one analyzes the case where the equation (22) no longer exhibits non-linearity with respect to g.

Theorem 6 *Suppose that the function g is constant, meaning*

$$\exists \zeta \in [l_1, l_2], \quad \forall T \in \mathbb{R}_+^*, \quad g(T) = \zeta.$$

If there exists $r^* \in \mathbb{R}_+^*$ such that $n_1(r^*) + n_2(r^*)\zeta = 0$, then:

1. *for all $r \in]r^*, +\infty[$, the associated equilibrium state is admissible for all $\mu \geq 0$;*
2. *for all $r \in]0, r^*]$, the associated equilibrium state is admissible if and only if $\mu\zeta > 0$.*

Otherwise,

- *if $\mu > 0$, then for all $r \in]0; +\infty[$, the associated equilibrium state is admissible.*
- *if $\mu = 0$, then the associated equilibrium state is not admissible.*

Proof Taking into account the previous assumptions, equation (22) becomes

$$i(r)T^2 + (n_1(r) + n_2(r)\zeta)T - m(r)\zeta = 0. \tag{25}$$

Due to $(n_1(r) + n_2(r)\zeta)^2 + 4i(r)m(r)\zeta \geq 0$, it is easy to show that it has two real roots (not necessarily distinct), namely:

$$T_\mu^{(1)} = \frac{-(n_1(r) + n_2(r)\zeta) + \sqrt{(n_1(r) + n_2(r)\zeta)^2 + 4i(r)m(r)\zeta}}{2i(r)}$$

and

$$T_\mu^{(2)} = \frac{-(n_1(r) + n_2(r)\zeta) - \sqrt{(n_1(r) + n_2(r)\zeta)^2 + 4i(r)m(r)\zeta}}{2i(r)}.$$

It is clear that $T_\mu^{(2)} \leq T_\mu^{(1)}$.

The assumption $a > \alpha$ implies that $n_1(r) < 0$ for all $r \in \mathbb{R}_+^*$. Since one is looking for a positive root, study the sign of $T_\mu^{(1)}$ will be considered.

1. If there exists $r^* \in \mathbb{R}_+^*$ such that $n_1(r^*) + n_2(r^*)\zeta = 0$, and since n is increasing on $\mathbb{R}_+^*$, then:

   a. for any $r \in]r^*; +\infty[$, one has: $0 \leq \zeta = n(r^*) < n(r)$ (that is, $n_1(r) + n_2(r)\zeta < 0$), from which one deduces that

   $$\forall r \in]r^*, +\infty[, \quad -(n_1(r) + n_2(r)\zeta) + \sqrt{(n_1(r) + n_2(r)\zeta)^2 + 4i(r)m(r)\zeta} > 0$$

   and in this case, there exists a strictly positive solution $T_\mu^{(1)}$ that satisfies (25).

b. for any $r \in]0, r^*]$, one has: $\zeta = n(r^*) \geq n(r) > 0$ (i.e., $n_1(r) + n_2(r)\zeta \geq 0$), and consequently:

$$-(n_1(r) + n_2(r)\zeta) + \sqrt{(n_1(r) + n_2(r)\zeta)^2 + 4i(r)m(r)\zeta} > 0$$

if

$$\sqrt{(n_1(r) + n_2(r)\zeta)^2 + 4i(r)m(r)\zeta} > (n_1(r) + n_2(r)\zeta) \geq 0.$$

This inequality is true if
$(n_1(r) + n_2(r)\zeta)^2 + 4i(r)m(r)\zeta > (n_1(r) + n_2(r)\zeta)^2$

$$\Leftrightarrow i(r)m(r)\zeta = \alpha\lambda I(r + \lambda + \mu)(r + \mu + \nu + \delta)^2\mu\zeta > 0,$$

which is satisfied if and only if $\mu\zeta > 0$.

2. Otherwise,

- if $\mu > 0$, then the equation $n_1(r) + n_2(r)\zeta = 0$ has no solution in $\mathbb{R}_+^*$. Since $n_1(r) + n_2(r)\zeta$ is a second-degree trinomial with a strictly negative leading coefficient, then for any $r \in \mathbb{R}_+^*$, one has $n_1(r) + n_2(r)\zeta < 0$. To conclude, one revisits the proof of 1 with $r^* = 0$.
- if $\mu = 0$, then the equation $n_1(r) + n_2(r)\zeta = 0$ becomes

$$-[(a - \alpha)(\nu + \delta) - I\zeta]r + I(\nu + \delta)\zeta = 0.$$

It can only have a solution if $(a - \alpha)(\nu + \delta) \neq I\zeta$.

And in this case, the solution is $r = \dfrac{I(\nu + \delta)\zeta}{(a - \alpha)(\nu + \delta) - I\zeta}$.
Since the solution does not belong to $\mathbb{R}_+^*$, necessarily $(a - \alpha)(\nu + \delta) < I\zeta$, and for any $r > 0$, $n_1(r) + n_2(r)\zeta < 0$. One deduces that:

$$\forall r \in \mathbb{R}_+^*, \quad 0 < \frac{(a - \alpha)(\nu + \delta)}{I} < \zeta \leq n_0 = 0.$$

This is absurd.

The existence of a solution $T_\mu^{(1)} > 0$ to equation (22) implies the attainment of an admissible equilibrium state:

$$\begin{cases} D^* = \dfrac{I}{\lambda+\mu} + \dfrac{2\alpha i(r)}{(\lambda+\mu)\left[-(n_1(r)+n_2(r)\zeta)+\sqrt{(n_1(r)+n_2(r)\zeta)^2+4i(r)m(r)\zeta}\right]} \\ C^* = \dfrac{\lambda I}{(\mu+\nu+\delta)(\lambda+\mu)} + \dfrac{(n_1(r)+n_2(r)\zeta)+\sqrt{(n_1(r)+n_2(r)\zeta)^2+4i(r)m(r)\zeta}}{2(\mu+\nu+\delta)(\lambda+\mu)(r+\mu+\nu+\delta)\zeta} \\ P_D^* = \dfrac{-n_1(r)+n_2(r)\zeta+\sqrt{(n_1(r)+n_2(r)\zeta)^2+4i(r)m(r)\zeta}}{2\lambda I(r+\mu+\nu+\delta)(r+\mu)} \\ \qquad\qquad + \dfrac{(n_1(r)+n_2(r)\zeta)-\sqrt{(n_1(r)+n_2(r)\zeta)^2+4i(r)m(r)\zeta}}{2i} \\ P_C^* = \dfrac{-n_1(r)+n_2(r)\zeta+\sqrt{(n_1(r)+n_2(r)\zeta)^2+4i(r)m(r)\zeta}}{2\lambda I(r+\mu+\nu+\delta)(r+\mu)}, \end{cases}$$

with

$$\gamma^* = \begin{cases} \dfrac{2\alpha\mu(r+\lambda+\mu)(r+\mu+\nu+\delta)+(n_1(r)+n_2(r)\zeta)}{2\mu(r+\mu)(a-\alpha)} \\ \qquad + \dfrac{\sqrt{(n_1(r)+n_2(r)\zeta)^2+4i(r)m(r)\zeta}}{2\mu(r+\mu)(a-\alpha)}, & \text{if } \mu > 0, \\ \dfrac{\alpha(r+\lambda)(r+\nu+\delta)(\nu+\delta)}{r(a-\alpha)(\nu+\delta)-I(r+\nu+\delta)\zeta}, & \text{if } \mu = 0. \end{cases}$$

□

Remark 5 1. Theorem 6 generalizes some results from [1, 3]. Specifically, the case where $K(D) = D$, i.e., $\zeta = 1$, and $\mu = 0$. If $\alpha < a - I(\nu+\delta)^{-1}$ then there exists $r^* = \dfrac{I(\nu+\delta)}{(a-\alpha)(\nu+\delta)-I} > 0$ such that $n_1(r^*) + n_2(r^*)\zeta = 0$. Since $\mu = 0$, by applying Theorem 6, the equilibrium state is admissible for all $r \in]r^*, +\infty[$.
2. If $\mu > 0$, the equation $n_1(r) + n_2(r)\zeta = 0$ implies

$$-\alpha\mu r^2 + (-a\rho\sigma - \alpha\mu\rho - \alpha\mu\sigma + \alpha\rho\sigma + \lambda I\zeta)r - a\mu\rho\sigma + \lambda I\zeta\sigma = 0 \tag{26}$$

and its discriminant Δ is such that

$$\Delta = (-a\rho\sigma - \alpha\mu\rho - \alpha\mu\sigma + \alpha\rho\sigma + \lambda I\zeta)^2 + 4\alpha\mu(-a\mu\rho\sigma + \lambda\sigma I\zeta) = B_0\lambda^2 + B_1\lambda + B_2$$

where $B_0 = (I\zeta)^2$, $B_1 = 2(-a\rho\sigma - \alpha\mu\rho - \alpha\mu\sigma + \alpha\rho\sigma)I\zeta + 4\alpha\mu\sigma I\zeta$ and $B_2 = (-a\rho\sigma - \alpha\mu\rho - \alpha\mu\sigma + \alpha\rho\sigma)^2 - 4a\alpha\mu^2\rho\sigma$.

Now, the discriminant Δ yields $-16\alpha\mu\rho\sigma I^2\zeta^2(\nu+\delta)(a-\alpha) \leq 0$. This guarantees that Δ is positive for all $\lambda > 0$. Consequently, equation (26) has two real solutions

$$r_1^* = \frac{-a\rho\sigma - \alpha\mu\rho - \alpha\mu\sigma + \alpha\rho\sigma + \lambda I l_2 - \sqrt{B_0\lambda^2 + B_1\lambda + B_2}}{2\alpha\mu}$$

and

$$r_2^* = \frac{-\alpha\rho\sigma - \alpha\mu\rho - \alpha\mu\sigma + \alpha\rho\sigma + \lambda I l_2 + \sqrt{B_0\lambda^2 + B_1\lambda + B_2}}{2\alpha\mu}.$$

According to Theorem 6,

1. for all $r \in]r_2^*, +\infty[$ and $r_2^* > 0$, the associated equilibrium state is admissible for all $\mu > 0$;
2. for all $r \in]0, r_1^*]$ and $r_1^* > 0$, the associated equilibrium state is admissible if and only if $\mu\zeta > 0$
3. if $r_1^* \leq r_2^* < 0$, then for all $r \in]0; +\infty[$, the associated equilibrium state is admissible.

5.2 *Case Where g is Strictly Increasing*

Since g is not constant, solving equation (22) is transformed into a fixed-point problem. For $h \in \mathbb{R}_+^*$, one introduces the following auxiliary problem:

$$i(r)T_{h,\mu}^2 + [n_1(r) + n_2(r)g(h)]T_{h,\mu} - m(r)g(h) = 0. \tag{27}$$

For fixed h in $\mathbb{R}_+^*$, this auxiliary problem is a particular case in the previous section.

Thus, since $g(h) \in \mathbb{R}_+^*$ and $(n_1(r) + n_2(r)g(h))^2 + 4i(r)m(r)g(h) \geq 0$, by applying the Schauder fixed-point theorem (see [21]), one can assert that it has two real roots (not necessarily distinct), namely:

$$T_{h,\mu}^{(1)} = \frac{-(n_1(r) + n_2(r)g(h)) + \sqrt{(n_1(r) + n_2(r)g(h))^2 + 4i(r)m(r)g(h)}}{2i(r)}$$

and

$$T_{h,\mu}^{(2)} = \frac{-(n_1(r) + n_2(r)g(h)) - \sqrt{(n_1(r) + n_2(r)g(h))^2 + 4i(r)m(r)g(h)}}{2i(r)}.$$

It is clear that $T_{h,\mu}^{(2)} \leq T_{h,\mu}^{(1)}$.

Proposition 1 *Let $h \in \mathbb{R}_+^*$ be fixed. Then:*
If there exists $r_h^* \in \mathbb{R}_+^*$ such that $n_1(r_h^*) + n_2(r_h^*)g(h) = 0$, then:

1. *for all $r \in]r_h^*, +\infty[$, the associated equilibrium state is admissible for all $\mu \geq 0$;*
2. *for all $r \in]0, r_h^*]$, the associated equilibrium state is admissible if and only if $\mu g(h) > 0$.*

Otherwise,

- *if* $\mu > 0$, *then for all* $r \in]0; +\infty[$, *the equilibrium state associated with h is admissible.*
- *if* $\mu = 0$, *then the equilibrium state associated with h is not admissible.*

Proof It suffices to apply Theorem 6 with $\zeta = g(h)$. □

Proposition 2 *Suppose* $h \in \mathbb{R}_+^*$ *is fixed. Then:*

1. *if there exists* $r_{2,h}^* \in \mathbb{R}_+^*$ *such that* $n_1(r_{2,h}^*) + n_2(r_{2,h}^*)l_2 = 0$, *then for all* $r \in]r_{2,h}^*, +\infty[$, *the equilibrium state associated with h is admissible for all* $\mu \geq 0$;
2. *if there exists* $r_{1,h}^* \in \mathbb{R}_+^*$ *such that* $n_1(r_{1,h}^*) + n_2(r_{1,h}^*)l_1 = 0$, *then for all* $r \in]0, r_{1,h}^*]$, *the equilibrium state associated with h is admissible if and only if* $\mu > 0$.

Otherwise,

- *if* $\mu > 0$, *then for all* $r \in]0; +\infty[$, *the equilibrium state associated with h is admissible;*
- *if* $\mu = 0$, *then the equilibrium state associated with h is never admissible.*

Proof Since n increases over $\mathbb{R}_+^*$, then:

1. If there exists $r_2^* \in \mathbb{R}_+^*$ such that $n_1(r_2^*) + n_2(r_2^*)l_2 = 0$, then for all $r \in]r_2^*; +\infty[$, one has
 $g(h) \leq l_2 = n(r_2^*) < n(r)$ (meaning $n_1(r) + n_2(r)g(h) < 0$), from which one deduces that,
 $$\forall r \in]r_2^*; +\infty[,$$
 $$-(n_1(r) + n_2(r)g(h)) + \sqrt{(n_1(r) + n_2(r)g(h))^2 + 4i(r)m(r)g(h)} > 0$$
 and in this case, there indeed exists a strictly positive solution $T_{h,\mu}^{(1)}$ satisfying (27).
2. If there exists $r_1^* \in \mathbb{R}_+^*$ such that $n_1(r_1^*) + n_2(r_1^*)l_1 = 0$, then for all $r \in]0; r_1^*[$, one has
 $n(r) < n(r_1^*) = l_1 \leq g(h)$ (meaning $n_1(r) + n_2(r)g(h) \geq 0$), and consequently:
 $$-(n_1(r) + n_2(r)g(h)) + \sqrt{(n_1(r) + n_2(r)g(h))^2 + 4i(r)m(r)g(h)} > 0$$
 if and only if
 $$\sqrt{(n_1(r) + n_2(r)g(h))^2 + 4i(r)m(r)g(h)} > (n_1(r) + n_2(r)g(h)) \geq 0.$$
 The latter inequality holds true if and only if
 $(n_1(r) + n_2(r)g(h))^2 + 4i(r)m(r)g(h) > (n_1(r) + n_2(r)g(h))^2$

$$\Leftrightarrow i(r)m(r)g(h) = \alpha\lambda I(r+\lambda+\mu)(r+\mu+\nu+\delta)^2\mu g(h) > 0,$$

which is satisfied if and only if $\mu > 0$.

3. Otherwise,

- If $\mu > 0$, the equations $n_1(r) + n_2(r)l_2 = 0$ have no solution in $\mathbb{R}_+^*$. Since $n_1(r) + n_2(r)l_2$ is a quadratic polynomial with a strictly negative leading coefficient, then for all $r \in \mathbb{R}_+^*$, one has $n_1(r) + n_2(r)l_2 < 0$. To conclude, one revisits the proof of 2 with $r_2^* = 0$.
- If $\mu = 0$, then the equation $n_1(r) + n_2(r)l_2 = 0$ becomes

$$-[(a-\alpha)(\nu+\delta) - Il_2]r + I(\nu+\delta)l_2 = 0.$$

It can only have a solution if $(a-\alpha)(\nu+\delta) \neq Il_2$.

And in this case, the solution is $r = \dfrac{I(\nu+\delta)l_2}{(a-\alpha)(\nu+\delta) - Il_2}$.
Since the solution does not belong to $\mathbb{R}_+^*$, necessarily $(a-\alpha)(\nu+\delta) < Il_2$, and for all $r > 0$, $n_1(r) + n_2(r)l_2 < 0$. Thus, one deduces that:

$$\forall r \in \mathbb{R}_+^*, \quad 0 < \frac{(a-\alpha)(\nu+\delta)}{I} < l_2 \leq n_0 = 0.$$

Absurd.

□

Remark 6 If $\mu = 0$, according to 2. of Proposition 2, for all $r \in]r_{2,h}^*; +\infty[$ such that $n_1(r) + n_2(r)g(h) < 0$, one has: $T_{h,0} = -\dfrac{n_1(r) + n_2(r)g(h)}{i(r)} > 0$.

Theorem 7 *Consider the function defined for $h \in \mathbb{R}_+^*$, by*

$$\tau_{r,\mu}(h) = \frac{-(n_1(r) + n_2(r)g(h)) + \sqrt{(n_1(r) + n_2(r)g(h))^2 + 4i(r)m(r)g(h)}}{2i(r)}.$$

Then, there exists an interval $I_{\mu,\lambda,\alpha,a}^f \subset \mathbb{R}_+^*$ such that

$$\forall r \in I_{\mu,\lambda,\alpha,a}^f, \quad \forall h \in \mathbb{R}_+^*, \quad \tau_{r,\mu}(h) > 0.$$

Moreover, for all $r \in I_{\mu,\lambda,\alpha,a}^f$, the function $\tau_{r,\mu}$ is continuous, strictly decreasing, and has a unique fixed point $T^*(r)$ satisfying

$$g(T^*(r)) = \zeta^*(r).$$

Proof Let $h \in \mathbb{R}_+^*$.

1. if $\mu = 0$, according to 1. of Proposition 2, there exists $r^*_{2,h} > 0$ such that for all $r \in]r^*_{2,h}; +\infty[$, one has: $n_1(r) + n_2(r)g(h) < 0$, and consequently $\tau_{r,0}(h) = -\dfrac{n_1(r) + n_2(r)g(h)}{i(r)} > 0.$
2. if $\mu > 0$, the hypotheses $(\mathbf{H}_1)$ and $(\mathbf{H}_2)$ imply that $4i(r)m(r)g(h) > 0$, and thus for all $r \in]0, +\infty[$, one has: $\tau_{r,\mu}(h) > 0$,

This justifies the existence of $I^f_{\mu,\lambda,\alpha,a}$.

Moreover, $\tau_{r,\mu}$ is differentiable on $I^f_{\mu,\lambda,\alpha,a}$ as a composition of differentiable functions on $\mathbb{R}^*_+$.

Thus, its derivative is given by the following relations

1. if $\mu = 0$, according to the above, for all $r \in I^f_{\mu,\lambda,\alpha,a} =]r^*_{2,h}, +\infty[$,

$$\tau'_{r,\mu}(h) = \frac{-n_2(r)}{i(r)} g'(h) < 0$$

2. if $\mu > 0$, then

$$\tau'_{r,\mu}(h) = \frac{n_2(r)}{i(r)} \left[\frac{(f(r) - n(r))\, g'(h) w}{\sqrt{(g(h) - n(r))^2 + 2\,(n(r) - f(r))\, g(h)}} \right] \cdot \left[\frac{1}{\sqrt{(g(h) - n(r))^2 + 2\,(n(r) - f(r))\, g(h)} + (g(h) - f(r))} \right]$$

with $w = \dfrac{1}{2}(n + f)(r)$. Therefore, according to Lemma 3, for $r \in I^f_{\mu,\lambda,\alpha,a}$, $\tau'_{r,\mu}$ is strictly negative, and $\tau_{r,\mu}$ is strictly decreasing with respect to h.

Moreover, relations (24) and (23) lead to respectively:

1. $\theta_\mu(r) = \lim\limits_{h \longrightarrow +\infty} \tau_{r,\mu}(h)$

$$= \begin{cases} \dfrac{-(n_1(r) + n_2(r)l_2) + \sqrt{(n_1(r) + n_2(r)l_2)^2 + 4i(r)m(r)l_2}}{2i(r)}, & \text{if } \mu > 0 \\ \dfrac{-(n_1(r) + n_2(r)l_2)}{i(r)}, & \text{if } \mu = 0, \end{cases}$$

where $\theta_\mu(r) > 0$ since $l_2 > 0$.

2. $\kappa_\mu(r) = \lim\limits_{h \longrightarrow 0^+} \tau_{r,\mu}(h) =$

$$\begin{cases} \dfrac{-(n_1(r)+n_2(r)l_1)+\sqrt{(n_1(r)+n_2(r)l_1)^2+4i(r)m(r)l_1}}{2i(r)}, \text{ if } l_1>0 \text{ and } \mu>0 \\ \dfrac{-(n_1(r)+n_2(r)l_1)}{i(r)}, \text{ if } l_1>0 \text{ and } \mu=0 \\ \dfrac{-n_1(r)}{i(r)}, \text{ if } l_1=0 \text{ and } \mu=0. \end{cases}$$

Since $\tau_{r,\mu}$ is continuous and strictly decreasing, this implies that $0<\theta_\mu(\cdot)<\kappa_\mu(\cdot)$ and $\tau_{r,\mu}(\,]0;+\infty[)$ is included in $X_r=\left[\theta_\mu(r)\,;\,\kappa_\mu(r)\right]$. Let $\tilde{\tau}_{r,\mu}$ be the restriction of $\tau_{r,\mu}$ on X_r, defined as

$$\begin{aligned} \tilde{\tau}_{r,\mu} : X_r &\longrightarrow X_r \\ h &\longmapsto \tau_{r,\mu}(h) \end{aligned}$$

where X_r is non-empty, convex, closed, and bounded in $\mathbb{R}$. According to the Schauder Fixed Point Theorem (see [21]), $\tilde{\tau}_{r,\mu}$ has a unique fixed point as well as $\tau_{r,\mu}$. The uniqueness of the fixed point is ensured by the strict monotonicity of $\tau_{r,\mu}$ with respect to h.

Let T^* denote this unique fixed point: the equality $\tau_{r,\mu}(T^*(r))=T^*(r)$ implies that $g(T^*(r))=\zeta^*(r)$ with

$$\zeta^*(r)=\frac{(i(r)T^*(r)+n_1(r))T^*(r)}{m(r)-n_2(r)T^*(r)}.$$

□

Furthermore, one has

$$\tau_{r,\mu}(T^*(r))=\frac{-(n_1(r)+n_2(r)\zeta^*(r))+\sqrt{(n_1(r)+n_2(r)\zeta^*(r))^2+4i(r)m(r)\zeta^*(r)}}{2i(r)}.$$

It is now important to specify the following two points:

Remark 7 If $\mu=0$, according to Theorem 7, the existence of the interval $\mathcal{I}_{0,\lambda,\alpha,a}\subset\mathbb{R}_+^*$ such that

$$\forall r\in\mathcal{I}_{0,\lambda,\alpha,a},\quad \forall h\in\mathbb{R}_+^*,\quad \tau_{r,0}(h)>0$$

is assured if and only if $n_1(r^*)+n_2(r^*)\zeta(r^*)<0$ (that is, if and only if $\zeta(r^*)<n(r^*)$).

Remark 8 Initially, solving equation $n_1(r)+n_2(r)\zeta^*(r)=0$ for $r\in\mathbb{R}_+^*$ was considered. To our knowledge, it is not possible to do so. This led us to propose the following theorem.

One draws inspiration from the case where g is constant, which allows us to state the following theorem.

Theorem 8 *The following properties are verified:*

1. *If there exists $r_2^* \in \mathbb{R}_+^*$ such that $n_1(r_2^*) + n_2(r_2^*)l_2 = 0$ and $]r_2^*, +\infty[\cap \mathcal{I}^f_{\mu,\lambda,\alpha,a}$ is non-empty, then for all $r \in]r_2^*, +\infty[\cap \mathcal{I}^f_{\mu,\lambda,\alpha,a}$, the equilibrium state associated with $\zeta^*(r)$ is admissible for all $\mu \geq 0$;*
2. *If there exists $r_1^* \in \mathbb{R}_+^*$ such that $n_1(r_1^*) + n_2(r_1^*)l_1 = 0$ and $]0, r_1^*] \cap \mathcal{I}^f_{\mu,\lambda,\alpha,a}$ is non-empty, then for all $r \in]0, r_1^*] \cap \mathcal{I}^f_{\mu,\lambda,\alpha,a}$, the equilibrium state associated with $\zeta^*(r)$ is admissible if and only if $\mu > 0$.*

Otherwise,

- *if $\mu > 0$, then for all $r \in \mathcal{I}^f_{\mu,\lambda,\alpha,a}$, the equilibrium state associated with $\zeta^*(r)$ is admissible.*
- *if $\mu = 0$, then for all $r \in \mathcal{I}^f_{\mu,\lambda,\alpha,a}$, the equilibrium state associated with $\zeta^*(r)$ is not admissible.*

Proof Since n increases on $\mathbb{R}_+^*$, then:

1. If there exists $r_2^* \in \mathbb{R}_+^*$ such that $n_1(r_2^*) + n_2(r_2^*)l_2 = 0$, then for all $r \in]r_2^*; +\infty[\cap \mathcal{I}^f_{\mu,\lambda,\alpha,a}$, one has $\zeta^*(r) \leq l_2 = n(r_2^*) < n(r)$ (i.e., $n_1(r) + n_2(r)\zeta^*(r) < 0$), hence one deduces that $\forall r \in]r_2^*; +\infty[\cap \mathcal{I}^f_{\mu,\lambda,\alpha,a}$,
$$-(n_1(r) + n_2(r)\zeta^*(r)) + \sqrt{(n_1(r) + n_2(r)\zeta^*(r))^2 + 4i(r)m(r)\zeta^*(r)} > 0$$
and in this case, there exists a strictly positive solution T^* which satisfies (27).
2. If there exists $r_1^* \in \mathbb{R}_+^*$ such that $n_1(r_1^*) + n_2(r_1^*)l_1 = 0$, then for all $r \in]0; r_1^*[\cap \mathcal{I}^f_{\mu,\lambda,\alpha,a}$, one has $n(r) < n(r_1^*) = l_1 \leq \zeta^*(r)$ (i.e., $n_1(r) + n_2(r)\zeta^*(r) \geq 0$), and consequently:
$$-(n_1(r) + n_2(r)\zeta^*(r)) + \sqrt{(n_1(r) + n_2(r)\zeta^*(r))^2 + 4i(r)m(r)\zeta^*(r)} > 0$$
if and only if
$$\sqrt{(n_1(r) + n_2(r)\zeta^*(r))^2 + 4i(r)m(r)\zeta^*(r)} > (n_1(r) + n_2(r)\zeta^*(r)) \geq 0.$$
This last inequality is true if and only if $(n_1(r) + n_2(r)\zeta^*(r))^2 + 4i(r)m(r)\zeta^*(r) > (n_1(r) + n_2(r)\zeta^*(r))^2$
$$\Leftrightarrow i(r)m(r)\zeta^*(r) = \alpha\lambda I(r + \lambda + \mu)(r + \mu + \nu + \delta)^2\mu\zeta^*(r) > 0,$$
which is verified if and only if $\mu > 0$.

Otherwise,

- If $\mu > 0$, the equations $n_1(r) + n_2(r)l_2 = 0$ admit no solution in $\mathbb{R}_+^*$. Since $n_1(r) + n_2(r)l_2$ is a second-degree trinomial with strictly negative leading coefficient, then for all $r \in \mathbb{R}_+^*$, $n_1(r) + n_2(r)l_2 < 0$. To conclude, one uses the demonstration from 1. with $r_2^* = 0$.
- If $\mu = 0$, then the equation $n_1(r) + n_2(r)l_2 = 0$ becomes

$$-\left((a-\alpha)(\nu+\delta) - Il_2\right)r + I(\nu+\delta)l_2 = 0.$$

It can only have a solution if $(a-\alpha)(\nu+\delta) \neq Il_2$.
And this case, the solution is $r = \dfrac{I(\nu+\delta)l_2}{(a-\alpha)(\nu+\delta) - Il_2}$. Since the solution does not belong to $\mathbb{R}_+^*$, necessarily $(a-\alpha)(\nu+\delta) < Il_2$, and for all $r > 0$, $n_1(r) + n_2(r)\zeta^*(r) < n_1(r) + n_2(r)l_2 < 0$. Thus:

$$\forall r \in \mathbb{R}_+^*, \quad 0 < \zeta^*(r) < l_2 \leq n_0 = 0.$$

Absurd.

□

Remark 9 At this stage, it is important to note that for $k \in \{1, 2\}$, $r_{k,h}^* = r_k^*$.

Proposition 3 *The following properties hold:*

1. *if* $\max\{l_1, l_2\} = l_2 \leq \dfrac{a\mu(\lambda+\mu)}{\lambda I}$, *then the equation* $n_1(r) + n_2(r)l_k = 0$ *does not have a solution in* $\mathbb{R}_+^*$ *for* $k \in \{1, 2\}$.
2. *if* $0 \leq l_1 < l_2 < \dfrac{(a-\alpha)(\nu+\delta)}{I}$ *and* $\mu = 0$ *or* $\mu > 0$, *then the intersection* $]r_2^*, +\infty[\cap I_{\mu,\lambda,\alpha,a}^f$ *is nonempty when*
   a. $I_{\mu,\lambda,\alpha,a}^f =]0, +\infty[$
   b. *if* $\mu > 0$, *then*
      a. $I_{\mu,\lambda,\alpha,a}^f = [r_f^{(1)}, +\infty[$,
      b. $I_{\mu,\lambda,\alpha,a}^f =]0, +\infty[$
      c. $I_{\mu,\lambda,\alpha,a}^f =]0, r_f^{(?)}] \cup [r_f^{(3)}, +\infty[$
3. *if* $0 < l_1 < \dfrac{(a-\alpha)(\nu+\delta)}{I} \leq l_2$ *and* $\mu = 0$ *or* $\dfrac{a\mu(\lambda+\mu)}{\lambda I} < l_1$ *and* $\mu > 0$, *then the intersection* $]0, r_1^*] \cap I_{\mu,\lambda,\alpha,a}^f$ *is non-empty when*
   a. $I_{\mu,\lambda,\alpha,a}^f = [r_f^{(1)}, +\infty[$ *with* $r_f^{(1)} \leq r_1^*$.
   b. $I_{\mu,\lambda,\alpha,a}^f =]0, +\infty[$ *or* $I_{\mu,\lambda,\alpha,a}^f =]0, r_f^{(2)}] \cup [r_f^{(3)}, +\infty[$.

See Appendix 10 for the proof.

Remark 10 When equation $n_1(r) + n_2(r)l_k = 0$ has a solution for all $k \in \{1, 2\}$, with $a - \alpha > 0$, then it suffices to use the part concerning $k = 2$ which provides admissibility for all possible values of μ. In this case, the latter equation is equivalent to

$$-\alpha\mu r^2 + (-a\rho\sigma - \alpha\mu\rho - \alpha\mu\sigma + \alpha\rho\sigma + \lambda I l_2)r - a\mu\rho\sigma + \lambda I l_2\sigma = 0 \quad (28)$$

and its discriminant Δ is such that

$$\Delta = (-a\rho\sigma - \alpha\mu\rho - \alpha\mu\sigma + \alpha\rho\sigma + \lambda I l_2)^2 + 4\alpha\mu(-a\mu\rho\sigma + \lambda\sigma I l_2) = B_0\lambda^2 + B_1\lambda + B_2$$

with $B_0 = (Il_2)^2$, $B_1 = 2(-a\rho\sigma - \alpha\mu\rho - \alpha\mu\sigma + \alpha\rho\sigma)Il_2 + 4\alpha\mu\sigma Il_2$ and $B_2 = (-a\rho\sigma - \alpha\mu\rho - \alpha\mu\sigma + \alpha\rho\sigma)^2 - 4a\alpha\mu^2\rho\sigma$.

Now, calculating the discriminant gives $\Delta = -16\alpha\mu\rho\sigma I^2 l_2^2(\nu + \delta)(a - \alpha) \leq 0$, which guarantees that Δ is positive for all $\lambda > 0$. Consequently, equation (28) has two real solutions.

$$r_1^* = \frac{-a\rho\sigma - \alpha\mu\rho - \alpha\mu\sigma + \alpha\rho\sigma + \lambda I l_2 - \sqrt{B_0\lambda^2 + B_1\lambda + B_2}}{2\alpha\mu} \quad (29)$$

and

$$r_2^* = \frac{-a\rho\sigma - \alpha\mu\rho - \alpha\mu\sigma + \alpha\rho\sigma + \lambda I l_2 + \sqrt{B_0\lambda^2 + B_1\lambda + B_2}}{2\alpha\mu} \quad (30)$$

when $\mu > 0$, and a real solution $r_0^* = \dfrac{Il_2\sigma}{(a-\alpha)\sigma - Il_2}$ when $\mu = 0$ with $a - \alpha > \dfrac{Il_2}{\sigma}$. Let's examine the part concerning $k = 2$ which gives the admissibility for all possible values of μ. So, r_2^* is strictly positive if and only if one of the four following properties holds:

1. if $\mu < \sqrt{\lambda(\nu + \delta)}$ and if there exists α^* and a that satisfy one of the two following conditions
   a. $\max\left\{0; \dfrac{a(\lambda + \mu)\sigma - \lambda I l_2}{\lambda(\nu + \delta) - \mu^2}\right\} < \alpha^* < a$
   b. $\min\left\{a; \dfrac{a(\lambda + \mu)\sigma - \lambda I l_2}{\lambda(\nu + \delta) - \mu^2}\right\} > \alpha^* > 0$ and
   $\dfrac{\lambda I l_2}{\mu(\lambda + \mu)} > a > \max\left\{\alpha^*, \dfrac{\lambda I l_2}{(\lambda + \mu)\sigma}\right\}$
2. if $\mu > \sqrt{\lambda(\nu + \delta)}$ and if there exists α^* and a that satisfy one of the two following conditions

a. $\min\left\{a;\ \dfrac{a(\lambda+\mu)\sigma-\lambda I l_2}{\lambda(\nu+\delta)-\mu^2}\right\} > \alpha^*$ and $a < \dfrac{\lambda I l_2}{(\lambda+\mu)\sigma}$

b. $\max\left\{0;\ \dfrac{a(\lambda+\mu)\sigma-\lambda I l_2}{\lambda(\nu+\delta)-\mu^2}\right\} < \alpha^*$ and $\dfrac{\lambda I l_2}{\mu(\lambda+\mu)} > a > \alpha^*$.

3. if $\mu = \sqrt{\lambda(\nu+\delta)}$ and for any a such that $0 < \alpha^* < a$, one of the two following conditions is satisfied.

a $a < \dfrac{\lambda I l_2}{(\lambda+\mu)\sigma}$

b $\dfrac{\lambda I l_2}{\mu(\lambda+\mu)} > a \geq \dfrac{\lambda I l_2}{(\lambda+\mu)\sigma}$.

4. if $\mu = 0$ and if there exists α^* that satisfies $a - \alpha^* > \dfrac{I l_2}{\sigma}$.

Example 1 Let's suppose that $K(D) = \ln D$. By the definition of the function g in Lemma 1, one has

$$g(T) = K'\left(\frac{IT+\alpha}{(\lambda+\mu)T}\right) = \frac{(\lambda+\mu)T}{IT+\alpha},$$

which is such that

$$l_2 = \lim_{T\to+\infty} g(T) = \frac{\lambda+\mu}{I}.$$

According to Theorem 8 and Remark 10, the associated equilibrium state is admissible if and only if one of the following four properties holds:

1. if $\mu < \sqrt{\lambda(\nu+\delta)}$ and if there exist α^* and a satisfying one of the following two conditions:

a. $\max\left\{0;\ \dfrac{(\lambda+\mu)(a\sigma-\lambda)}{\lambda(\nu+\delta)-\mu^2}\right\} < \alpha^* < a$

b. $\min\left\{a;\ \dfrac{(\lambda+\mu)(a\sigma-\lambda)}{\lambda(\nu+\delta)-\mu^2}\right\} > \alpha^* > 0$ and $\dfrac{\lambda}{\mu} > a > \max\left\{\alpha^*, \dfrac{\lambda}{\sigma}\right\}$

2. if $\mu > \sqrt{\lambda(\nu+\delta)}$ and if there exists α^* and a satisfying one of the following two conditions:

a. $\min\left\{a;\ \dfrac{(\lambda+\mu)(a\sigma-\lambda)}{\lambda(\nu+\delta)-\mu^2}\right\} > \alpha^*$ and $a < \dfrac{\lambda}{\sigma}$

b. $\max\left\{0;\ \dfrac{(\lambda+\mu)(a\sigma-\lambda)}{\lambda(\nu+\delta)-\mu^2}\right\} < \alpha^*$ and $\dfrac{\lambda}{\mu} > a > \alpha^*$.

3. if $\mu = \sqrt{\lambda(\nu+\delta)}$ and for all a such that $0 < \alpha^* < a$, one of the following two conditions are checked

a. $a < \dfrac{\lambda}{\sigma}$

b. $\dfrac{\lambda}{\mu} > a \geq \dfrac{\lambda}{\sigma}$.

4. if $\mu = 0$ and if there exists α^* satisfying $a - \alpha^* > \dfrac{\lambda}{\sigma}$.

Example 2 Let's assume that $K(D) = \psi(1 - \exp(-D))$, where $\psi > 0$. According to the function g from Lemma 1, one has

$$g(T) = K'\left(\frac{IT + \alpha}{(\lambda + \mu)T}\right) = \psi \exp\left(-\frac{IT + \alpha}{(\lambda + \mu)T}\right),$$

which is such that

$$l_2 = \lim_{T \to +\infty} g(T) = \psi \exp\left(-\frac{I}{\lambda + \mu}\right) = \phi.$$

According to Theorem 8 and Remark 10, the associated equilibrium state is admissible if one of the following four properties holds:

1. if $\mu < \sqrt{\lambda(\nu + \delta)}$ and if there exists α^* that satisfies one of the two conditions below.

   a. if $\mu < \sqrt{\lambda(\nu + \delta)}$ and if there exist α^* and a that satisfy one of the two conditions below.

      i. $\max\left\{0; \dfrac{a(\lambda + \mu)\sigma - \lambda I\phi}{\lambda(\nu + \delta) - \mu^2}\right\} < \alpha^* < a$

      ii. $\min\left\{a; \dfrac{a(\lambda + \mu)\sigma - \lambda I\phi}{\lambda(\nu + \delta) - \mu^2}\right\} > \alpha^* > 0$ and $\dfrac{\lambda I\phi}{\mu(\lambda + \mu)} > a > \max\left\{\alpha^*, \dfrac{\lambda I\phi}{(\lambda + \mu)\sigma}\right\}$

   b. if $\mu > \sqrt{\lambda(\nu + \delta)}$ and if there exists α^* and a that satisfy one of the two conditions below

      i. $\min\left\{a; \dfrac{a(\lambda + \mu)\sigma - \lambda I\phi}{\lambda(\nu + \delta) - \mu^2}\right\} > \alpha^*$ and $a < \dfrac{\lambda I\phi}{(\lambda + \mu)\sigma}$

      ii. $\max\left\{0; \dfrac{a(\lambda + \mu)\sigma - \lambda I\phi}{\lambda(\nu + \delta) - \mu^2}\right\} < \alpha^*$ and $\dfrac{\lambda I\phi}{\mu(\lambda + \mu)} > a > \alpha^*$.

   c. if $\mu = \sqrt{\lambda(\nu + \delta)}$ and for any a such that $0 < \alpha^* < a$, one of the two conditions below is satisfied

      i. $a < \dfrac{\lambda I\phi}{(\lambda + \mu)\sigma}$

      ii. $\dfrac{\lambda I\phi}{\mu(\lambda + \mu)} > a \geq \dfrac{\lambda I\phi}{(\lambda + \mu)\sigma}$.

   d. if $\mu = 0$ and if there exists α^* that satisfies $a - \alpha^* > \dfrac{I\phi}{\sigma}$.

Table 1 List of model parameters used for simulations

Parameters	Descriptions	Values
I	Diabetes incidence	2644
γ	Recovery rate	0.06
λ	Probability of developing a complication	0.08061
ν	Rate at which patients with complications become severely disabled	0.01
δ	Mortality rate due to complications	0.001

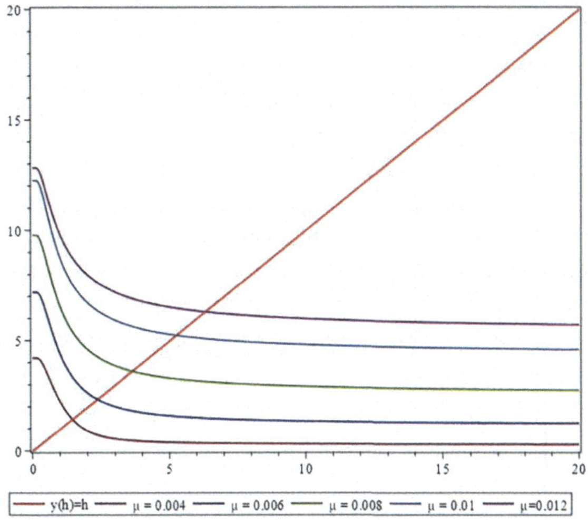

Fig. 2 The curve of $\tau_{r,\mu}$ and $y(h) = h$

Through the following remark, one presents the qualitative behavior of $\tau_{r,\mu}$ from Theorem 7. One could choose to work with the data from Example 1, which would yield a similar result.

Remark 11 Using the results of Example 2 and the parameter values calculated from the data in [39] presented in Table 1, and choosing the constants a and α as follows: $a = 2$ and $\alpha = \frac{1}{10}$, one presents the graph of the function $\tau_{r,\mu}$ in Fig. 2. This graphical representation allows us to visualize the behavior of $\tau_{r,\mu}$ as a function of μ. The various curves show the monotonic properties of the decreasing function $\tau_{r,\mu}$. The fixed point of $\tau_{r,\mu}$ is clearly identified on the graph.

Remark 12 In the case where $\beta > 0$, the equation (20) becomes:

$$TY^3(T) + 2\sigma[g(T) - (1-\rho)T]Y^2(T) \\ + 2\sigma[(AI + 2\sigma\rho)T + Aa - 2\rho\sigma g(T)]Y(T) - 4A\rho\sigma^2(a-\alpha) = 0 \quad (31)$$

with $A = \dfrac{2\sigma\beta\,[r - (\lambda - \sigma)]}{(r+\sigma)^2}$.
This equation involves a transcendental variable and is doubly non-linear. Its solution, which is initially more complex, is left open for further developments.

6 Formulation of the Stochastic Model

The current environmental fluctuations have a significant impact on all aspects of real life. With this in mind, one will examine the influence of these fluctuations on the diffusion rate β. To do this, one will transform the deterministic model (1) by adding a diffusion term that depends on useful insights into the disease and models random perturbations of parameters related to the development of diabetic complications. This diffusion term will be a measurable function $\mathcal{J}$ depending notably on C and D, aimed at encouraging individuals to return to the compartment of diabetics without complications, and a Brownian process $B(t) = (B_1(t), B_2(t))$ defined on a filtered probability space $\left(\Omega, \mathcal{F}, \{\mathcal{F}_t\}_{t\geq 0}, \mathbb{P}\right)$, where the filtration $\{\mathcal{F}_t\}_{t\geq 0}$ satisfies the usual conditions (growth, continuity, with $\mathcal{F}_0$ that contains all $\mathbb{P}$ null sets). The other parameters are assumed to be defined as previously. Thus, one obtains the following new stochastic model:

$$\begin{cases} dD(t) = & (I - (\lambda(t) + \mu(t))\, D(t) + \gamma(t)C(t) - \beta(t)D(t)C(t))dt \\ & \qquad + \mathcal{J}_1(D(t), C(t))dB_1(t) \\ dC(t) = & (\lambda(t)D(t) - (\gamma(t) + \mu(t) + \nu(t) + \delta(t))\, C(t) \\ & \qquad + \beta(t)D(t)C(t))dt \; - \; \mathcal{J}_2(D(t), C(t))dB_2(t), \end{cases} \tag{32}$$

where $\mathcal{J}_1(D, C) \in L^2$ and $\mathcal{J}_2(D, C) \in L^2$ are two locally Lipschitz continuous functions with positive values, such that

1. $\exists\, \varepsilon_1, \mathcal{E}_1 \in \mathbb{R}_+, \quad \varepsilon_1 \leq \left| \dfrac{\mathcal{J}_1(D(t), C(t))}{D(t)C(t)} \right| \leq \mathcal{E}_1$
2. $\exists\, \varepsilon_2, \mathcal{E}_2 \in \mathbb{R}_+, \quad \varepsilon_2 \leq \left| \dfrac{\mathcal{J}_2(D(t), C(t))}{D(t)C(t)} \right| \leq \mathcal{E}_2,$

where $\max\{\mathcal{E}_1, \mathcal{E}_2\} > 0$.

Example 3 In (32), a specific form of stochastic perturbation that can be considered is the addition of white noise to the parameters β as follows:

$$\beta \mapsto \beta + \sigma_i \dot{B}_i(t) \tag{33}$$

where B_i is a white noise, $\dot{B}_i(t)$ its derivative and σ_i its intensity. This leads to the case where

$$\mathcal{J}_i(D(t), C(t)) = -\sigma_i D(t)C(t). \tag{34}$$

Remark 13 Our model encompasses media coverage; however, it does not fully encapsulate the scenario analyzed in [13]. To achieve this, the model must be extended by generalizing the selection of the β parameter as a function of (D, C).

In the following analysis, it is assumed that the population size $N(t)$ and all the parameters in the proposed model remain invariant over time. Assuming $i = \frac{I}{N}$, $b = \frac{D}{N}$ and $c = \frac{C}{N}$, the system (32) becomes:

$$\begin{cases} db(t) = & (i - (\lambda + \mu)\, b(t) + \gamma c(t) - \beta b(t)c(t)N)dt \\ & \qquad + \mathcal{J}_1(Nb(t), Nc(t))N^{-1}dB_1(t) \\ dc(t) = & (\lambda b(t) - (\gamma + \mu + \nu + \delta)\, c(t) + \beta b(t)c(t)N)dt \\ & \qquad - \mathcal{J}_2(Nb(t), Nc(t))N^{-1}dB_2(t), \end{cases} \tag{35}$$

with

$$\varepsilon \begin{cases} 1 \le \left| \dfrac{\mathcal{J}_1(Nb(t), Nc(t))}{N^2 b(t)c(t)} \right| \le \mathcal{E}_1 & \text{(36a)} \\ \varepsilon_2 \le \left| \dfrac{\mathcal{J}_2(Nb(t), Nc(t))}{N^2 b(t)c(t)} \right| \le \mathcal{E}_2. & \text{(36b)} \end{cases}$$

In addition, considering

$$Y(t) = \begin{pmatrix} b(t) \\ c(t) \end{pmatrix},$$

$$M(t, Y(t)) = N^{-1} \begin{pmatrix} \mathcal{J}_1(Nb(t), Nc(t)) & 0 \\ 0 & -\mathcal{J}_2(Nb(t), Nc(t)) \end{pmatrix}$$

and

$$F(t, Y(t)) = \begin{pmatrix} i - (\lambda + \mu)\, b(t) + \gamma c(t) - \beta b(t)c(t)N \\ \lambda b(t) - (\gamma + \mu + \nu + \delta)\, c(t) + \beta b(t)c(t)N \end{pmatrix},$$

leads to the following EDS:

$$dY(t) = F(t, Y(t))dt + M(t, Y(t))dB(t), \tag{37}$$

of initial condition $Y(t_0) = Y_0$ with $0 \le t_0 < T < +\infty$ and X_0 a $\mathcal{F}_{t_0}$-measurable random variable taking values in $\mathbb{R}^2$ such that $E\left[\,|X_0|^2\right] < +\infty$. Thus, $F(t, \cdot)$ and $M(t, \cdot)$ are respectively called the drift and diffusion functions. In the following, one will assume

$$\jmath_1(b(t), c(t)) = \mathcal{J}_1(Nb(t), Nc(t))N^{-1}, \quad \jmath_2(b(t), c(t)) = \mathcal{J}_2(Nb(t), Nc(t))N^{-1}$$

and one draws on some of K. Valmont's [44] and [30, 34] approaches to prove some of our results.

7 Study of Certain Fundamental Properties

In this section, one studies the existence, uniqueness, and positivity of the solution of the model (32), as well as the persistence of diabetes complications. Based on the seminal contributions of B. Øksendal [37], one declines the set C^f, which encompasses continuous mappings from the interval $[0, T]$ to the Euclidean space $\mathbb{R}^2$.

7.1 *Existence, Uniqueness of Global Solution Between 0 and 1, and Limits of Solutions*

In this subsection, the existence and uniqueness of local and global solutions between 0 and 1, as well as the limits of the solutions, are proved.

Theorem 9 *For any initial condition* $Y(0) = Y_0 \in (0; 1)^2$, *the stochastic differential equation* (37) *admits a unique local solution in* C^f.

See Appendix 10 for the demonstration.

Theorem 10 *For any given initial value* $(b(0); c(0)) \in (0; 1)^2$ *of* (37)*, there exists a unique global solution* $(b(t); c(t)) \in (0; 1)^2$ *for all* $t \geq 0$*, and this solution remains almost surely in* $(0; 1)^2$*, that is*

$$\mathbb{P}\left\{(b(t); c(t)) \in (0; 1)^2; t \in \mathbb{R}_+\right\} = 1.$$

Proof Let τ_e be the explosion time defined as in [31]. The existence of a local solution $(b(t); c(t))$ of (35) for $(t \in [0; \tau_e))$ has been proven by Theorem 9. To prove that this solution is global, let us show that $\tau_e = +\infty$ almost surely (a.s.). To do so, one assumes that there exists $n_0 > 0$ sufficiently large such that $b_0, c_0 \in \left(\frac{1}{n_0}, 1 - \frac{1}{n_0}\right)$, and for each integer $n > n_0$, one defines the stopping time:

$$\tau_n = \inf\left\{t \in [0, \tau_e); \max\{b(t); c(t)\} \leq \frac{1}{n} \quad \text{or} \quad \min\{b(t); c(t)\} \geq 1 - \frac{1}{n}\right\}.$$

Let $\inf \phi = +\infty$ (with ϕ being the empty set). Since τ_n is increasing, one sets $\tau_{+\infty} = \lim_{n \longrightarrow +\infty} \tau_n$. It follows that $\tau_{+\infty} \leq \tau_e$ (because $\tau_n \leq \tau_e$) a.s. At this stage, our goal is to show that $\tau_{+\infty} = +\infty$ a.s. To do this, assume that $\tau_{+\infty} \neq +\infty$. Thus, there exist two constants $T > 0$ and $\varsigma \in (0, 1)$ such that $\mathbb{P}(\tau_{+\infty} \leq T) > \varsigma$, and an integer $n_1 \geq n_0$ such that

$$\mathbb{P}(\tau_n \leq T) \geq \varsigma, \quad \forall n \geq n_1. \tag{38}$$

Let's consider the set E defined by $E = \{(x, y) \in (0, 1)^2 : x + y < 1\}$ and the function V defined as follows:

$$V : (x, y) \in E \longmapsto (x^{-1} + y^{-1}) \in \mathbb{R}_+$$

Using the Itô formula, one obtains

$$\begin{aligned} dV(b(t), c(t)) &= \frac{\partial V}{\partial t}dt + \frac{\partial V}{\partial b}db + \frac{\partial V}{\partial c}dc \\ &+ \frac{1}{2}\left(\frac{\partial^2 V}{\partial b^2}(db)^2 + \frac{\partial^2 V}{\partial c^2}(dc)^2 + \frac{\partial^2 V}{\partial b \partial c}dbdc + \frac{\partial^2 V}{\partial c \partial b}dcdb\right). \end{aligned} \tag{39}$$

Given that: $dt \times dt = dt \times dB_t = dB_t \times dt = 0$, $dB_t \times dB_t = dt$, one has:

$$(db)^2(t) = \jmath_1^2(b(t), c(t))dt, \ (dc)^2(t) = \jmath_2^2(b(t), c(t))dt, \quad \text{and} \quad dbdc = dcdb = 0.$$

Moreover, one notices that

$$\frac{dV}{\partial t} \cdot dt = 0,$$

$$\begin{aligned} \frac{dV}{\partial b} \cdot db(t) &= -\frac{(i - (\lambda + \mu)\, b(t) + \gamma c(t) - \beta b(t)c(t)N)dt + \jmath_1(b(t), c(t))dB_1(t)}{b^2(t)}, \\ \frac{dV}{\partial c} \cdot dc(t) &= -\frac{(\lambda b(t) - (\gamma + \sigma)\, c(t) + \beta b(t)c(t)N)dt - \jmath_2(b(t), c(t))dB_2(t)}{c^2(t)}, \end{aligned}$$

$$\frac{d^2 V}{\partial b^2} \cdot (db(t))^2 = \left(\frac{2}{b^3(t)}\right) \cdot (db(t))^2 = \frac{(db(t))^2}{b^3(t)} = \frac{2\jmath_1^2(b(t), c(t))\, dt}{b^3(t)} \leq \frac{2\mathcal{E}_1^2 N^2 c^2(t)}{b(t)}$$

and

$$\frac{d^2 V}{\partial c^2} \cdot (dc(t))^2 = \left(\frac{2}{c^3(t)}\right) \cdot (dc(t))^2 = \frac{(dc(t))^2}{c^3(t)} = \frac{2\jmath_2^2(b(t), c(t))\, dt}{c^3(t)} \leq \frac{2\mathcal{E}_2^2 N^2 b^2(t)}{c(t)}.$$

Based on the previous observations, equation (39) becomes

$$\begin{aligned} dV(b(t), c(t)) &\leq LV(b(t), c(t))\, dt \\ &+ \left(\frac{\jmath_2(b(t), c(t))}{c^2}dB_2(t) - \frac{\jmath_1(b(t), c(t))}{b^2}dB_1(t)\right) \end{aligned} \tag{40}$$

with

$$LV(b, c) = -\frac{i}{b^2} + \frac{\lambda + \mu}{b} + \frac{\beta c N}{b} - \frac{\gamma c}{b^2} - \frac{\lambda b}{c^2} + \frac{\gamma + \mu + \nu + \delta}{c} - \frac{\beta b N}{c} + \frac{2\mathcal{E}_1^2 N^2 c^2}{b} + \frac{2\mathcal{E}_2^2 N^2 b^2}{c},$$

and by bounding, one obtains

$$LV(b(t), c(t)) \leq M(t)\left(\frac{1}{b(t)} + \frac{1}{c(t)}\right) = M(t)V(b(t), c(t)),$$

with

$$M(t) = \max\{(\lambda + \mu + \beta N + 2\mathcal{E}_1^2 N^2 \mathcal{M}_t); (\gamma + \mu + \nu + \delta + 2\mathcal{E}_2^2 N^2 \mathcal{M}_t)\},$$

where $\mathcal{M}_t = \max\left\{b^2(t), c^2(t)\right\}$, which is locally bounded since one is working in C^f. With this last inequality, equation (40) becomes

$$dV\,(b(t), c(t)) \leq M(t)\cdot V(b(t), c(t)) + N\left(\frac{\mathcal{E}_2 b(t)}{c(t)}dB_2(t) - \frac{\varepsilon_1 c(t)}{b(t)}dB_1(t)\right). \tag{41}$$

Integrating inequality (41) term by term over $[0, \tau_m)$, with $\tau_m = \min\{\tau_n; t\}$, and taking the expectation of both sides of the obtained inequality, one has:

$$E\left[V\,(b(\tau_m), c(\tau_m))\right] \leq E\left[V\,(b(0), c(0))\right] + E\left[\int_0^{\tau_m} M(s)V(b(s), c(s))ds\right]$$

since

$$E\left[\mathcal{E}_2 N\int_0^{\tau_m}\frac{b(s)}{c(s)}dB_1(s)\right] = 0 \quad\text{and}\quad E\left[\varepsilon_1 N\int_0^{\tau_m}\frac{c(s)}{b(s)}dB_2(s)\right] = 0.$$

Using Gronwall's lemma, one obtains

$$E\left[V\,(b(\tau_m), c(\tau_m))\right] \leq V\,(b(0), c(0))\, e^{\int_0^T M(s)ds}. \tag{42}$$

Now, let's define the set Ω_n as $\Omega_n = \{\tau_n \leq T\}$ for $n \geq n_1$. Therefore, according to (38), $\mathbb{P}(\Omega_n) \geq \varsigma$ for $n \geq n_1$. Consequently, for any $\omega \in \Omega_n$, at least one of $b(\tau_n, \omega), c(\tau_n, \omega)$ is equal to $1 - \frac{1}{n}$ or $\frac{1}{n}$.

If $b(\tau_n, \omega) = 1 - \frac{1}{n}$ or $\frac{1}{n}$ then $V\,(b(\tau_n, \omega), c(\tau_n, \omega)) \geq n$. If $c(\tau_n, \omega) = 1 - \frac{1}{n}$ or $\frac{1}{n}$ then $V\,(b(\tau_n, \omega), c(\tau_n, \omega)) \geq n$. According to (38) and (42), one has:

$$V(b_0, c_0)e^{\int_0^T M(s)ds} \geq E\left[\mathbb{I}_{\Omega_n(\omega)}V\,(b(\tau_n, \omega), c(\tau_n, \omega))\right] \geq \varsigma n$$

with $\mathbb{I}_{\Omega_n(\omega)}$ being the indicator function of Ω_n. Since ς is fixed in $(0, 1)$, as for $n \longrightarrow +\infty$, one gets $\infty > V(b_0, c_0)e^{\int_0^T M(s)ds} = \infty$. This is not possible. Therefore, $\tau_\infty = \infty$ a.s., concluding the proof. □

Remark 14 In the context of the example 3, one can conjecture that the asymptotic behavior of the stochastic model around (b^*, c^*), the steady state of the deterministic model, is such that

$$\limsup_{t\to\infty} \frac{1}{t}\int_0^t \left((b(s)-b^*)^2+(c(s)-c^*)^2\right)ds \le (K_1\sigma_1^2+K_2\sigma_2^2),$$

where $K_1, K_2 \in \mathbb{R}_+$. This analysis can be carried out using the strong law of large numbers for continuous local martingales.

7.2 *Reducing the Number of Diabetics with Complications*

In this subsection, the results are presented regarding the reduction of the number of diabetics with complications.

Remark 15 Let $(b(\cdot), c(\cdot))$ be a solution of (35), with the initial condition (b_0, c_0) in $(0, 1)^2$. Almost surely, one has

$$\lim_{t\to\infty}\frac{b(t)}{t}=0 \text{ and } \lim_{t\to\infty}\frac{c(t)}{t}=0.$$

Lemma 4 *For any solution* $(b(\cdot), c(\cdot))$ *of* (35)*, with the initial condition* (b_0, c_0)*, one has*

$$\lim_{t\to\infty}\frac{\int_0^t b(\zeta)dB_2(\zeta)}{t}=\lim_{t\to\infty}\frac{\int_0^t c(\zeta)dB_1(\zeta)}{t}=\lim_{t\to\infty}\frac{\int_0^t b(\zeta)c(\zeta)dB_1(\zeta)}{t}$$
$$=\lim_{t\to\infty}\frac{\int_0^t b(\zeta)c(\zeta)dB_2(\zeta)}{t}=0.$$

Proof Let
$X(t)=\int_0^t b(\zeta)dB_2(\zeta)$, $Y(t)=\int_0^t c(\zeta)B_1(\zeta)$, $H_1(t)=\int_0^t b(\zeta)c(\zeta)dB_1(\zeta)$, and $H_2(t)=\int_0^t b(\zeta)c(\zeta)dB_2(\zeta)$.
According to the Burkholder-Davis-Gundy inequality for any $p>2$, there exists a constant C_p such that

$$E\left[\sup_{\zeta\in[0,t]}|H_1(\zeta)|^p\right]\le C_p\mathcal{E}_1^p t^{\frac{p}{2}}E\left[\sup_{\zeta\in[0,t]}b^p(\zeta)\sup_{\zeta\in[0,t]}c^p(\zeta)\right]\le 2\mathcal{E}_1^p C_p t^{\frac{p}{2}}M^2.$$

For $\vartheta>0$ sufficiently small and k a non-zero natural integer, as t tends to $(k+1)\vartheta$, one has

$$E\left[|H_1((k+1)\vartheta)|^p\right]\le 2\mathcal{E}_1^p C_p[(k+1)\vartheta]^{\frac{p}{2}}M^2. \tag{43}$$

For any choice of $\xi_{H_1} \geq 0$ and by applying Doob's inequality to $H_1(t)$, one has

$$\mathbb{P}\left\{\omega : \sup_{\zeta\in[k\vartheta,(k+1)\vartheta]} |H_1(\zeta)|^p > (k\vartheta)^{1+\xi_{H_1}+\frac{p}{2}}\right\} \leq \frac{E\left[|H_1((k+1)\vartheta)|^p\right]}{(k\vartheta)^{1+\xi_{H_1}+\frac{p}{2}}}. \quad (44)$$

Applying the upper bound from inequality (43) to inequality (44), one obtains

$$\mathbb{P}\left\{\omega : \sup_{\zeta\in[k\vartheta,(k+1)\vartheta]} |H_1(\zeta)|^p > (k\vartheta)^{1+\xi_{H_1}+\frac{p}{2}}\right\} \leq \frac{2\mathcal{E}_1^2 M^2 C_p\,(k\vartheta+\vartheta)^{\frac{p}{2}}}{(k\vartheta)^{1+\xi_{H_1}+\frac{p}{2}}} < +\infty. \quad (45)$$

Then, by using the Borel-Cantelli Lemma, for almost every $\omega \in \Omega$, one has

$$\sup_{\zeta\in[k\vartheta,(k+1)\vartheta]} |H_1(t)|^p \leq (k\vartheta)^{1+\xi_{H_1}+\frac{p}{2}}, \quad k \in \mathbb{N}^*.$$

This proves that there exists a positive $k_{H_{1,0}}(\omega)$ such that for almost every $\omega \in \Omega$, for $k \geq k_{H_{1,0}}$, the relation remains true. For almost every $\omega \in \Omega$, if $k \geq k_{H_{1,0}}$ and $k\vartheta \leq t \leq (k+1)\vartheta$, one has

$$\lim_{t\to+\infty} \sup \frac{\ln|H_1(t)|}{\ln t} \leq \frac{1+\xi_{H_1}+\frac{p}{2}}{p}.$$

And when ξ_{H_1} tends toward 0, one has: $\lim_{t\to+\infty} \sup \dfrac{\ln|H_1(t)|}{\ln t} \leq \dfrac{2+p}{2p}$. So for $0 < \eta < \dfrac{1}{2} - \dfrac{1}{p}$ sufficiently small, there is a constant $\bar{T} = \bar{T}(\omega) > 0$ and a set Ω such that $\mathbb{P}(\Omega) \geq 1-\eta$, pour $t \geq \bar{T}$ and $\omega \in \Omega_n$,

$$\ln|H_1(t)| \leq (\frac{1}{2} + \frac{1}{p} + \eta)\ln t$$

and therefore

$$\lim_{t\to+\infty} \sup \frac{H_1(t)}{t} \leq \lim_{t\to+\infty} \frac{t^{\frac{1}{2}+\frac{1}{p}+\eta}}{t} = \lim_{t\to+\infty} \frac{1}{t^{1-\frac{1}{2}-\frac{1}{p}-\eta}} = 0.$$

Given that $\lim_{t\to+\infty} \inf \left(\dfrac{H_1(t)}{t}\right) \geq 0$, this means that $\lim_{t\to+\infty} \dfrac{|H_1(t)|}{t} = 0$ p.s. Hence $\lim_{t\to+\infty} \dfrac{H_1(t)}{t} = 0$. An analogous demonstration shows that

$$\lim_{t\to+\infty} \frac{H_2(t)}{t} = 0, \quad \lim_{t\to+\infty} \frac{X(t)}{t} = 0, \quad \lim_{t\to+\infty} \frac{Y(t)}{t} = 0.$$

Which proves the advertised result. □

One states the following assumption:
($\mathbf{H}_5$) : the set of points, where $c(t)$ vanishes, is negligible and

$$\int_0^t b(s)c^{-1}(s)ds = O(t)$$

in the neighborhood of $+\infty$, meaning that, for all $r \in \mathbb{R}_+$, there exists a constant $\alpha_{c,r} \geq 0$ such that

$$\forall t \in [r, +\infty[, \int_0^t b(s)c^{-1}(s)ds \leq \alpha_{c,r}t. \tag{46}$$

Theorem 11 *One assumes that hypothesis* ($\boldsymbol{H}_5$) *is satisfied and that* $(b(\cdot), c(\cdot))$ *is a solution of* (35) *with the initial condition* $(b_0, c_0) \in (0,1)^2$. *If, for* $r \in \mathbb{R}_+$ *large enough, one has :* $\alpha_{c,r} < \dfrac{\gamma+\mu+\nu+\delta}{\lambda}$ *and* $\beta < \dfrac{\mu(\gamma+\mu+\nu+\delta-\lambda\alpha_{c,r})}{I}$*, then*

$$\lim_{t\to+\infty} \sup \frac{\ln c(t)}{t} \leq 0 \ \ a.s.,$$

with $\mu \neq 0$. *In other words,* $c(\cdot)$ *almost surely tends exponentially toward zero, meaning that complications decrease with probability* 1.

Proof One considers $r \in [0, +\infty[$ to be sufficiently large and $t \in [r, +\infty[$. By integrating the equations of system (35) over the interval $[0, t]$, one obtains

$$\begin{aligned} b(t) - b_0 + c(t) - c_0 \leq it - \mu \int_0^t b(t)dt - (\mu+\nu+\delta)\int_0^t c(t)dt \\ + \mathcal{E}_1 N \int_0^t b(t)c(t)dB_1(t) - \varepsilon_2 N \int_0^t b(t)c(t)dB_2(t) \end{aligned} \tag{47}$$

and

$$\begin{aligned} b(t) - b_0 + c(t) - c_0 \geq It - \mu \int_0^t b(t)dt - (\mu+\nu+\delta)\int_0^t c(t)dt \\ + \varepsilon_1 N \int_0^t b(t)c(t)dB_1(t) - \mathcal{E}_2 N \int_0^t b(t)c(t)dB_2(t). \end{aligned} \tag{48}$$

Thus, from equation (47), one deduces that

$$\int_0^t b(t)dt \leq \frac{i}{\mu}t - \frac{\mu+\nu+\delta}{\mu}\int_0^t c(t)dt + \frac{1}{\mu}\varphi_1(t) \tag{49}$$

and from equation (48), one deduces that

$$\int_0^t b(t)dt \geq \frac{i}{\mu}t - \frac{\mu+\nu+\delta}{\mu}\int_0^t c(t)dt + \frac{1}{\mu}\varphi_2(t) \tag{50}$$

where

$$\varphi_1(t) = -b(t) - b_0 + c(t) - c_0 - \mathcal{E}_1 N\int_0^t b(t)c(t)dB_1(t) + \varepsilon_2 N\int_0^t b(t)c(t)dB_2(t)$$

and

$$\varphi_2(t) = -b(t) - b_0 + c(t) - c_0 - \varepsilon_1 N\int_0^t b(t)c(t)dB_1(t) + \mathcal{E}_2 N\int_0^t b(t)c(t)dB_2(t).$$

Applying Itô's formula to $\ln c(t)$ with $c(t) > 0$ for almost every t, one has

$$\begin{aligned} d(\ln c(t)) = \frac{1}{c(t)}&\left[(\lambda b(t) - (\gamma+\mu+\nu+\delta)\,c(t) + \beta b(t)c(t)N)dt\right.\\ &\left.-\jmath_2(b(t),c(t))dB_2(t)\right] - \frac{1}{2}\frac{\jmath_2^2(b(t),c(t))dt}{c^2(t)}. \end{aligned} \tag{51}$$

Using equation (46) from assumption $\mathbf{H}_5$, equation (51), and relation (36b), one obtains the following two inequalities:

$$\begin{aligned} \ln c(t) \leq &-(\gamma+\mu+\nu+\delta)\,t + \beta N\int_0^t b(\zeta)d\zeta + \lambda\alpha_{c,r}t\\ &-\varepsilon_2\int_0^t b(\zeta)dB_2(\zeta) - \frac{1}{2}\varepsilon_2^2N^2\int_0^t b^2(\zeta)d\zeta \end{aligned} \tag{52}$$

$$\begin{aligned} \text{and}\quad \ln c(t) \geq &-(\gamma+\mu+\nu+\delta)\,t + \beta N\int_0^t b(\zeta)d\zeta\\ &-\mathcal{E}_2\int_0^t b(\zeta)dB_2(\zeta) - \frac{1}{2}\mathcal{E}_2^2N^2\int_0^t b^2(\zeta)d\zeta \end{aligned} \tag{53}$$

since $b \leq 1$ and $c \leq 1$.
Next, using relation (49) and bounding equation (52), one has

$$\ln c(t) \leq \left[\frac{\beta I}{\mu} - (\gamma+\mu+\nu+\delta) + \lambda\alpha_{c,r}\right]t + \frac{\beta N}{\mu}\varphi_1(t) - \varepsilon_2\int_0^t b(\zeta)dB_2(\zeta).$$

Thus, by dividing both sides of the above inequality by t, one gets

$$\frac{\ln c(t)}{t} \leq (\gamma+\mu+\nu+\delta)\left(\frac{\beta I+\mu\lambda\alpha_{c,r}}{\mu(\gamma+\mu+\nu+\delta)}-1\right) + \frac{\beta N}{\mu}\frac{\varphi_1(t)}{t} - \varepsilon_2\frac{\int_0^t b(\zeta)dB_2(\zeta)}{t}. \tag{54}$$

By assumption, one has: $(\gamma+\mu+\nu+\delta)\left(\frac{\beta I+\mu\lambda\alpha_{c,r}}{\mu(\gamma+\mu+\nu+\delta)}-1\right) < 0$. As t tends to ∞, one has: $\lim_{t\to+\infty}\sup\frac{\ln c(t)}{t} \leq 0$. This completes the proof. □

7.3 *Persistence of Diabetes Complications in the Considered Population*

This subsection focuses on the persistence of diabetic complications, which leads us to recall the following definition.

Definition 1 ([30, 34]). The model (35) is said to be persistent if $\lim_{t\to\infty}\inf\langle c(t)\rangle > 0$ almost surely (a.s.), where $\langle c(t)\rangle = \frac{1}{t}\int_0^t c(\zeta)d\zeta$.

Theorem 12 *Let* $(b(\cdot); c(\cdot))$ *be the solution of system* (35) *with the initial condition* $(b_0; c_0) \in (0; 1)^2$. *If* $\frac{2\beta I-\mu\mathcal{E}_2^2N^2}{2\mu(\gamma+\mu+\nu+\delta)} > 1$, *then*

$$\lim_{t\to\infty}\inf\langle c(t)\rangle \geq \frac{\mu(\gamma+\mu+\nu+\delta)}{(\mu+\nu+\delta)\beta N}\left(\frac{2\beta I-\mu\mathcal{E}_2^2N^2}{2\mu(\gamma+\mu+\nu+\delta)}-1\right) > 0 \quad a.s. \tag{55}$$

Remark 16 This inequality implies that there are almost surely a certain number of diabetics with complications in the population, which means that complications persist on average with probability 1.

Proof Returning to the inequality (53) and using the inequality (50), one has

$$\ln c(t) \geq (\gamma+\mu+\nu+\delta)\left(\frac{2\beta I-\mu\mathcal{E}_2^2N^2}{2\mu(\gamma+\mu+\nu+\delta)}-1\right)t - \frac{\sigma\beta N}{\mu}\int_0^t c(\zeta)d\zeta + \Psi^{(2)}(t)$$

where $\Psi^{(2)}(t) = \frac{\beta N}{\mu}\varphi_2(t) - \mathcal{E}_2\int_0^t b(\zeta)dB_2(\zeta)$.

Dividing both sides of the above inequality by t, one gets

$$\frac{\ln c(t)}{t} \geq (\gamma + \mu + \nu + \delta)\left(\frac{2\beta I - \mu \mathcal{E}_2^2 N^2}{2\mu(\gamma + \mu + \nu + \delta)} - 1\right) - \frac{\sigma\beta N}{\mu}\frac{\int_0^t c(\zeta)d\zeta}{t} + \frac{\Psi^{(2)}(t)}{t}.$$

Applying Remark 15 and Lemma 4, one has: $\lim\limits_{t\to+\infty} \frac{\Psi^{(2)}(t)}{t} = 0.$
Setting
$\wp_1 = (\gamma + \mu + \nu + \delta)\left(\frac{2\beta I - \mu \mathcal{E}_2^2 N^2}{2\mu(\gamma + \mu + \nu + \delta)} - 1\right), \quad \wp_0 = \frac{\sigma\beta N}{\mu},$
and applying the general lemmas of [27], see its appendix, one obtains

$$\liminf_{t\to\infty} \langle c(t)\rangle \geq \frac{\mu\,(\gamma + \mu + \nu + \delta)}{(\mu + \nu + \delta)\beta N}\left(\mathcal{T}_0^{(2)} - 1\right),$$

where $\mathcal{T}_0^{(2)} = \frac{2\beta I - \mu \mathcal{E}_2^2 N^2}{2\mu(\gamma + \mu + \nu + \delta)}$. This completes the proof. □

8 Numerical Simulations

In this section, one uses both the explicit Euler method and the Euler-Maruyama method, with the assistance of Maple, to illustrate our results. One chooses $r = r^* + \epsilon$ so that r is strictly greater than r^* from equation (30), so $\epsilon = 0.001$ is suitable. One assumes $\varepsilon_1 = \mathcal{E}_1$, $\varepsilon_2 = \mathcal{E}_2$, and these values are arbitrarily chosen. The time interval $[t_0, T]$ is discretized with a step size h such that $t_i = t_0 + ih, i = 0, 1, \cdots, n, t_n = T$,

$$\bar{\gamma}(t_0) = 0 \quad \text{and} \quad \bar{\gamma}(t_{i+1}) = \min\left\{1; \max\left\{0; \frac{\alpha}{(P_C - P_D)(t_i)\cdot C(t_i)}\right\}\right\}.$$

To apply control theory results, one uses data from Table 1 and those drawn from [39], which are listed in Table 2, covering a period of 365 days.

Figure 3(a_1) and (a_2) depict graphical representations of simulated trajectories of model (1) for different values of β with $\bar{\gamma}(t) = 0$. Observe that as the parameter β increases, the curves representing diabetics with complications increase significantly, while the curves representing diabetics without complications decrease.

Table 2 Initial numbers of diabetic patients with and without complications

Diabetic Population	Description	Value
D_0	Diabetics without complications	29734
C_0	Diabetics with complications	2607

Figure 3(b_1) and (b_2) present graphical representations of the simulated trajectories of model (1) for different scenarios. In the absence of control and when the parameter β is zero, the number of diabetics, with or without complications, appears to remain constant. With control and a higher or looner rate of β, the number of diabetics without complications increases at the expense of those with complications. However, even with control, a high rate of β promotes the transfer of individuals from the compartment of diabetics without complications to that of diabetics with complications.

The graphs in Fig. 3(c_1) and (c_2) illustrate trajectories of the solution of model (32) for the following cases: $\mathcal{T}_0^{(2)} = -0.8664597170$ with $\beta = 0$, $\mathcal{T}_0^{(2)} = 0.7481849870$ with $\beta = 0.25$, $\mathcal{T}_0^{(2)} = 2.039900751 > 1$ with $\beta = 0.45$, and the case $\mathcal{T}_0^{(2)} = 2.362829691 > 1$ with $\beta = 0.50$, when $\gamma = 0.0082$, $\varepsilon_1 = \varepsilon_2 = 0.6$ ie $\sigma_1 = \sigma_2 = 0.6$ in example 3.

By comparing the graphs generated by the deterministic model and the stochastic model, one notices marked differences in the temporal dynamics of the data. For example, the results of the deterministic model for diabetes simulation appear to follow a more regular and predictable distribution, while those of the stochastic model exhibit a more random distribution. Furthermore, these graphs reveal that the stochastic model produces more extreme results than the deterministic model.

9 Conclusion and Future Perspectives

In this chapter, one has presented a deterministic model for a diabetic population, considering the influence that one person can exert on another's behavior, either through education or through misleading advice. In addition, a stochastic model has been proposed to assess the impact of health education on the progression of diabetes complications. This stochastic model provides a better understanding of the progression of complications by incorporating random effects. The population was divided into two groups: people with diabetes without complications (D) and those with complications (C). After establishing the existence and characterization of an equilibrium state for the deterministic model, one demonstrated that there is no limit cycle between the number of diabetics with complications, the number of diabetics without complications, and the chosen control parameter under certain conditions. A study on eligibility for control was also conducted. For the stochastic model, the existence, uniqueness, and positivity of the solution were proved. In addition, the thresholds for persistence and reduction of diabetes complications were determined. Finally, numerical simulations were performed that highlighted the effectiveness of the proposed control strategies, particularly in reducing complications. These simulations also demonstrate that the low quality of education in the region has undoubtedly contributed to the increase in complications.

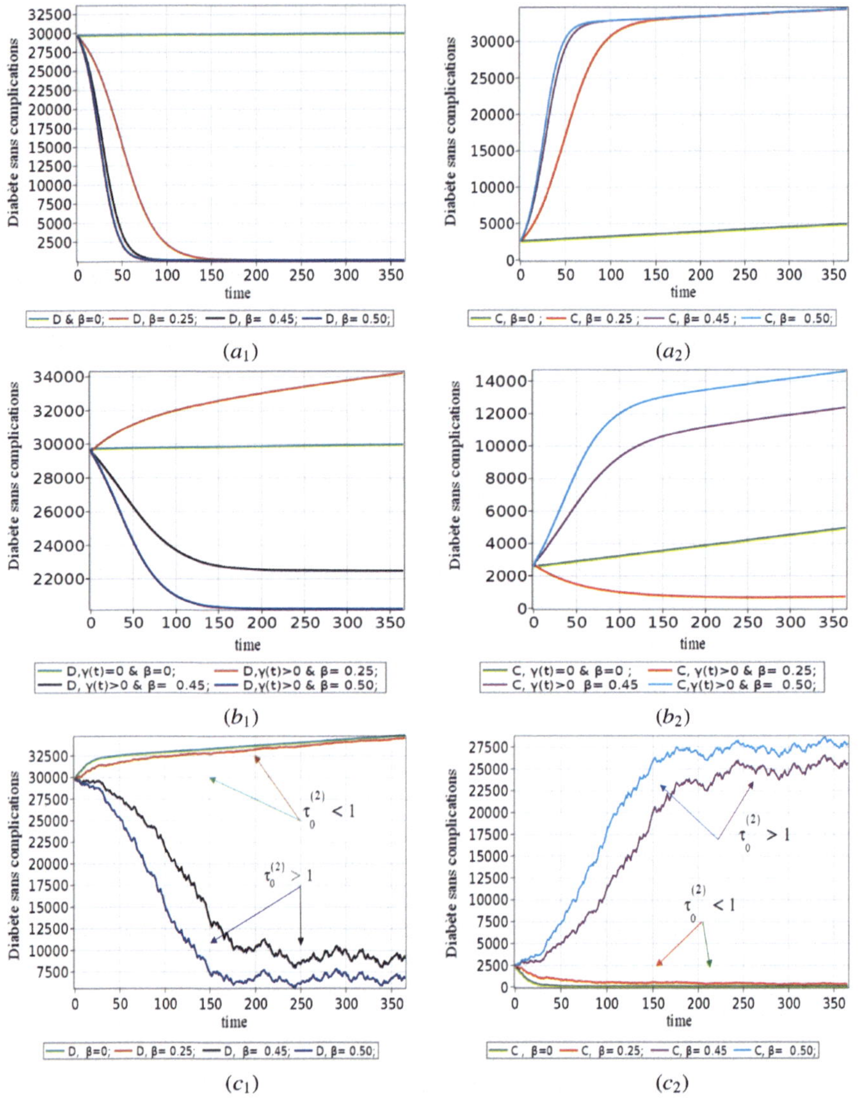

Fig. 3 (a_1, b_1, c_1): The evolution of the number of diabetics without complications; (a_2, b_2, c_2): The evolution of the number of diabetics with complications

Competing Interests The authors have no conflicts of interest to declare that are relevant to the content of this chapter.

Acknowledgements The Bank of the Republic of Haiti (BRH) and the French Embassy in Haiti are thanked for funding Woody Merat's research.

10 Appendix

Proof Lemma 3. Note that

$$f(r) = \frac{-\alpha\mu r^2 + [\lambda(a-\alpha)(\mu+\nu+\delta) + \mu(a-2\alpha)(\mu+\nu+\delta) - (\lambda+\mu)\alpha\mu]r}{\lambda I(r+\mu+\nu+\delta)} + \frac{(a-2\alpha)(\lambda+\mu)(\mu+\nu+\delta)\mu}{\lambda I(r+\mu+\nu+\delta)}.$$

A standard calculation yields the stated limits. In addition, the derivative of f is given by

$$f'(r) = \frac{-\alpha\mu r^2 - 2\alpha\mu(\mu+\nu+\delta)r + (a-\alpha)(\nu+\delta)(\mu+\lambda)(\mu+\nu+\delta) - \alpha\mu\sigma^2}{\lambda I(r+\sigma)^2}.$$

1. if $\mu = 0$, then $f'(r) = \dfrac{(a-\alpha)(\mu+\nu+\delta)^2}{I(r+\sigma)^2} > 0$
2. if not, because $[-2\alpha\mu(\mu+\nu+\delta)]^2 - 4(-\alpha\mu)[(a-\alpha)(\nu+\delta)(\mu+\lambda)(\mu+\nu+\delta) - \alpha\mu(\mu+\nu+\delta)^2] = 4\alpha\mu(a-\alpha)(\nu+\delta)(\mu+\lambda)(\mu+\nu+\delta) > 0$

   f' has the following two real roots:

$$r_{f_1}^{(0)} = \frac{-\alpha\mu(\mu+\nu+\delta) - \sqrt{\alpha\mu(a-\alpha)(\nu+\delta)(\mu+\lambda)(\mu+\nu+\delta)}}{\alpha\mu} \le 0$$

   and

$$r_{f_2}^{(0)} = \frac{-\alpha\mu(\mu+\nu+\delta) + \sqrt{\alpha\mu(a-\alpha)(\nu+\delta)(\mu+\lambda)(\mu+\nu+\delta)}}{\alpha\mu}.$$

   In addition

$$f(r_{f_2}^{(0)}) = \frac{(a\lambda + a\mu - \alpha\lambda)(\mu+\nu+\delta) - \alpha(\lambda+\mu)\mu}{\lambda I} - \frac{2\sqrt{\alpha\mu(a-\alpha)(\nu+\delta)(\mu+\lambda)\sigma}}{\lambda I} > 0$$

   if and only if

$$[(a\lambda + a\mu - \alpha\lambda)(\mu+\nu+\delta) - \alpha\lambda\mu - \alpha\mu^2]^2 - [2\sqrt{\alpha\mu(a-\alpha)(\nu+\delta)(\mu+\lambda)\sigma}]^2 > 0,$$

Let's define

$$(a\delta + a\mu + av - \alpha\delta - 2\alpha\mu - \alpha v)^2\lambda^2 + 2\mu\left[(a-2\alpha)(\mu+v+\delta)^2(a-\alpha)\right.$$
$$\left.-\mu\alpha^2(v+\delta)\right]\lambda + \mu^2(a\delta + a\mu + av - 2\alpha\delta - \alpha\mu - 2\alpha v)^2 = 0 \quad (56)$$

Now, since the discriminant of equation (56) is equal to $-16\alpha^2\mu^3(v+\delta)(\mu+v+\delta)^2(a-\alpha)(a-2\alpha)$,

a. if $a > 2\alpha$, the discriminant is strictly negative, and $f(r_{f_2}^{(0)})$ is strictly positive. By the intermediate value theorem, there exists $r_f^{(1)} > r_{f_2}^{(0)}$ such that $f(r_f^{(1)}) = 0$, since $f_\infty = -\infty$.
b. Otherwise, if not, the discriminant is positive, and in this case, there exists $\lambda_f^{(1)}$ and $\lambda_f^{(2)}$ solutions of (56) such that
   i. for all $\lambda_f^{(1)} \le \lambda \le \lambda_f^{(2)}$, $f(r_{f_2}^{(0)})$ is negative.
   ii. for all $\lambda < \lambda_f^{(1)}$ or $\lambda_f^{(2)} < \lambda$, $f(r_{f_2}^{(0)})$ is strictly positive, with

$$\lambda_f^{(1)} = \frac{\Theta^* - \sqrt{16\alpha^2\mu^3(v+\delta)(\mu+v+\delta)^2(a-\alpha)(2\alpha-a)}}{2(a\delta + a\mu + av - \alpha\delta - 2\alpha\mu - \alpha v)^2} \quad \text{and}$$

$$\lambda_f^{(2)} = \frac{\Theta^* + \sqrt{16\alpha^2\mu^3(v+\delta)(\mu+v+\delta)^2(a-\alpha)(2\alpha-a)}}{2(a\delta + a\mu + av - \alpha\delta - 2\alpha\mu - \alpha v)^2}$$

where $\Theta^* = 2\mu[(2\alpha - a)(\mu+v+\delta)^2(a-\alpha) + \mu\alpha^2(v+\delta)] > 0$

Moreover, when $f(r_{f_2}^{(0)}) > 0$, by the intermediate value theorem, there exist

- $r_f^{(2)} \in]0, r_{f_2}^{(0)}[$ such that $f(r_f^{(2)}) = 0$, since $f_0 < 0$
- $r_f^{(3)} > r_{f_2}^{(0)}$ such that $f(r_f^{(3)}) = 0$, since $f_\infty = -\infty$.

□

Proof Proposition 3. One first notice that solving the equation $n_1(r) + n_2(r)l_k = 0$, for $k \in \{1, 2\}$ is equivalent to solving the equation $n(r) = l_k$, arguing that $n_2 > 0$. Firstly, if for all $k \in \{1, 2\}$ the equation $n(r) = l_k$ has no solution in $\mathbb{R}_+^*$, then according to Theorem 8, for all $r \in I_{\mu,\lambda,\alpha,a}^f$, the equilibrium state associated with $\zeta^*(r)$ is admissible when $\mu > 0$. For the equation $n(r) = l_k$ to have no solution in $\mathbb{R}_+^*$, for all $k \in \{1, 2\}$, it suffices that

$$\max\{l_1, l_2\} = l_2 \le \frac{a\mu(\lambda+\mu)}{\lambda I}.$$

Secondly, one will seek to solve the equation $n_1(r) + n_2(r)l_k = 0$, for $k \in \{1, 2\}$, on $r \in \mathbb{R}_+^*$. This is equivalent to solving the equation $n(r) = l_k$. According to Lemma

1, the existence of a solution in $\mathbb{R}_+^*$ is equivalent to $l_k \in]n_0, n_\infty[$ for at least one index k, meaning:

- if $\mu = 0 : 0 \leq l_1 < l_2 < \frac{(a-\alpha)(\nu+\delta)}{I}$ or $0 < l_1 < \frac{(a-\alpha)(\nu+\delta)}{I} \leq l_2$
- if $\mu > 0 : \frac{a\mu(\lambda+\mu)}{\lambda I} < l_1$.

The satisfaction of one of the above conditions induces the existence of $r_k^* \in \mathbb{R}_+^*$ satisfying

$$n_1(r_k^*) + n_2(r_k^*)l_k = 0.$$

One will examine the two possibilities

1. if $k = 1$ then according to Theorem 8, One can only consider the case where $\mu > 0$, and it suffices for the interval $]0, r_1^*] \cap \mathcal{I}^f_{\mu,\lambda,\alpha,a}$ to be nonempty. There are three possibilities:
   a. $\mathcal{I}^f_{\mu,\lambda,\alpha,a} = [r_f^{(1)}, +\infty[$ and the intersection $]0, r_1^*] \cap \mathcal{I}^f_{\mu,\lambda,\alpha,a} = [r_f^{(1)}, r_1^*]$ is non-empty for $r_f^{(1)} \leq r_1^*$.
   b. $\mathcal{I}^f_{\mu,\lambda,\alpha,a} =]0, +\infty[$ and thus $]0, r_1^*] \cap \mathcal{I}^f_{\mu,\lambda,\alpha,a} =]0, r_1^*]$ which is non-empty.
   c. $\mathcal{I}^f_{\mu,\lambda,\alpha,a} =]0, r_f^{(2)}] \cup [r_f^{(3)}, +\infty[$ and the intersection $]0, r_1^*] \cap \mathcal{I}^f_{\mu,\lambda,\alpha,a} =]0, \min\{r_1^*, r_f^{(2)}\}] \cup A_{r_1^*, r_f^{(3)}}$ is non-empty, with

$$A_{r_1^*, r_f^{(3)}} = \begin{cases} [r_f^{(3)}, r_1^*] \text{ if } & r_1^* \geq r_f^{(3)} \\ \phi & \text{otherwise.} \end{cases}$$

2. if $k = 2$, then it suffices for the interval $]r_2^*, +\infty[\cap \mathcal{I}^f_{\mu,\lambda,\alpha,a}$ to be non-empty:
   a. if $\mu = 0$, then $\mathcal{I}^f_{\mu,\lambda,\alpha,a} =]0, +\infty[$ and thus $]r_2^*, +\infty[\cap \mathcal{I}^f_{\mu,\lambda,\alpha,a} =]r_2^*, +\infty[$ which is non-empty.
   b. if $\mu > 0$, then
      i. $\mathcal{I}^f_{\mu,\lambda,\alpha,a} = [r_f^{(1)}, +\infty[$, and the intersection $]r_2^*, +\infty[\cap \mathcal{I}^f_{\mu,\lambda,\alpha,a} =]\max\{r_f^{(1)}, r_2^*\}, +\infty[$ which is non-empty
      ii. $\mathcal{I}^f_{\mu,\lambda,\alpha,a} =]0, +\infty[$ and thus $]r_2^*, +\infty[\cap \mathcal{I}_\mu =]r_2^*, +\infty[$ which is non-empty.
      iii. $\mathcal{I}^f_{\mu,\lambda,\alpha,a} =]0, r_f^{(2)}] \cup [r_f^{(3)}, +\infty[$ and the intersection $]r_2^*, +\infty[\cap \mathcal{I}^f_{\mu,\lambda,\alpha,a} =]\max\{r_f^{(3)}, r_2^*\}, +\infty[\cup]\min\{r_f^{(2)}, r_2^*\}, r_f^{(2)}]$ which is non-empty.

□

Proof Theorem 9. Since the function F is composed of measurable functions, it is measurable. The same applies to M. Verify the assumptions of Theorem 5.2.1 in B. Oksendal's book (sixth edition, page 68) in order to be able to apply it.

1. Let $Y(t) = (b(t), c(t))$ and $t \in [0, T]$. Then

$$\|F(t, Y(t))\| + |||M(t, Y(t))||| = \sqrt{(J_1(b(t), c(t)))^2 + (J_2(b(t), c(t)))^2} + \sqrt{\left[i - (\lambda + \mu)\, b(t) + \gamma c(t) - \beta b(t)c(t)N\right]^2 + \left[\lambda b(t) - (\gamma + \sigma)\, c(t) + \beta b(t)c(t)N\right]^2}.$$

Using the relation $(a + b)^2 \leq 2a^2 + 2b^2$, one has

$$(i - (\lambda + \mu)\, b(t) + \gamma c(t) - \beta b(t)c(t)N)^2 \leq 2i^2 + \mathcal{M}_t \left(2\gamma^2 + \beta^2 b^2(t)N^2\right) + \mathcal{M}_t \left(4(\lambda^2 + \mu^2) + \beta^2 c^2(t)N^2\right)$$

with $\mathcal{M}_t = \max \left\{b^2(t), c^2(t)\right\}$, which is locally bounded since one is working in C^f. Similarly,

$$(\lambda b(t) - (\gamma + \sigma)\, c(t) + \beta b(t)c(t)N)^2 \leq \mathcal{M}_t \left(2\lambda^2 + \beta^2 c^2(t)N^2\right) + \mathcal{M}_t \left((\gamma + \sigma)^2 + \beta^2 b^2(t)N^2\right)$$

and $\sqrt{(J_1(b(t), c(t)))^2 + (J_2(b(t), c(t)))^2} \leq N\mathcal{M}_t\sqrt{\mathcal{E}_1^2 + \mathcal{E}_2^2}.$

With the three previous estimates, one finds

$$\|F(t, Y(t))\| + |||M(t, Y(t))||| \leq N\mathcal{M}_t\sqrt{\mathcal{E}_1^2 + \mathcal{E}_2^2} + \sqrt{2i^2 + \mathcal{M}_t[2\gamma^2 + 4(\lambda^2 + \mu^2) + 2\lambda^2 + (\gamma + \sigma)^2 + 2\beta^2 N^2 \left(b^2(t) + c^2(t)\right)]}.$$

Since $(b(t) + c(t))^2 = 1$, one obtains

$$\|F(t, Y(t))\| + |||M(t, Y(t))||| \leq l_1 \,(1 + ||Y(t)||)$$

with

$$l_1^2 = \max \left\{2\left[2i^2 + \mathcal{M}_t \left(2\gamma^2 + 4(\lambda^2 + \mu^2) + 2\lambda^2 + (\gamma + \sigma)^2 + \beta^2 N^2\right)\right];\; N^2\mathcal{M}_t^2(\mathcal{E}_1^2 + \mathcal{E}_2^2)\right\}.$$

2. Let $Y_1(t) = (b_1(t), c_1(t))$, $Y_2(t) = (b_2(t), c_2(t))$, and $t \in [0, T]$. Then

$$\|F(t, Y_2(t)) - F(t, Y_1(t))\| + |||M(t, Y_2(t)) - M(t, Y_1(t))||| = \sqrt{A_1^2 + A_2^2} + \sqrt{A_3^2}$$

where

$$
\begin{aligned}
A_1^2 &= (-(\lambda+\mu)(b_2(t)-b_1(t)) + \gamma\,(c_2(t)-c_1(t)) - \beta N\,(b_2(t)c_2(t)-b_1(t)c_1(t)))^2\,,\\
A_2^2 &= (\lambda\,(b_2(t)-b_1(t)) - (\gamma+\sigma)\,(c_2(t)-c_1(t))\\
&\quad + N\beta\,(b_2(t)c_2(t)-b_1(t)c_1(t)))^2\,,\\
A_3^2 &= (\jmath_1(b_1(t),c_1(t)) - \jmath_1(b_2(t),c_2(t)))^2 + (\jmath_2(b_1(t),c_1(t)) - \jmath_2(b_2(t),c_2(t)))^2\,.
\end{aligned}
$$

Thus,

$$
A_1^2 \le 4((\lambda+\mu)^2 + \beta^2 M_t N^2)(b_2(t)-b_1(t))^2 + 4(\gamma^2+\beta^2 M_t N^2)\,(c_2(t)-c_1(t))^2
$$

and

$$
A_2^2 \le 4(\lambda^2 + M_t N^2\beta^2)(b_2(t)-b_1(t))^2 + 4((\gamma+\sigma)^2 + M_t N^2\beta^2)(c_2(t)-c_1(t))^2.
$$

Moreover, considering the equality

$$
A_3^2 = (\jmath_1(b_1(t),c_1(t)) - \jmath_1(b_2(t),c_2(t)))^2 + (\jmath_2(b_1(t),c_1(t)) - \jmath_2(b_2(t),c_2(t)))^2
$$

and the fact that $\mathcal{J}_1$ and $\mathcal{J}_2$ are locally Lipschitz functions, there exists a constant $k^{J_{1,2}} \in \mathbb{R}^+$ such that

$$
A_3^2 \le k^{J_{1,2}}\left[(b_2(t)-b_1(t))^2 + (c_2(t)-c_1(t))^2\right].
$$

Therefore, one has

$$
\|F(t,Y_2(t)) - F(t,Y_1(t))\| + |||M(t,Y_2(t)) - M(t,Y_1(t))||| \le
l_2\sqrt{(b_2(t)-b_1(t))^2 + (c_2(t)-c_1(t))^2}
$$

where

$$
l_2^2 = 4\max\left\{4((\lambda+\mu)^2 + 2\beta^2 M_t N^2 + \lambda^2);\, 4((\lambda+\mu)^2 + 2\beta^2 M_t N^2 + \lambda^2);\, k^{J_{1,2}}\right\}.
$$

Finally, one obtains

$$
\|F(t,Y_2(t)) - F(t,Y_1(t))\| + |||M(t,Y_2(t)) - M(t,Y_1(t))||| \le l_2\|Y_2(t) - Y_1(t)\|.
$$

With the conditions satisfied, according to Theorem 5.2.1 in B. Oksendal's book (sixth edition, page 68), the stochastic differential equation (SDE) (37) admits a unique local solution, which leads to the announced result.

□

References

1. Bernard, S., Nuiro, S.P., Pietrus, A.: Diabetes, complications and limit cycles. Appl. Math. E-Notes **15**, 197–206 (2015)
2. Bernard, S., Pietrus, A.: Optimal glucose modelling for diabetes. e-Journal Caribbean Acad. Sci. **8**(1), 1–8, (2015)
3. Bernard, S., César, T., Nuiro, S.P., Piétrus, A.: Unexistence of limit cycle in an optimal control problem of a population of diabetics. Revista de Matemática: Teoría y Aplicaciones **25**(2), 239–259 (2018)
4. Bennett, N.R., Francis, D.K., Ferguson, T.S., Hennis, A.J.M., Wilks, R.J., Harris, E.N., MacLeish, M.M.Y., Sullivan, L.W.: Disparities in diabetes mellitus among Caribbean populations: a scoping review. Int. J. Equity Health **14**(1), 1–17 (2015)
5. Bihan, H., Choleau, C., Cohen, R., Reach, G.: Functional insulin therapy for type 1 diabetes and precariousness. Med. Metabolic Dis. **3**(1), 55–58, Elsevier (2009)
6. Blauw, L.L., Aziz, N.A., Tannemaat, M.R., Blauw, C.A., de Craen, A.J., Pijl, H., Rensen, P.C.N.: Diabetes incidence and glucose intolerance prevalence increase with higher outdoor temperature. BMJ Open Diabet. Res. Care **5**(1), e000317 (2017)
7. Bodet, C., Radermecker, R., Philips, J.-C., Rorive, M., Benmouna, K., Forthomme, B., Wang, F.-C., Kaux, J.-F.: The musculoskeletal complications of diabetes. Revue Médicale de Liège **74**(11), 572–579 (2019)
8. Bolie, V.W.: Coefficients of normal blood glucose regulation. J. Appl. Physiol. **16**(5), 783–788 (1961)
9. Bordier, L., Sollier, M., Dolz, M., Garcia, C., Allain, J., Bauduceau, B.: Does improving glycemic control reduce the macroangiopathic complications of diabetes? Med. Metabolic Dis. **13**(1), 36–41, Elsevier (2019)
10. Boutayeb, A., Derouich, M.: Age structured models for diabetes in East Morocco. Math. Comput. Simul. **58**(3), 215–229. Elsevier (2002)
11. Boutayeb, A., Twizell, E.H., Achouayb, K., Chetouani, A.: A mathematical model for the burden of diabetes and its complications. Biomed. Eng. Online **3**, 1–8, Springer (2004)
12. Bouchardat, A.: About Glycosuria or Diabetes Mellitus: Its Hygienic Treatment with Notes Documents on the Nature and Treatment of Gout, Uric Gravel; on Oliguria, Diabetes Insipidus with Excess Urea, Hippuric Acid, Pimelorrhea, etc. In: Germer Bailliére Bookshop (1875)
13. Cai, Y., Kang, Y., Banerjee, M., Wang, W.: A stochastic epidemic model incorporating media coverage. Commun. Math. Sci. **14**, 893–910 (2016)
14. Carl, J., Caspersen, G., Thomas, D., Boseman, L.A., Gloria, L.A., Beckles, A., Albright, L.: Aging, diabetes, and the public health system in the United States. Am. J. Publ. Health **102**(8), 1482–1497 (2012)
15. Cole, J.B., Florez, J.C.: Genetics of diabetes mellitus and diabetes complications. Nat. Rev. Nephrol. **16**(7), 377–390 (2020)
16. Derouich, M., Boutayeb, A.: The effect of physical exercise on the dynamics of glucose and insulin. J. Biomech. **35** (7), 911–917, Elsevier (2002)
17. Deshpande, A.D., Harris-Hayes, M., Schootman, M.: Epidemiology of diabetes and diabetes-related complications. Phys. Therapy **88**(11), 1254–1264 (2008)
18. Diop, S.N., Djrolo, F., Traoré Sidibé, A., Baldé, N.M., Monabeka, H.G., Epaka, M.E., Drabo, Y.J., Abodo, J.R., Kasiam, J.B., Mbaye, N.M.: Consensus for the management of hyperglycemia in type 2 diabetes in sub-Saharan Africa. Written by a group of African diabetes experts. Med. Metabolic Dis. **13** (2), 210–216, Elsevier (2019)
19. France-Antilles: Diabetes, a major problem in the Caribbean. https://www.guadeloupe.franceantilles.fr/actualite/sante/le-diabete-un-gros-probleme-pour-la-caraibe-703339.php. Cited 17 Jul 2023 (2023)
20. Geslot, A., Bertrand, M., Galinier, A., Hanaire, H.: Vitamin B6 deficiency, a factor associated with diabetic complications? Clin. Nutr. Metabol. **36**(1), S42 (2022)
21. Granas, A., Dugundji, J.: Elementary fixed point theorems. In: Fixed Point Theory, pp. 9–84. Springer (2003)

22. Hanssen Mark, J.W., van der Lans Anouk, A.J.J., Brans, B., Hoeks, J., Jardon, K.M.C., Schaart, G., Mottaghy, F.M., Schrauwen, P., van Marken Lichtenbelt, W.D.: Short-term cold acclimation recruits brown adipose tissue in obese humans. Diabetes **65**(5), 1179–1189, Am. Diabetes Assoc. (2016) https://doi.org/10.2337/db15-1372.
23. Holt, R.I.G., De Groot, M., Golden, S.H..: Diabetes and depression. Curr. Diabetes Rep. **14**, 1–9, Springer (2014)
24. Huang, J., Jin, X.: Preventing rumor spreading on small-world networks. J. Syst. Sci. Complex. **24**(3), 449–456, Springer (2011)
25. Hussein, F., Hejase, H.J.: Artificial intelligence and fake news. Žurnalistikos tyrimai **16**, 39–71 (2022)
26. IRIS: Toolbox to fight NCD misinformation. In: Forum to Fight Health and NCD Misinformation. https://iris.who.int/handle/10665/363688. Cited 05 Apr 2024 (2024)
27. Ji, C., Jiang, D.: Threshold behaviour of a stochastic SIR model. Appl. Math. Model. **38**(21–22), 5067–5079 (2014)
28. Kaux, J.-F., Bodet, C.: Musculoskeletal complications of diabetes. In: 41st Seminar on Sports Traumatology and Cardiology (2022)
29. Khiari, M., Zribi, S., Zahra, H., Boukhayatia, F., Mizouri, R., Temessek, A., Ben Mami, F.: Degenerative complications: type 1 diabetes vs. type 2 diabetes, Annals of Endocrinology, p. 494 (2018)
30. Liu, P., Huang, L., Din, A., Huang, X.: Impact of information Lévy noise on stochastic COVID-19 epidemic model under real statistical data. J. Biologic. Dyn. **16**(1), 236–253 (2022)
31. Mao, X.: Stochastic Differential Equations and Applications. Elsevier (2007)
32. Mealey, B.L., Ocampo, G.L.: Diabetes mellitus and periodontal disease. Periodontology 2000 **44**(1), 127–153 (2007)
33. Menció-Padrón, D., Bouza-Allende, G.: Simulating the propagation of rumours with spreaders of distinct characters. Oper. Res. Decis. **33**(3), 75–87. Oficyna Wydawnicza Politechniki Wroc ıawskiej (2023)
34. Mohamed, E.F., Sekkak, I.: Lévy noise impact on a stochastic delayed epidemic model with Crowly-Martin incidence crowding effect. Phys. A: Stat. Mech. Appl. **541**, 123315 (2020)
35. Mumtaz, N., Green, C., Duggan, J.: Exploring the effect of misinformation on infectious disease transmission. Systems **10**(2), 50, Multidisciplinary Digital Publishing Institute (2022)
36. Nazar, C.M.J., Bojerenu, M.M., Safdar, M., Marwat, J.: Effectiveness of diabetes education and awareness of diabetes mellitus in combating diabetes in the United Kingdom; a literature review. J. Nephropharmacol. **5**(2), 110. Society of Diabetic Nephropathy Prevention (2016)
37. Oksendal, B.: Stochastic Differential Equations: An Introduction with Applications. Springer Science & Business Media (2013)
38. OMS: Combating online misinformation about non-communicable diseases requires collaboration. https://www.who.int/europe/fr/news/item/20-10-2022-collaboration-is-key-to-countering-online-misinformation-about-noncommunicable-diseases--new-who-europe-toolkit-shows-how. Cited 05 Apr 2024 (2022)
39. ORSAG: Chronic disease: Diabetes in Guadeloupe https://orsag.fr/wp-content/uploads/2021/11/. Cited 17 Jul 2023 (2021)
40. Philips, J., Vandelaer, A., Quintana, C., Grandfils, S.: Metabolic complications after gestational diabetes: who to follow up first? Gynecol. Obs. Fertility Senol. **50**(5), 431 (2022)
41. Ricci, P., Blotière, P.-O., Weill, A., Simon, D., Tuppin, P., Ricordeau, P., Allemand, H.: Treated diabetes: what changes between 2000 and 2009 in France. Bull. Epidemiol. Hebd. **42**(43), 425–431 (2010)
42. Salathé, M., Khandelwal, S.: Assessing vaccination sentiments with online social media: implications for infectious disease dynamics and control. PLoS Comput. Biol. **7** (10), e1002199. Public Library of Science (2011) https://doi.org/10.1371/journal.pcbi.1002199
43. Scheen, A., Paquot, N.: A revolution in the treatment of type 2 diabetes: antidiabetic drugs with cardio-renal protection independent of glycemic control! Revue Médicale de Liège **75**(5–6), 392–398 (2020)

44. Valmont, K.: Stochastic optimal control with applications to e-rumeur propagation, PhD thesis, Université des Antilles (2019)
45. Wang, J., Zhao, L., Huang, R.: SIR a rumor spreading model in complex networks. Phys. A: Stat. Mech. Appl. **398**, 43–55, Elsevier (2014)
46. World Health Organization: Global strategic directions for strengthening nursing and midwifery 2016-2020. https://iris.who.int/bitstream/handle/10665/275453/9789241510455-eng.pdf?sequence=1. Cited 17 Jul 2023 (2016)
47. World Health Organization: Global Diabetes Report. https://apps.who.int/iris/handle/10665/254648. Cited 17 Jul 2023 (2023)

A Compartmental Mathematical Model for the Evolution from a Population of Healthy Men to Population of Men Affected by MetS, BPH and CVD/PCa

Mohamed Derouich, Mohamed E. N. Lamlili, Abdesslam Boutayeb, and Fatima Zahra Benassila

Abstract Benign prostatic hyperplasia (BPH) is a highly prevalent disease affecting a large proportion of men especially by the age of 60 years and above. Defined as a non-malignant enlargement of the prostate gland, it is frequently associated with lower urinary tract symptoms (LUTS) and it ultimately obstructs the bladder outflow. Sometimes, this disease can also prelude the prostate cancer (PCa) which is the second more prevalent men cancer in developed countries. While for a long period of time, benign prostatic hyperplasia was considered as an irreversible age-dependent disease, recent clinical and epidemiological studies brought out evidence on the correlation between BPH and diabetes homeostasis (hyperglycaemia, hyperinsulenemia) and more generally with metabolism syndrome (MetS). Following the interesting findings of a large number of studies in terms of correlation between benign prostatic hyperplasia and the components of the metabolism syndrome, we propose a mathematical model that illustrates the men population dynamics based on the flow out of and in compartments of men with MetS, BPH, CVDs/PCa. Carrying out simulations using and varying values of some parameters of our model shows how the number of men affected by BPH can be controlled through action on the main risk factors of MetS (healthy diet, physical activity, weight control, ...). Moreover, our model indicates that action on these risk factors can reduce the mixed burden of MetS, BPH and CVDs/PCa.

M. Derouich · M. E. N. Lamlili
Mohammed First University, Oujda, Morocco
e-mail: m.derouich@ump.ac.ma

A. Boutayeb (✉)
Emirates Aviation University, Dubaï, UAE
e-mail: x.boutayeb@gmail.com

F. Z. Benassila
Endocrinologist, Endocrinology Unit, Al Farabi Hospital, Oujda, Morocco
e-mail: fzahra.ben@hotmail.fr

A. Boutayeb and M. E. N. Lamlili (eds.), *Computational Mathematics and Modelling for Diabetes*, Industrial and Applied Mathematics,
https://doi.org/10.1007/978-981-96-1925-2_7

1 Introduction

Characterised by an age-dependent non-malignant enlargement of the prostate gland, benign prostatic hyperplasia (BPH) is a highly prevalent disease affecting around half of the men population at the age 60 and more than 80% of men aged 75 years and above [5, 12, 38, 53]. Frequently associated with lower urinary tract symptoms (LUTS) and ultimately obstructing bladder outflow, this disease can also prelude the prostate cancer (PCa) which is the second more prevalent men cancer in developed countries and which affected globally nearly 1.5 million new cases and caused more than 375,000 deaths in 2020 [55]. However, it should be stressed that association (which may be due to common risk factors or pathophysiological mechanisms) between BPH and PCa remains a controversial issue [16].

For a long period of time (until the 1980s), BPH and LUTS were considered as irreversible processes of age leading to resection or enucleating of the prostate adenoma as the most common treatment of LUTS caused by BPH [5, 40, 52]. Following the papers published by Huggins et al in the 1940s [32], indicating that prostate cancer may be controlled through castration while administration of testosterone can cause prostate growth, research focussed on symptoms of BPH/LUTS and the role of hormones like testosterone [28, 51]. In recent decades, however, it was shown that prostate volume continues to increase while testosterone levels start decreasing regularly around the age of 40 years [28, 39, 51, 63]. In parallel, some urologists have raised the question about benefits/risks of surgical intervention for BPH and researchers have particularly highlighted the exponentially increasing number of prostate ablations [39, 53, 58].

In the mean time, clinical and epidemiological studies brought out evidence on the correlation between BPH and diabetes homeostasis (hyperglycaemia, hyperinsulenemia) and more generally with metabolism syndrome (MetS), defined as the presence of three or more criteria among the following: (1) abdominal obesity (waist circumference >102 cm), (2) hypertriglyceridemia (>150 mg/dl) or medications, (3) low high-density lipoprotein (HDL) cholesterol (<40 mg/dl) or medications, (4) hypertension ($>130/85$ mmHg) or medication, (5) high fasting glucose (>110 mg/dl) or medication [31]. Indeed, a multitude of papers were devoted to the association of benign prostatic hyperplasia with type 2 diabetes [37, 43] or particularly with insulin resistance and/or hyperglycaemia [3, 10, 54, 56, 71]. Although the pathways underlying the relationship between MetS and BPH are still needing acute research, the association between BPH and metabolism syndrome or separately with its components is supported by a large number of studies as indicated by many recent reviews and meta-analyses [1, 13, 19, 25, 41, 44, 45, 61, 64, 73]. It should be stressed that recently, Dobrowolski et al. [23] proposed a new definition of MetS based on the fact that obesity coexists with hypertension, disordered glucose and lipid metabolism. Consequently, the new definition proposed for MetS assumes the presence of obesity besides two of the three following conditions: impaired glucose metabolism, high blood pressure and elevated non-high-density lipoprotein (non-HDL) cholesterol level.

In a recent book edited by Boutayeb, a whole chapter is dedicated to "The association between Benign Prostatic Hyperplasia and Metabolic Syndrome/Type2 Diabetes" [6]. According to a comprehensive review dealing with this theme, the quasi-totality of studies in which authors analysed the effect of MetS on the prostate enlargement concluded that prostatic volume was greater in people with MetS than in their counterpart non-MetS. The results are illustrated in Fig. 1 which shows that 30 out of 33 studies revealed that the mean prostatic volume was greater or equal in people with MetS than in their counterparts with non-MetS [1, 2, 11, 14, 15, 17, 18, 20–22, 24–27, 29, 30, 34–36, 42, 46–50, 62, 65–70, 72].

Taking into account the epidemiological and clinical finding on the association between BPH and MetS or diabetes and following previous mathematical models dealing with diabetes population dynamics (see Chap. 1 in this book), we propose a new mathematical model illustrating the link between MetS and BPH and how they contribute to the increasing incidence and prevalence of non-communicable diseases like cardiovascular diseases and cancer.

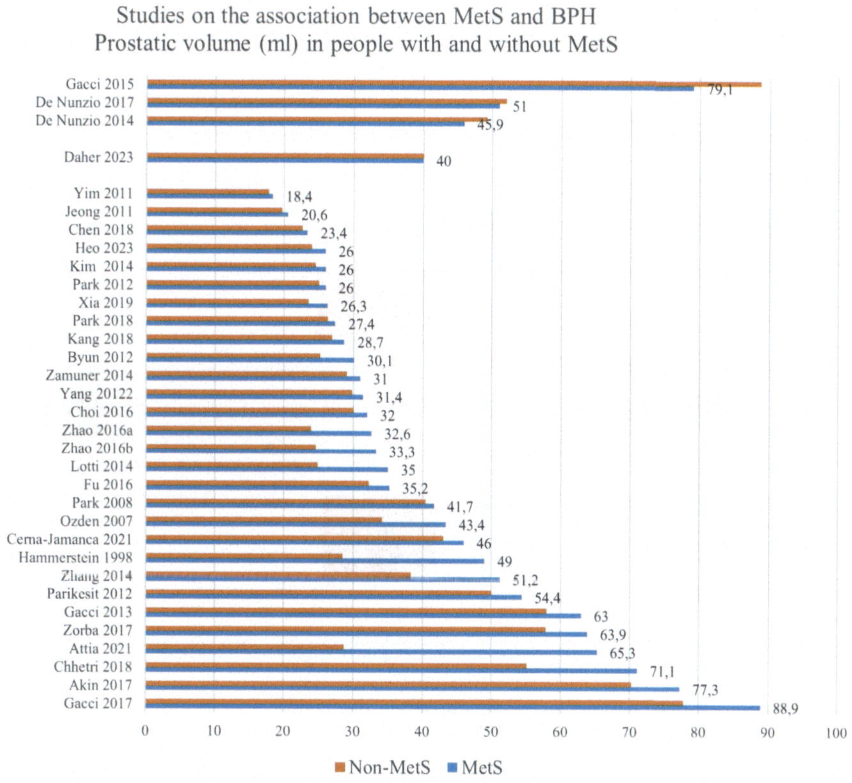

Fig. 1 Studies that analysed the correlation between MetS and BPH

2 The Mathematical Model

Following the multitude of published mathematical models on diabetes population dynamics cited in Chap. 1 and in particular the models published by the team of Boutayeb et al. [7–9], we assume that the present mathematical model is governed by four differential equations translating the flow in and out of the four compartments indicated in Fig. 2.

Here:

- P: Population of men over 20 years old not affected by MetS, BPH, CVD/PCa.
- Ms: Population with metabolic syndrome.
- Bp: Population with benign prostatic hyperplasia.
- Cc: Population with cardiovascular diseases and prostate cancer.

While the other parameters and their values used in simulation are given in Table 1, the mortality rates μ_1, μ_2, μ_3, μ_4 associated to each compartment are assumed to have the same value μ.

The mathematical model is given by the following system of ordinary differential equations:

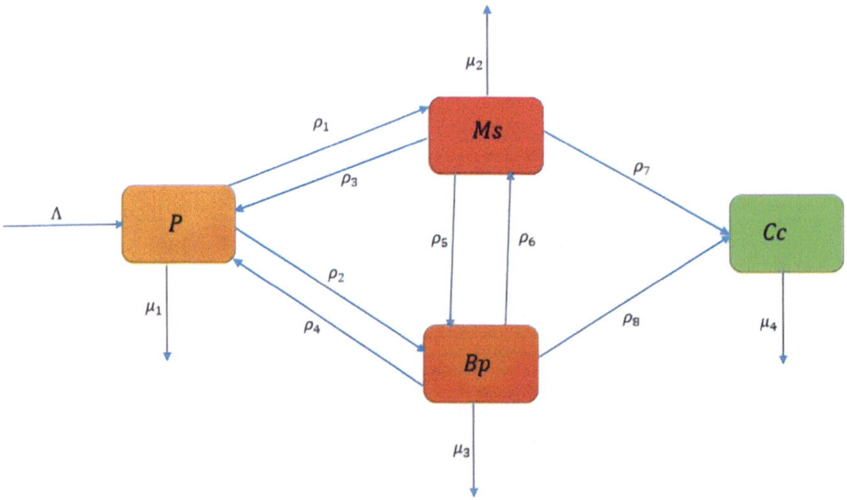

Fig. 2 The flow in and out of the four compartments of the model

Table 1 Parameters used in the mathematical model

Parameter	Signification	Value	Source
ρ_1	Probability of evolution from compartment P to Ms	0.01	[6, 33, 60]
ρ_2	Probability of evolution from compartment P to Bp	0.008	[6, 60]
ρ_3	Probability of evolution from compartment Ms to P	0.005	[4, 6, 60]
ρ_4	Probability of evolution from compartment Bp to P	0.001	[60]
ρ_5	Probability of evolution from compartment Ms to Bp	0.018	[6, 60]
ρ_6	Probability of evolution from compartment Bp to Ms	0.001	[6, 60]
ρ_7	Probability of evolution from compartment Ms to Cc	0.001	[33, 60]
ρ_8	Probability of evolution from compartment Bp to Cc	0.002	[60]
μ	Mortality rate	0.013	[59, 60]
N	Total population = P + Ms + Bp + Cc	40×10^6	
Λ	New men free of MetS, BPH, CVD/PCa	μxN	

$$\begin{cases} \dfrac{dP(t)}{dt} = \Lambda - (\rho_1 + \rho_2 + \mu_1)P(t) + \rho_3 Ms(t) + \rho_4 Bp(t) \\ \dfrac{dMs(t)}{dt} = \rho_1 P(t) - (\rho_3 + \rho_5 + \rho_7 + \mu_2)Ms(t) + \rho_6 Bp(t) \\ \dfrac{dBp(t)}{dt} = \rho_2 P(t) + \rho_5 Ms(t) - (\rho_4 + \rho_6 + \rho_8 + \mu_3)Bp(t) \\ \dfrac{dCc(t)}{dt} = \rho_7 Ms(t) + \rho_8 Bp(t) - \mu_4 Cc(t) \end{cases} \tag{1}$$

Let $\alpha = \rho_1 + \rho_2 + \mu_1$, $\beta = \rho_3 + \rho_5 + \rho_7 + \mu_2$ and $\gamma = \rho_4 + \rho_6 + \rho_8 + \mu_3$. The previous system becomes

$$\begin{cases} \dfrac{dP(t)}{dt} = \Lambda - \alpha P(t) + \rho_3 Ms(t) + \rho_4 Bp(t) \\ \dfrac{dMs(t)}{dt} = \rho_1 P(t) - \beta Ms(t) + \rho_6 Bp(t) \\ \dfrac{dBp(t)}{dt} = \rho_2 P(t) + \rho_5 Ms(t) - \gamma Bp(t) \\ \dfrac{dCc(t)}{dt} = \rho_7 Ms(t) + \rho_8 Bp(t) - \mu_4 Cc(t). \end{cases} \tag{2}$$

2.1 *Stability Analysis*

2.1.1 Equilibrium Points

Theorem 1 *The system admits one equilibrium point* $E = (P^*, Ms^*, Bp^*, Cc^*)$ *which is positive where*

$$\begin{aligned}
P^* &= \frac{\Lambda\gamma(\beta\gamma - \rho_6\rho_5)}{\alpha\beta\gamma - \rho_1\rho_3\gamma - \alpha\rho_5\rho_6 - \beta\rho_4\rho_2 - \rho_2\rho_3\rho_6 - \rho_1\rho_4\rho_5} \\
Ms^* &= \left(\frac{\rho_1\gamma + \rho_6\rho_2}{\beta\gamma - \rho_6\rho_5}\right) P^* \\
Bp^* &= \frac{1}{\gamma}\left(\rho_2 + \rho_5\frac{\rho_1\gamma + \rho_6\rho_2}{\beta\gamma - \rho_6\rho_5}\right) P^* \\
C_c^* &= \frac{1}{\mu_4}\left[\rho_7\frac{\rho_1\gamma + \rho_6\rho_2}{\beta\gamma - \rho_6\rho_5} + \frac{\rho_8}{\gamma}\left(\rho_2 + \rho_5\frac{\rho_1\gamma + \rho_6\rho_2}{\beta\gamma - \rho_6\rho_5}\right)\right] P^*.
\end{aligned}$$

Proof For the model above, equilibrium points are defined such that there is no variations in P, Ms, Bp, Cc with respect to t:

$$\Lambda - \alpha P + \rho_3 Ms + \rho_4 Bp = 0 \tag{3}$$

$$\rho_1 P - \beta Ms + \rho_6 Bp = 0 \tag{4}$$

$$\rho_2 P + \rho_5 Ms - \gamma Bp = 0 \tag{5}$$

$$\rho_7 Ms + \rho_8 Bp - \mu_4 Cc = 0. \tag{6}$$

From (6) we obtain

$$C_c = \frac{1}{\mu_4}(\rho_7 Ms + \rho_8 Bp). \tag{7}$$

From (5) we obtain

$$Bp = \frac{1}{\gamma}(\rho_2 P + \rho_5 Ms). \tag{8}$$

In Eq. (4) we substitue Bp, therefore

$$\begin{aligned}
\rho_1 P - \beta Ms + \frac{\rho_6}{\rho_4 + \rho_6 + \rho_8 + \mu_3}(\rho_2 P + \rho_5 Ms) = 0 \\
\Rightarrow \left(\rho_1 + \frac{\rho_6\rho_2}{\gamma}\right) P + \left(\frac{\rho_6\rho_5}{\gamma} - (\rho_3 + \rho_5 + \mu_2)\right) Ms = 0 \\
\Rightarrow Ms = \left(+\frac{\rho_1(\gamma) + \rho_6\rho_2}{\beta\gamma - \rho_6\rho_5}\right) P
\end{aligned}$$

$\Longrightarrow$

$$Ms = \left(\frac{\rho_1\gamma + \rho_6\rho_2}{\beta\gamma - \rho_6\rho_5}\right) P. \tag{9}$$

From (8) we obtain

$$Bp = \frac{1}{\gamma}\left(\rho_2 + \rho_5\frac{\rho_1\gamma + \rho_6\rho_2}{\beta\gamma - \rho_6\rho_5}\right) P \tag{10}$$

So

$$C_c = \frac{1}{\mu_4\gamma}\left[\rho_2\rho_8 + \frac{(\rho_7\gamma + \rho_5\rho_8)(\rho_1\gamma + \rho_6\rho_2)}{\beta\gamma - \rho_6\rho_5}\right] P. \tag{11}$$

In Eq. (3) we substitue Bp, therefore

$$\Lambda - \alpha P + \rho_3 Ms + \rho_4 Bp = 0$$

$$\Rightarrow \Lambda + \left(-\alpha + \rho_3\frac{\rho_1\gamma + \rho_6\rho_2}{\beta\gamma - \rho_6\rho_5} + \frac{\rho_4}{\gamma}\left(\rho_2 + \rho_5\frac{\rho_1\gamma + \rho_6\rho_2}{\beta\gamma - \rho_6\rho_5}\right)\right) P = 0$$

$$\Rightarrow \Lambda\gamma(\beta\gamma - \rho_6\rho_5) - \big[\alpha\gamma(\beta\gamma - \rho_6\rho_5) - \gamma\rho_3(\rho_1\gamma + \rho_6\rho_2) - \rho_4\rho_2(\beta\gamma - \rho_6\rho_5)$$
$$+ \rho_4\rho_5\rho_1\gamma + \rho_4\rho_5\rho_6\rho_2\big] P = 0$$

$$\Rightarrow P = \frac{\Lambda\gamma(\beta\gamma - \rho_6\rho_5)}{(\alpha\gamma - \rho_4\rho_2)(\beta\gamma - \rho_6\rho_5) - (\gamma\rho_3 + \rho_4\rho_5)(\rho_1\gamma + \rho_6\rho_2)}$$

$$\begin{aligned} P &= \frac{\Lambda\gamma(\beta\gamma - \rho_6\rho_5)}{(\alpha\gamma - \rho_4\rho_2)(\beta\gamma - \rho_6\rho_5) - (\gamma\rho_3 + \rho_4\rho_5)(\rho_1\gamma + \rho_6\rho_2)} \\ &= \frac{\Lambda\gamma(\beta\gamma - \rho_6\rho_5)}{\alpha\beta\gamma - \rho_1\rho_3\gamma - \alpha\rho_5\rho_6 - \beta\rho_4\rho_2 - \rho_2\rho_3\rho_6 - \rho_1\rho_4\rho_5}. \end{aligned}$$

For any values of the parameters, the disease-free equilibrium point is given by $E - (P^*, Ms^*, Bp^*, Cc^*)$,
where

$$\begin{aligned} P^* &= \frac{\Lambda\gamma(\beta\gamma - \rho_6\rho_5)}{\alpha\beta\gamma - \rho_1\rho_3\gamma - \alpha\rho_5\rho_6 - \beta\rho_4\rho_2 - \rho_2\rho_3\rho_6 - \rho_1\rho_4\rho_5} \\ Ms^* &= \left(\frac{\rho_1\gamma + \rho_6\rho_2}{\beta\gamma - \rho_6\rho_5}\right) P^* \\ Bp^* &= \frac{1}{\gamma}\left(\rho_2 + \rho_5\frac{\rho_1\gamma + \rho_6\rho_2}{\beta\gamma - \rho_6\rho_5}\right) P^* \\ C_c^* &= \frac{1}{\mu_4}\left[\rho_7\frac{\rho_1\gamma + \rho_6\rho_2}{\beta\gamma - \rho_6\rho_5} + \frac{\rho_8}{\gamma}\left(\rho_2 + \rho_5\frac{\rho_1\gamma + \rho_6\rho_2}{\beta\gamma - \rho_6\rho_5}\right)\right] P^*. \end{aligned}$$

On the one hand:

$$\beta = \rho_3 + \rho_5 + \rho_7 + \mu_2 > \rho_5,\ \gamma = \rho_4 + \rho_6 + \rho_8 + \mu_3 > \rho_6 \Longrightarrow \beta\gamma - \rho_6\rho_5 > 0$$

$$\begin{aligned}
\alpha\beta\gamma - \rho_1\rho_3\gamma - \alpha\rho_5\rho_6 = {} & \rho_1\rho_5\rho_8 + \rho_1\rho_5\mu_3 + \rho_1\rho_4\mu_7 + \rho_1\rho_6\rho_7 + \rho_1\rho_7\rho_8 \\
-\beta\rho_4\rho_2 - \rho_2\rho_3\rho_6 - \rho_1\rho_4\rho_5 \quad & +\rho_1\rho_7\mu_3 + \rho_1\rho_4\mu_2 + \rho_1\rho_6\mu_2 + \rho_1\rho_8\mu_2 \\
& +\rho_1\mu_1\mu_2 + \rho_2\rho_3\rho_8 + \rho_2\rho_3\mu_3 + \rho_2\rho_5\mu_8 + \rho_2\rho_5\mu_3 \\
& +\rho_2\rho_4\rho_6 + \rho_2\rho_7\rho_8 + \rho_2\rho_7\mu_3 + \rho_2\rho_6\mu_2 + \rho_2\rho_8\mu_1 \\
& +\rho_2\mu_2\mu_3 + \rho_3\rho_4\mu_1 + \rho_3\rho_6\mu_1 + \rho_3\rho_8\mu_1 + \rho_3\mu_1\mu_3 \\
& +\rho_4\rho_5\mu_1 + \rho_5\rho_8\mu_1 + \rho_5\mu_1\mu_3 + \rho_4\rho_7\mu_1 + \rho_6\rho_7\mu_1 \\
& +\rho_7\rho_8\mu_1 + \rho_4\mu_1\mu_2 + \rho_6\mu_1\mu_2 + \rho_8\mu_1\mu_2 + \mu_1\mu_2\mu_3 \\
& > 0.
\end{aligned}$$

So $P^* > 0$, $Ms^* > 0$, $Bp^* > 0$ and $Cc^* > 0$.

2.1.2 Local Stability of the Equilibria

Theorem 2 *The equilibrium point $E = (P^*, Ms^*, Bp^*, Cc^*)$ is locally asymptotically stable.*

Proof The local stability analysis based on variational principle is used. The Jacobian of the system (1) at point E is written as

$$A(E) = \begin{pmatrix} -\alpha & \rho_3 & \rho_4 & 0 \\ \rho_1 & -\beta & \rho_6 & 0 \\ \rho_2 & \rho_5 & -\gamma & 0 \\ 0 & \rho_7 & \rho_8 & -\mu_4 \end{pmatrix}.$$

Then the characteristic polynomial is given by

$$\begin{aligned}
P(\lambda) &= \det(A - \lambda I) \\
&= \begin{vmatrix} -\alpha-\lambda & \rho_3 & \rho_4 & 0 \\ \rho_1 & -\beta-\lambda & \rho_6 & 0 \\ \rho_2 & \rho_5 & -\gamma-\lambda & 0 \\ 0 & \rho_7 & \rho_8 & -\mu_4-\lambda \end{vmatrix} \\
&= -(\mu_4+\lambda)\left[(-\alpha-\lambda)((-\beta-\lambda)(-\gamma-\lambda) - \rho_5\rho_6) - \rho_1(\rho_3(-\gamma-\lambda) - \rho_4\rho_5)\right. \\
&\quad \left. +\rho_2(\rho_3\rho_6 + \rho_4(\beta-\lambda))\right] \\
&= (\mu_4+\lambda)[(\alpha+\lambda)(\beta+\lambda)(\gamma+\lambda) - \rho_5\rho_6(\alpha+\lambda) - \rho_1\rho_3(\gamma+\lambda) - \rho_1\rho_4\rho_5 \\
&\quad -\rho_2\rho_3\rho_6 - \rho_2\rho_4(\beta-\lambda)] \\
&= (\mu_4+\lambda)[\lambda^3 + (\alpha+\beta+\gamma)\lambda^2 + (\alpha\beta+\alpha\gamma+\beta\gamma)\lambda + \alpha\beta\gamma - \alpha\rho_5\rho_6 - \rho_5\rho_6\lambda
\end{aligned}$$

$$
\begin{aligned}
&-\gamma\rho_1\rho_3 - \rho_1\rho_3\lambda) - \rho_1\rho_4\rho_5 - \rho_2\rho_3\rho_6 - \beta\rho_2\rho_4 - \rho_2\rho_4\lambda] \\
&= (\mu_4 + \lambda)[\lambda^3 + (\alpha + \beta + \gamma)\lambda^2 + (\alpha\beta + \alpha\gamma + \beta\gamma - \rho_5\rho_6 - \rho_1\rho_3 - \rho_2\rho_4)\lambda \\
&+\alpha\beta\gamma - \alpha\rho_5\rho_6 - \gamma\rho_1\rho_3 - \rho_1\rho_4\rho_5 - \rho_2\rho_3\rho_6 - \beta\rho_2\rho_4] \\
&= (\mu_4 + \lambda)[\lambda^3 + A\lambda^2 + B\lambda + C]
\end{aligned}
$$

where

$$
\begin{aligned}
A &= \alpha + \beta + \gamma > 0 \\
B &= \alpha(\beta + \gamma) + \beta\gamma + \rho_2\rho_4 - \rho_1\rho_3 - \rho_5\rho_6 \\
C &= \alpha\beta\gamma - \alpha\rho_5\rho_6 - \rho_1\rho_3\gamma - \rho_2\rho_4\beta - \rho_1\rho_4\rho_5 - \rho_2\rho_3\rho_6
\end{aligned}
$$

$$
\begin{aligned}
B &= \alpha(\beta + \gamma) + \beta\gamma + \rho_2\rho_4 - \rho_1\rho_3 - \rho_5\rho_6 \\
&= \alpha\beta - \rho_1\rho_3 + \alpha\gamma - \rho_2\rho_4 + \beta\gamma - \rho_5\rho_6
\end{aligned}
$$

Or $\alpha = \rho_1 + \rho_2 + \mu_1 \geq \rho_1$, $\beta = \rho_3 + \rho_5 + \rho_7 + \mu_2 \geq \rho_3$, $\beta = \rho_3 + \rho_5 + \rho_7 + \mu_2 \geq \rho_5$ and $\gamma = \rho_4 + \rho_6 + \rho_8 + \mu_3 \geq \rho_4$ so $\alpha\beta - \rho_1\rho_3 \geq 0$ and $\beta\gamma - \rho_5\rho_6 \geq 0$ then $B > 0$

$$
\begin{aligned}
C &= \alpha\beta\gamma - \alpha\rho_5\rho_6 - \rho_1\rho_3\gamma - \rho_2\rho_4\beta - \rho_1\rho_4\rho_5 - \rho_2\rho_3\rho_6 \\
&= (\rho_1 + \rho_2 + \mu_1)(\rho_3 + \rho_5 + \rho_7 + \mu_2)(\rho_4 + \rho_6 + \rho_8 + \mu_3) - \rho_5\rho_6(\rho_1 + \rho_2 + \mu_1) \\
&\quad -\rho_1\rho_3(\rho_4 + \rho_6 + \rho_8 + \mu_3) - \rho_1\rho_4\rho_5 - \rho_2\rho_3\rho_6 - \rho_2\rho_4(\rho_3 + \rho_5 + \rho_7 + \mu_2) \\
&= \rho_1\rho_4\rho_7 + \rho_1\rho_4\mu_2 + \rho_3\rho_4\mu_1 + \rho_4\rho_5\mu_1 + \rho_4\rho_7\mu_1 + \rho_4\mu_1\mu_2 + \rho_1\rho_6\rho_7 + \rho_1\rho_6\mu_2 \\
&\quad +\rho_2\rho_6\rho_7 + \rho_2\rho_6\mu_2 + \rho_3\rho_6\mu_1 + \rho_6\rho_7\mu_1 + \rho_6\mu_1\mu_2 + \rho_1\rho_5\rho_8 + \rho_1\rho_7\rho_8 \\
&\quad +\rho_1\rho_8\mu_2 + \rho_2\rho_5\rho_8 + \rho_2\rho_3\rho_8 + \rho_2\rho_7\rho_8 + \rho_2\rho_8\mu_2 + \rho_5\rho_8\mu_1 + \rho_7\rho_8\mu_1 \\
&\quad +\rho_8\mu_1\mu_2 + \rho_1\rho_5\mu_3 + \rho_1\rho_7\mu_3 + \rho_1\mu_2\mu_3 + \rho_2\rho_3\mu_3 + \rho_2\rho_5\mu_3 \\
&\quad +\rho_2\rho_7\mu_3 + \rho_2\mu_2\mu_3 + \rho_3\mu_1\mu_3 + \rho_5\mu_1\mu_3 + \rho_7\mu_1\mu_3 + \mu_1\mu_2\mu_3 \\
&> 0.
\end{aligned}
$$

So $A > 0$, $B > 0$ and $C > 0$.

On the other hand

$$
\begin{aligned}
AB - C &= (\alpha + \beta + \gamma)[\alpha(\beta + \gamma) + \beta\gamma - \rho_2\rho_4 - \rho_1\rho_3 - \rho_5\rho_6] - \alpha\beta\gamma - \alpha\rho_5\rho_6 \\
&\quad -\rho_1\rho_3\gamma - \rho_2\rho_4\beta - \rho_1\rho_4\rho_5 - \rho_2\rho_3\rho_6 \\
&= \alpha^2(\beta + \gamma) + \alpha\beta\gamma - \alpha\rho_2\rho_4 - \alpha\rho_1\rho_3 - \alpha\rho_5\rho_6 + \alpha\beta^2 + \alpha\beta\gamma + \beta^2\gamma \\
&\quad -\beta\rho_2\rho_4 - \beta\rho_1\rho_3 - \beta\rho_5\rho_6 + \gamma\alpha\beta + \alpha\gamma^2 + \beta\gamma^2 - \gamma\rho_2\rho_4 - \gamma\rho_1\rho_3 \\
&\quad -\gamma\rho_5\rho_6 - \alpha\beta\gamma + \alpha\rho_5\rho_6 + \rho_1\rho_3\gamma + \rho_2\rho_4\beta + \rho_1\rho_4\rho_5 + \rho_2\rho_3\rho_6 \\
&= 2\alpha\beta\gamma + \alpha(\alpha\gamma\beta + \alpha\gamma - \rho_2\rho_4 - \rho_1\rho_3) + \beta(\alpha\beta + \beta\gamma - \rho_1\rho_3 - \rho_5\rho_6) \\
&\qquad +\gamma(\alpha\gamma + \beta\gamma - \rho_2\rho_4 - \rho_5\rho_6) + \rho_1\rho_4\rho_5 + \rho_2\rho_3\rho_6 \\
&= 2\alpha\beta\gamma + \rho_1\rho_4\rho_5 + \rho_2\rho_3\rho_6 + (\alpha + \beta)(\alpha\beta - \rho_1\rho_3) + (\alpha + \gamma)(\alpha\gamma - \rho_2\rho_4) \\
&\quad +(\beta + \gamma)(\beta\gamma - \rho_5\rho_6)
\end{aligned}
$$

Or $\alpha = \rho_1 + \rho_2 + \mu_1 \geq \rho_1, \quad \alpha = \rho_1 + \rho_2 + \mu_1 \geq \rho_2,$
$\beta = \rho_3 + \rho_5 + \mu_2 \geq \rho_3, \quad \beta = \rho_3 + \rho_5 + \mu_2 \geq \rho_5,$
$\gamma = \rho_4 + \rho_6 + \rho_8 + \mu_3 \geq \rho_4$ and $\gamma = \rho_4 + \rho_6 + \rho_8 + \mu_3 \geq \rho_6$.
So $\alpha\beta - \rho_1\rho_3 \geq 0, \quad \alpha\gamma - \rho_2\rho_4 \geq 0$ and $\beta\gamma - \rho_5\rho_6 \geq 0$.

Thus $AB - C > 0$.

We find $A > 0, \ B > 0, \ C > 0$ and $AB \geq C$. Using Routh–Hurwitz stability criterion, it can be concluded that E is locally asymptotically stable.

3 Simulation

3.1 Difficulties in Estimating the Values of Parameters Used in the Model

As indicated in Fig. 2, different parameters are used in the mathematical model to translate the flow in and out of each compartment. Significance, value and source of each parameter are indicated in Table 1. However, it should be stressed that the values of these parameters are used for the purpose of simulation while real values may depend on age, region, country, ethnicity and the definition of the disease. For instance, there are different definitions of MetS in the literature, including those given by WHO 1999, EGIR 1999, NCEP 2005 and IDF 2006, and consequently the estimate of the prevalence, incidence or death may vary significantly depending on the definition adopted as indicated by Boutayeb [6]. In 2021, Belete et al. conducted a meta-analysis on the global prevalence of MetS among patients with type 1 diabetes. They selected and analysed 27 studies carried out in 17 different countries with a total number of 45811 participants. Their analysis revealed a pooled prevalence of metabolic syndrome of 23.7% with a significant heterogeneity between countries, showing a prevalence of MetS in men varying from a minimum value of 3.7% in Italy and a maximum value of 80.7% in Finland (Table 2) [4].

3.2 Values of the Parameters Used for Simulation

In order to illustrate graphically the results yielded by our model, we carried out simulations with different values of the parameters indicated in Table 1.

3.3 Effect of Action on Parameters ρ_1 and ρ_5

We carried out simulations using different values of the parameters ρ_1 and ρ_5 separately and simultaneously.

Table 2 Prevalence of metabolism syndrome among patients with type 1 diabetes

Country	Sample size	Mean age (in years)	Diagnostic criteria	MetS prevalence in men (%)
Italy	412	17.3	IDF	3.7
Poland	163	14.8	IDF	6.6
China	754	16	IDF	7.5
Egypt	160	14.8	IDF	7.8
Brazil	261	34.8	NCEP	10.9
India	451	15.4	JIS	22.8
Australia	2120	44.3	WHO	31.8
Mexico	140	28	IS	44.0
Finland	791	45	JIS	80.7

Source Extracted from Belete et al. [4]

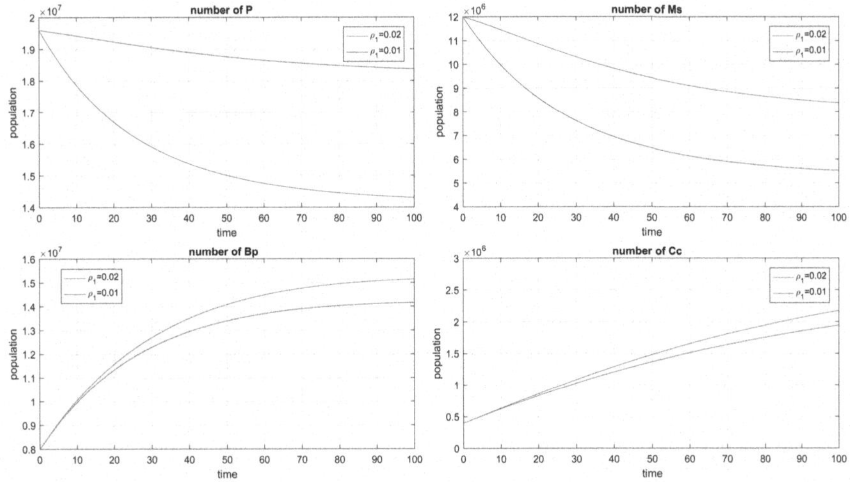

Fig. 3 Varying the parameter ρ_1

As indicated in Fig. 3, decreasing the value of the parameter ρ_1 from 0.02 to 0.01 affects the four compartments by decreasing simultaneously the number of men suffering from metabolism syndrome (Ms), benign prostatic hyperplasia (Pb) and cardiovascular diseases/cancer (Cc). Consequently, the population (P) of men free from these diseases/conditions increases significantly with time.

In order to illustrate the correlation between MetS and BPH, we varied the parameter ρ_5 which indicates the flow from the compartment Ms to the compartment Bp. Figure 4 shows that halving the value of the parameter ρ_5 from 0.018 to 0.009 engenders a lower number of men suffering from BPH and also those affected by CVDs/PCa.

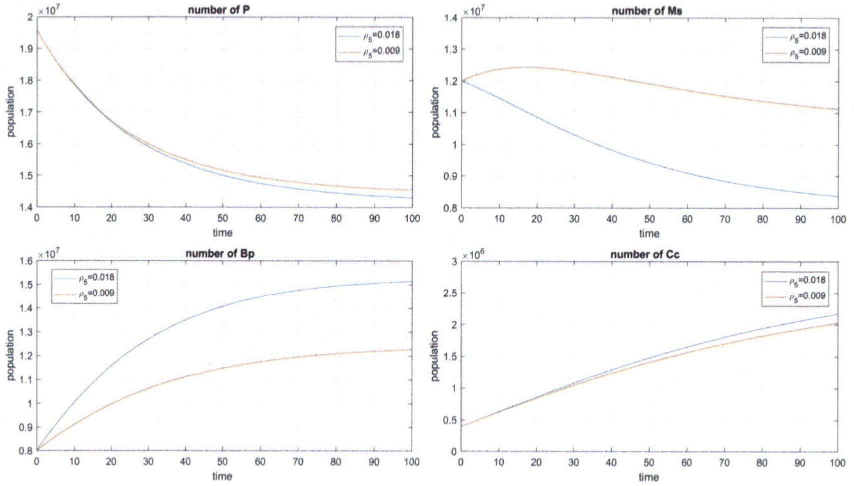

Fig. 4 Varying the parameter ρ_5

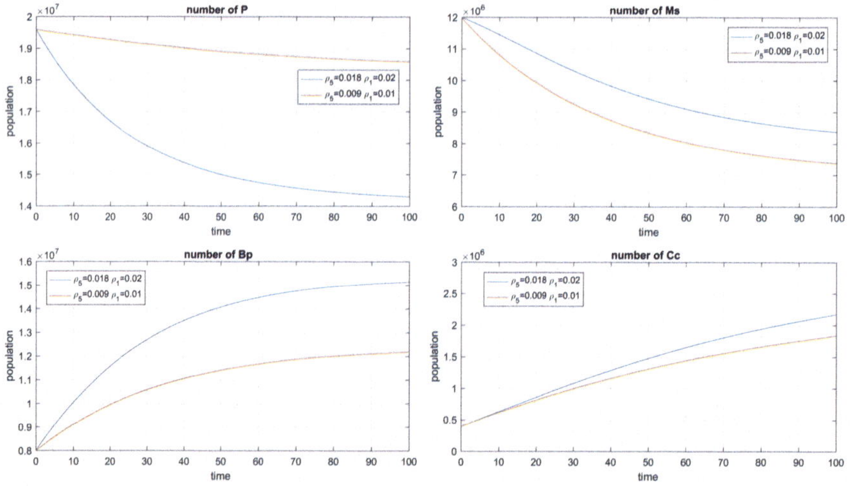

Fig. 5 Varying simultaneously the parameters ρ_1 and ρ_5

Finally, reducing simultaneously the values of ρ_1 and ρ_5 reduces further the numbers of men suffering from benign prostatic hyperplasia and especially the burden of CVDs/PCa (Fig. 5).

4 Discussion

Following the results of a large number of studies which revealed that people with MetS are more affected by benign prostatic hyperplasia than their counterparts with no MetS, our mathematical model is pioneer in illustrating the relation between MetS and BPH and indicating how the large number of men affected by a BPH can be controlled by reducing the number of men with metabolism syndrome. As indicated in the previous section, the results show that, with time, the number of men with BPH can be reduced through the flow from the compartment Ms to Bp by acting on the parameter ρ_5. Indeed, decreasing this parameter from 0.018 to 0.009 engenders a decrease in the number of men suffering from benign prostatic hyperplasia as well as the number of men affected by CVDs and PCa. Furthermore, reducing simultaneously values of the parameters ρ_1 and ρ_5 will improve significantly men's health by controlling the number of men affected by MetS, BPH and CVDs/PCa. In terms of pragmatic action, these results show that decision-makers can significantly reduce the mixed burden of MetS, BPH, CVDs/PCa by adopting efficient and regular strategies based on the promotion of healthy eating, physical activity and weight control.

In 2022, Xin et al. carried out a meta-analysis and analysed 18 selected articles comprising 1685 cases and 4653 controls. They found that prostate volume of BPH patients with diabetes was significantly higher than that of BPH patients without diabetes [64].

In 2024, Boutayeb edited a book dealing with different aspects and risk factors of diabetes. He stressed that "During the last decades clinical and epidemiological studies brought out evidence on the correlation between benign prostatic hyperplasia and diabetes homeostasis and more generally with metabolic syndrome or separately with its components (hypertension, obesity, hyperglycaemia, insulin resistance...)". He devoted Chap. 4 to the association of BPH with diabetes, hypertension, obesity and MetS and its components in general [6].

In the introduction of the present chapter, the increasing trends of both MetS and BPH, accompanied with a new evidence on the correlation between the two diseases, were analysed by a large number of authors who, generally, stressed the need for decision-makers to act through adapted strategies to limit the mixed burden of MetS and BPH.

The mechanism connecting BPH and metabolism syndrome is still unclear. It may be explained by increased oxidative stress in accumulated fat in obese patients, causing DNA damage in the prostate and promoting epithelial proliferation, stromal thickening and fibrosis, which interferes with the pathophysiology of BPH [57].

5 Conclusion

Although the mathematical model proposed in this chapter is based on the usual simple idea of compartmental models for non-communicable diseases, it illustrates a MetS-BHP relation which has not been given the attention it needs. Moreover, this model can be replicated and used in different countries and/or with different databases based on particular estimation of the parameters.

6 Limitation and Perspectives

The fact that all parameters used in our model are age dependent suggests the use of a model with age structure. It will also be interesting to introduce sensitivity analysis and optimal control to improve the results yielded by the present model.

References

1. Abd Allah, M.A., Yasien, H.A., Elgharbawy, M.S., AbdElaal, H.M., et al.: Correlation between metabolic syndrome and benign prostatic hyperplasia. Menoufia Medical Journal **34**(1), 43–46 (2021)
2. Akin, Y., Gulmez, H., Ates, E., Gulum, M., Savas, M.: Preliminary assessment of neck circumference in benign prostatic hyperplasia in patients with metabolic syndrome. Int Braz J Urol **43**(1), 95–103 (2017). https://doi.org/10.1590/S1677-5538.IBJU.2016.0139
3. AlFartosy, A.J.M., Ati, M.H.: The relationship of insulin resistance with some proinflammtory markers in prostate cancer and benign prostate hyperplasia patients. Annals of the Romanian Society for Cell Biology **25**(6), 14186–14194 (2021)
4. Belete, R., Ataro, Z., Abdu, A., Sheleme, M.: Global prevalence of metabolic syndrome among patients with type i diabetes mellitus: a systematic review and meta-analysis. Diabetology & Metabolic Syndrome **13**, 25 (2021). https://doi.org/10.1186/s13098-021-00641-8
5. Berry, S.J., Coffey, D.S., Walsh, P.C., Ewing, L.L.: The development of human benign prostatic hyperplasia with age. The Journal of urology **132**(3), 474–479 (1984)
6. Boutayeb, A.: Diabetes: From Prevention to Management; How to Deal with Its Different Aspects and Risk Factors, chap. 4, pp. 33–47. BOHR Publishers (2024https://doi.org/10.54646/DFPMDDARF.2024.02. http://doi.org/10.54646/DFPMDDARF.2024.02
7. Boutayeb, A., Chetouani, A.: A population model of diabetes and pre-diabetes. International Journal of Computer Mathematics **84**(1), 57–66 (2007)
8. Boutayeb, A., Chetouani, A., Achouyab, A., Twizell, E.: A non-linear population model of diabetes mellitus. Journal of Applied Mathematics and Computing **21**(1–2), 127–139 (2006)
9. Boutayeb, A., Twizell, E., Achouayb, K., Chetouani, A.: A mathematical model for the burden of diabetes and its complications. Biomedical Engineering Online **3**(1), 20 (2004)
10. Breyer, B.N., Sarma, A.V.: Hyperglycemia and insulin resistance and the risk of bph/luts: an update of recent literature. Current urology reports **15**(12), 1–6 (2014)
11. Byun, H., Sung, Y., Kim, W., Jung, J., Song, J., Chung, H.: Relationships between prostate-specific antigen, prostate volume, and components of metabolic syndrome in healthy korean men. Korean Journal of Urology **53**(11), 774–778 (2012)
12. Cannarella, R., Condorelli, R.A., Barbagallo, F., La Vignera, S., Calogero, A.E.: Endocrinology of the aging prostate: Current concepts. Frontiers in Endocrinology p. 19 (2021)
13. Castelli, T., Russo, G.I., Reale, G., Privitera, S., Chisari, M., Fragal'a, E., Favilla, V., Cimino, S., Morgia, G.: Metabolic syndrome and prostatic disease: potentially role of polyphenols in preventive strategies. a review. International braz j urol **42**, 422–430 (2016)
14. Cerna-Jamanca, J., Gamboa-Vicente, W.: Asociación entre síndrome metabólico y niveles de antígeno prostático específico en varones que acudieron a control prostático al servicio de urología del hospital belén de trujillo, perú. Rev Mex Urol **81**(2), 1–12 (2021)

15. Chen, Z., Yan, Y., Chen, C., et al.: The single nucleotide polymorphism sr700518 is an independent risk factor for mets and bph. Andrology **6**(4), 568–578 (2018)
16. Chetta, P., Zadra, G.: Metabolic reprogramming as an emerging mechanism of resistance to endocrine therapies in prostate cancer. Cancer Drug Resistance **4**(1), 143–162 (2021)
17. Chhetri, R., Baral, S., Thapa, N.: Metabolic syndrome and benign prostatic hyperplasia: A nepalese perspective. Journal of Lumbini Medical College **6**(2), 4 (2018) https://doi.org/10.22502/jlmc.v6i2.207
18. Choi, W., Heo, N., Paick, J., et al.: Prostate-specific antigen lowering effect of metabolic syndrome is influenced by prostate volume. Int J Urol **23**, 299–304 (2016). https://doi.org/10.1111/iju.13042
19. Corona, G., Vignozzi, L., Rastrelli, G., Lotti, F., Cipriani, S., Maggi, M.: Benign prostatic hyperplasia: a new metabolic disease of the aging male and its correlation with sexual dysfunctions. International journal of endocrinology **2014** (2014)
20. Daher, M., Saqer, T., Jabr, M., Al-Mousa, S.: Benign prostatic hyperplasia and metabolic syndrome; prevalence and association: a cross-sectional study in syria. BMC Urology **23**, 187 (2023). https://doi.org/10.1186/s12894-023-01365-9
21. De Nunzio, C., Aronson, W., Freedland, S., Giovannucci, E., Parsons, J.: The correlation between metabolic syndrome and prostatic diseases. Eur Urol **61**, 560–570 (2012)
22. De Nunzio, C., Lombardo, R., Gacci, M., et al.: Metabolic syndrome does not increase the risk of ejaculatory dysfunction in patients with lower urinary tract symptoms and benign prostatic enlargement: an italian single center cohort study. Urology **105**, 85–90 (2017)
23. Dobrowolski, P., Prejbisz, A., Kury lowicz, A., Baska, A., Burchardt, P., Chlebus, K., et al.: Metabolic syndrom-a new definition and management guidelines. a joint position paper by the polish society of hypertension, polish society for the treatment of obesity, polish lipid association, polish association for study of liver, polish society of family medicine, polish society of lifestyle medicine, division of prevention and epidemiology polish cardiac society, "club 30" polish cardiac society, and division of metabolic and bariatric surgery society of polish surgeons. Archives of Medical Science **18**(5), 1133–1156 (2022) https://doi.org/10.5114/aoms/152921
24. Fu, Y., Zhou, Z., Yang, B., et al.: The relationship between the clinical progression of benign prostatic hyperplasia and metabolic syndrome: a prospective study. Urol Int **97**(3), 330–335 (2016)
25. Gacci, M., Corona, G., Vignozzi, L., Salvi, M., Serni, S., De Nunzio, C., Tubaro, A., Oelke, M., Carini, M., Maggi, M.: Metabolic syndrome and benign prostatic enlargement: a systematic review and meta-analysis. BJU international **115**(1), 24–31 (2015)
26. Gacci, M., Sebastianelli, A., Salvi, M., et al.: Benign prostatic enlargement can be influenced by metabolic profile: results of a multicenter prospective study. BMC Urology **17**(22) (2017) https://doi.org/10.1186/s12894-017-0211-9
27. Gacci, M., Vignozzi, L., Sebastianelli, L., et al.: Metabolic syndrome and lower urinary tract symptoms: the role of inflammation. Prostate Cancer and Prostatic Diseases **16**(1), 101–106 (2013)
28. Hammarsten, J., Damber, J.E., Karlsson, M., Knutson, T., Ljunggren, Ö., Ohlsson, C., Peeker, R., Smith, U., Mellström, D.: Insulin and free oestradiol are independent risk factors for benign prostatic hyperplasia. Prostate cancer and prostatic diseases **12**(2), 160–165 (2009)
29. Hammarsten, J., Hogstedt, B., Holthunis, N., Mellstrom, D.: Components of the metabolic syndrome-risk factors for the development of benign prostatic hyperplasia. Cancer Prostatic Dis **1**(3), 157–162 (1998). https://doi.org/10.1038/sj.pcan.4500221
30. Heo, J., Kim, D., Yoo, J., Lee, K.: Metabolic syndrome-related factors as possible targets for lower urinary tract symptoms in korean males. The Aging Male **26**(1), 2166920 (2023). https://doi.org/10.1080/13685538.2023.2166920
31. Huang, P.: A comprehensive definition for metabolic syndrome. Dis. Model. Mech **2**(56), 231–237 (2009). https://doi.org/10.1242/dmm.001180
32. Huggins, C., Hodges, C.V.: Studies on prostatic cancer. i. the effect of castration, of estrogen and of androgen injection on serum phosphatases in metastatic carcinoma of the prostate. 1941. J. Urol. **167**(2 Pt 2), 948–51; discussion 952 (2002)

33. International Diabetes Federation: Idf diabetes atlas (2021). https://diabetesatlas.org. Accessed: 2024-06-20
34. Jeong, J., Kim, E., Kim, D.: Association of metabolic syndrome and benign prostate enlargement in young korean males. Korean Journal of Urology **52**(11), 757–762 (2011)
35. Kang, J., Kim, Y., Jeong, J.: Association between metabolic syndrome and lower urinary tract symptoms, prostate specific antigen, and prostate volume: Single center study. Journal of Radiological Science and Technology **41**(4), 313–319 (2018) https://doi.org/10.17946/JRST.2018.41.4.313
36. Kim, J., Doo, S., Yun, J., et al.: Lower likelihood of having moderate-to-severe lower urinary tract symptoms in middle-aged healthy korean men with metabolic syndrome. Urology **84**, 665–669 (2014)
37. Kim, W.T., Yun, S.J., Choi, Y.D., Kim, G.Y., Moon, S.K., Choi, Y.H., Kim, I.Y., Kim, W.J.: Prostate size correlates with fasting blood glucose in non-diabetic benign prostatic hyperplasia patients with normal testosterone levels. Journal of Korean medical science **26**(9), 1214–1218 (2011)
38. Kopp, W.: Diet-induced hyperinsulinemia as a key factor in the etiology of both benign prostatic hyperplasia and essential hypertension? Nutrition and Metabolic Insights **11**, 1178638818773072 (2018)
39. Kristal, A.R., Arnold, K.B., Schenk, J.M., Neuhouser, M.L., Goodman, P., Penson, D.F., Thompson, I.M.: Dietary patterns, supplement use, and the risk of symptomatic benign prostatic hyperplasia: results from the prostate cancer prevention trial. American journal of epidemiology **167**(8), 925–934 (2008)
40. Lepor, H.: Medical treatment of benign prostatic hyperplasia. Reviews in urology **13**(1), 20–33 (2011)
41. Li, J., Peng, L., Cao, D., Gou, H., Li, Y., Wei, Q.: The association between metabolic syndrome and benign prostatic hyperplasia: a systematic review and meta-analysis. The Aging Male **23**(5), 1388–1399 (2020)
42. Lotti, F., Corona, G., Vignozzi, L.: Metabolic syndrome and prostate abnormalities in male subjects of infertile couples. Asian Journal of Andrology **16**, 295–304 (2014). https://doi.org/10.4103/1008-682X.122341
43. Mubenga, L.E., Chimanuka, D., De Groote, P., Bwenge, E., Hermans, M., Tombal, B.: Comparison of prostate size and anthropometric parameters between diabetic and non-diabetic congolese patients who underwent transurethral prostate resection in the democratic republic of congo. African Journal of Urology **25**(1), 1–7 (2019)
44. Ngai, H.Y., Yuen, K.K.S., Ng, C.M., Cheng, C.H., Chu, S.K.P.: Metabolic syndrome and benign prostatic hyperplasia: an update. Asian journal of urology **4**(3), 164–173 (2017)
45. Omran, A., Leca, B.M., Oštarijaš, E., Graham, N., Da Silva, A.S., Za"ır, Z.M., Miras, A.D., le Roux, C.W., Vincent, R.P., Cardozo, L., et al.: Metabolic syndrome is associated with prostate enlargement: a systematic review, meta-analysis, and meta-regression on patients with lower urinary tract symptom factors. Therapeutic Advances in Endocrinology and Metabolism **12**, 20420188211066210 (2021)
46. Ozden, C., Ozdal, O., Urgancioglu, Koyuncu, G., Gokkaya, S., Memis, A.: The correlation between metabolic syndrome and prostatic growth in patients with benign prostatic hyperplasia. European Urology **51**(1), 199–206 (2007)
47. Parikesit, D., Witjaksono, F., Mochtar, C., Rasyid, N., Hamid, A.: Impact of metabolic syndrome on lower urinary tract symptoms in patients with benign prostate hyperplasia. Med J Indones **30**, 138–142 (2021) https://doi.org/10.13181/mji.oa.214256
48. Park, H., Lee, H., Lee, K., et al.: Relationship between lower urinary tract symptoms and metabolic syndrome in a community-based elderly population. Urology **72**(3), 556–560 (2008)
49. Park, S., Yeo, J., Cho, D., et al.: Impact of metabolic status on the association of serum vitamin d with hypogonadism and lower urinary tract symptoms/benign prostatic hyperplasia. Aging Male **21**, 55–59 (2018)
50. Park, Y., Min, S., Lee, J.: Relationship between lower urinary tract symptoms/benign prostatic hyperplasia and metabolic syndrome in korean men. World J Mens Health **30**, 183–188 (2012)

51. Rizzuti, A., Stocker, G., Santos, H.O.: Exploring the role of testosterone replacement therapy in benign prostatic hyperplasia and prostate cancer: A review of safety. Uro **2**(1), 30–39 (2022)
52. Shah, A., Shah, A.A., K, N., Lobo, R.: Mechanistic targets for bph and prostate cancer-a review. Reviews on Environmental Health **36**(2), 261–270 (2021) https://doi.org/10.1515/reveh-2020-0051
53. Shinohara, K.: Thermal ablation of prostate diseases: advantages and limitations. International journal of hyperthermia **20**(7), 679–697 (2004)
54. Son, H., Suh, B., Kim, S., Myong, J., Shon, Y., Kim, H., Jeong, H.: Association between insulin resistance and benign prostatic hyperplasia in healthy middle-aged korean men. Korean Journal of Health Promotion **15**(4), 202–208 (2015)
55. Sung, H., Ferlay, J., Siegel, R.L., Laversanne, M., Soerjomataram, I., Jemal, A., Bray, F.: Global cancer statistics 2020: Globocan estimates of incidence and mortality worldwide for 36 cancers in 185 countries. CA: A Cancer Journal for Clinicians **71**(3), 209–249 (2021). https://doi.org/10.3322/caac.21660
56. Vikram, A., Jena, G., Ramarao, P.: Insulin-resistance and benign prostatic hyperplasia: the connection. European journal of pharmacology **641**(2–3), 75–81 (2010)
57. Vital, P., Castro, P., Ittmann, M.: Oxidative stress promotes benign prostatic hyperplasia. Prostate **76**(1), 58–67 (2016)
58. Wilczynski, C., Lily Agrawal, L.: Testosterone effects on the prostate gland: Review of pathophysiology and considerations in prostate cancer. J Fam Med Dis Prev **1**(004) (2015)
59. World Bank: Population by gender and age. https://genderdata.worldbank.org/en/topics/population. Accessed: 2024-06-20
60. World Health Organisation: Non communicable diseases. https://www.who.int/news-room/fact-sheets/detail/noncommunicable-diseases. Accessed: 2024-06-20
61. Wu, S., He, H., Wang, Y., Xu, R., Zhu, B., Zhao, X.: Association between benign prostate hyperplasia and metabolic syndrome in men under 60 years old: a meta-analysis. Journal of International Medical Research **47**(11), 5389–5399 (2019)
62. Xia, B., Zhao, S., Chen, Z., et al.: The association of pathogenic factors of metabolic syndrome on serum prostate- specific antigen levels: a pilot study. BMC Urology **19**(119) (2019) https://doi.org/10.1186/s12894-019-0549-2
63. Xia, B.W., Zhao, S.C., Chen, Z.P., Chen, C., Liu, T.S., Yang, F., Yan, Y.: Relationship between serum total testosterone and prostate volume in aging men. Scientific Reports **11**(1), 1–7 (2021)
64. Xin, C., Fan, H., Xie, J., Hu, J., Sun, X., Liu, Q.: Impact of diabetes mellitus on lower urinary tract symptoms in benign prostatic hyperplasia patients: A meta-analysis. Frontiers in Endocrinology **12**, 741748 (2022). https://doi.org/10.3389/fendo.2021.741748
65. Yang, T., Hsieh, J., Chen, S., et al.: Metabolic syndrome associated with reduced lower urinary tract symptoms in middle-aged men receiving health checkup. Urology **80**, 1093–1097 (2012)
66. Yim, S., Cho, Y., Joo, K.: Relationship between metabolic syndrome and prostate volume in korean men under 50 years of age. Korean Journal of Urology **52**(6), 390–395 (2011). https://doi.org/10.4111/kju.2011.52.6.390
67. Zamuner, M., Laranja Walker, W., Alonso, J.C.C., et al.: Is metabolic syndrome truly a risk factor for male lower urinary tract symptoms or just an epiphenomenon? Advances in Urology **2014**, 1–5 (2014) https://doi.org/10.1155/2014/203854
68. Zhang, X., Zeng, X., Liu, Y., et al.: Impact of metabolic syndrome on benign prostatic hyperplasia in elderly chinese men. Urol Int **93**, 214–219 (2014)
69. Zhao, S., Tang, J., Shao, S., et al.: The relationship between benign prostatic hyperplasia/lower urinary tract symptoms and men platelet volume: the role of metabolic syndrome. Urol Int **66**(4), 449–458 (2016)
70. Zhao, S., Xia, M., Tang, J., et al.: Associations between metabolic syndrome and clinical benign prostatic hyperplasia in a northern urban han chinese population: a prospective cohort study. Sci Rep **6**, 33933 (2016)
71. Zhou, B., Wang, P., Xu, W., Li, Y., Tong, D., Jiang, J., Sun, Z.: Correlations of glucose metabolism, insulin resistance and inflammatory factors with symptom score of patients with benign prostatic hyperplasia. European Review for Medical and Pharmacological Sciences **22**(16), 5077–5081 (2018)

72. Zorba, O., Uzun, H., Akca, G., et al.: The effect of different metabolic syndrome: definitions on the relationship between metabolic syndrome and luts in men with bhp enlargement. Am J Mens Health **11**(1), 158–163 (2017)
73. Zou, C., Gong, D., Fang, N., Fan, Y.: Meta-analysis of metabolic syndrome and benign prostatic hyperplasia in chinese patients. World journal of urology **34**(2), 281–289 (2016)

A Compartmental Model for the Dynamics of a Population of Women with Gestational Diabetes

Wiam Boutayeb, Mohamed E. N. Lamlili, Abdesslam Boutayeb, and Fatima Zahra Benassila

Abstract Gestational diabetes is one of the three principal types of diabetes mellitus (type 1, type 2, gestational diabetes). It is due to the pancreatic β-cells' dysfunction or the delayed response of the β-cells to the glycaemic levels, and the marked insulin resistance following a potential hormonal release. It is diagnosed for the first time during pregnancy and may occur anytime during pregnancy. In our study, we are interested in predisposed women to gestational diabetes mellitus (GDM) who are considered as vulnerable subjects to diabetes. For this purpose, we propose a compartmental mathematical model, describing the passage of adult women from the stage pre-diabetes to gestational diabetes and diabetes. The local stability of the model's equilibrium point is carried out using the Routh–Hurwitz stability criterion while sensitivity analysis is based on the normal forward sensitivity index. A numerical simulation is carried out using different values of parameters. Our discussion indicates how decision-makers should pragmatically act to reduce both gestational diabetes and type 2 diabetes.

1 Introduction

First of all, it is necessary to distinguish between diabetes during pregnancy and gestational diabetes which occurs temporarily during pregnancy and contributes to maternal morbidity and mortality. For this purpose, the World Health Organization stressed that "*the diagnostic criteria for hyperglycaemia in pregnancy recommended by the World Health Organization (WHO) in 1999 were not evidence-based and needed to be updated in the light of previously unavailable data*". Consequently, the

W. Boutayeb · M. E. N. Lamlili
Mohammed First University, Oujda, Morocco

A. Boutayeb (✉)
Emirates Aviation University, Dubaï, UAE
e-mail: x.boutayeb@gmail.com

F. Z. Benassila
Endocrinologist, Endocrinology Unit, Al Farabi Hospital, Oujda, Morocco
e-mail: fzahra.ben@hotmail.fr

A. Boutayeb and M. E. N. Lamlili (eds.), *Computational Mathematics and Modelling for Diabetes*, Industrial and Applied Mathematics,
https://doi.org/10.1007/978-981-96-1925-2_8

Table 1 WHO criteria for diagnosis of diabetes in pregnancy and gestational diabetes

Diabetes in pregnancy should be diagnosed by the 2006 WHO criteria for diabetes if one or more of the following criteria are met	Gestational diabetes mellitus should be diagnosed at any time in pregnancy if one or more of the following criteria are met
* Fasting plasma glucose $\geq 7.0 mmol/l$ (126 mg/dl)	$*$ Fasting plasma glucose 5.1–6.9 mmol/l (92–125 mg/dl)
$*$ 2-hour plasma glucose $\geq 11.1 mmol/l$ (200 mg/dl) following a 75 g oral glucose load	$*$ 1-hour plasma glucose $\geq 10.0 mmol/l$ (180 mg/dl) following a 75 g oral glucose load
$*$ Random plasma glucose $\geq$ 11.1 mmol/l (200 mg/dl) in the presence of diabetes symptoms	$*$ 2-hour plasma glucose 8.5–11.0 mmol/l (153–199 mg/dl) following a 75 g oral glucose load

WHO proposed in 2013, different criteria for the diagnosis of diabetes and gestational diabetes as indicated in Table 1 [14].

In other words, gestational diabetes is defined as a glycaemic level higher than that of a non-diabetic woman and less than that of a woman diagnosed with diabetes [9]. However, the precise diagnostic criteria differ according to different health organizations. For instance, while the World Health Organization criteria suggest that gestational diabetes can be diagnosed at any time in pregnancy [12], the American Diabetes Association classify gestational diabetes mellitus as diabetes diagnosed in the second or third trimester of pregnancy that was not clearly overt diabetes prior to gestation [1]. Stressing that "*within and between countries there is disagreement between obstetric, medical and endocrine groups regarding the diagnosis and management of GDM*", Donaldson and McIntyre suggested consensus for the diagnosis of gestational diabetes [6].

As indicated by the International Diabetes Federation, despite its increasing prevalence, gestational diabetes remains a neglected issue, particularly in low- and middle-income countries. In 2019, it was estimated that 16% of live births had some form of hyperglycaemia in pregnancy (20 million women) of which nearly 85% were due to gestational diabetes [7]. The IDF Diabetes Atlas 2021 shows that the prevalence of gestational diabetes among pregnant women aged 20-49 years varies between 13% in Africa and 25.9% in Southeast Asia (see Fig. 1) [8].

In Australia, the women's royal hospital estimates that "*around 10–15% of pregnant people will get gestational diabetes (GDM) between the 24th and the 28th week of pregnancy, sometimes earlier. It usually goes away after the baby is born*" [13]. According to a survey carried out in 2019 on a cohort including 17145 women registered at 15 to 20 gestational weeks in Qingdao, China, the incidence of gestational diabetes was 17.4% [10]. In Canada, it is estimated that incidence of gestational diabetes is between 3 and 20% depending on risk factors [5].

Women who have gestational diabetes are exposed to diverse pregnancy-related complications like foetal anomalies, spontaneous abortion, obstructed labour, large birth weight, neonatal hypoglycaemia, obesity and high blood pressure. They are also at a high risk of developing gestational diabetes again in future pregnancies and type 2 diabetes after giving birth. Consequently, it is crucial for health decision-makers

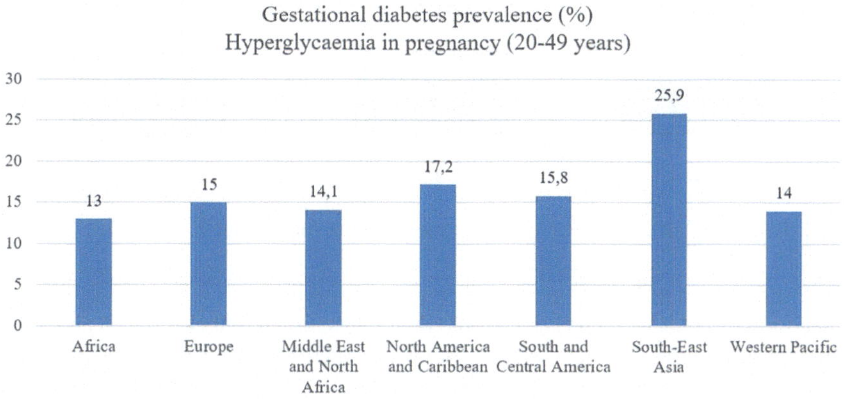

Fig. 1 Gestational diabetes prevalence by IDF regions in 2021. *Source* IDF Diabetes Atlas 2021

to adopt efficient and cost-effective strategies to reduce the incidence of gestational diabetes [3, 11].

As indicated in Chap. 4 of this book, many authors have dealt with the evolution from the stage of pre-diabetes or diabetes without complication to the stage of diabetes with complications. Obviously, the dynamic of populations is not regarded in the usual way of transmitted diseases but rather as a probabilistic evolution from one stage to another. To this end, different mathematical tools are used, including discrete mathematics, ordinary differential equations, partial differential equations, optimal control, numerical approximations and stochastic approaches [2]. To our knowledge, however, no compartmental mathematical model has been devoted to gestational diabetes. Although Mat Daud et al. published a paper entitled "*A mathematical model for the population dynamics of diabetes mellitus during pregnancy*", the authors of this model considered four compartments: (1) non-diabetic non-pregnant women of childbearing age, (2) diabetic non-pregnant women of childbearing age, (3) diabetic women during pregnancy and (4) pregnant women with complication of DM [4]. Consequently, gestational diabetes was not considered separately from diabetes as we indicated previously.

In this paper, we present a compartmental model dealing with gestational diabetes. The main objective is to describe and analyse the population dynamics related to gestational diabetes in order to suggest pragmatic actions that will reduce the incidence of gestational diabetes and avoid the evolution to type 2 diabetes.

2 Methods

Ordinary differential equations are used to express the population dynamics. The compartmental model considered is based on the flow in and out of each compartment

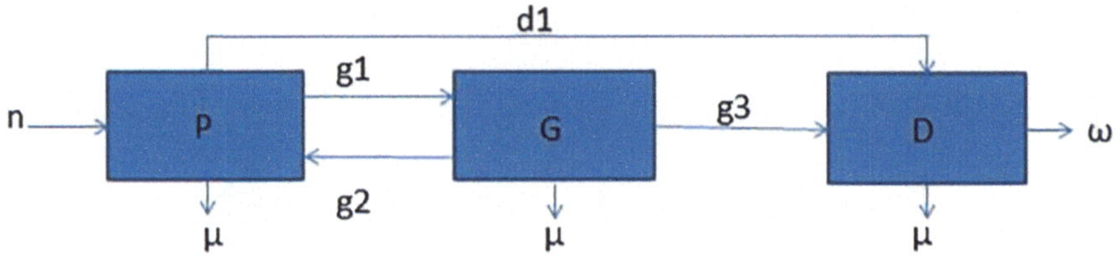

Fig. 2 Flowchart indicating the dynamics between different compartments

population of adult women (P(t)), women with gestational diabetes (G(t)) and adult women having type 2 diabetes (D(t)). The Routh–Hurwitz criterion is used for local stability analysis and the normal forward sensitivity index for sensitivity analysis.

Our compartmental mathematical model study is proposed in order to analyse the passage of adult women from the stages pre-diabetes to gestational diabetes and diabetes.

The proposed compartmental model is shown in the following figure (Fig. 2).

The different parameters used in our mathematical model are defined as follows:

- n: denotes incidence of worldwide women of childbearing age
- g_1: denotes the rate of healthy women developing gestational diabetes,
- g_2: denotes the rate of adult women recovering from gestational diabetes,
- g_3: denotes the rate of women with gestational diabetes becoming diabetics,
- d_1: denotes the rate of healthy women becoming diabetic,
- μ: natural mortality rate,
- ω: mortality rate related to complications.

The system of equations is

$$\begin{cases} \dfrac{dP(t)}{dt} = n - (g_1 + d_1 + \mu)\, P(t) + g_2 G(t) \\ \dfrac{dG(t)}{dt} = g_1 P(t) - (g_2 + g_3 + \mu)\, G(t) \\ \dfrac{dD(t)}{dt} = d_1 P(t) + g_3 G(t) - (\omega + \mu) D(t). \end{cases} \tag{1}$$

3 Results

Theorem 1 *Assuming that initial conditions* $P(0)$, $G(0)$ *and* $D(0)$ *are positive then the solution* $P(t)$, $G(t)$ *and* D(t) *will remain positive for all* $t > 0$.

Proof From the system of Eq. (1), it is easy to see that:

$$\mathrm{P}'(\mathrm{t}) > -\alpha \mathrm{P}(\mathrm{t}) \text{ or } \frac{dP(t)}{P(t)} > -\alpha \mathrm{dt}$$
$$\mathrm{G}'(\mathrm{t}) > -\beta \mathrm{G}(\mathrm{t}) \text{ or } \frac{dG(t)}{G(t)} > -\beta dt$$
$$\mathrm{D}'(\mathrm{t}) > -\theta \mathrm{D}(\mathrm{t}) \text{ or } \frac{dD(t)}{D(t)} > -\theta \mathrm{dt}.$$

From which we deduce after integration that

$$\mathrm{P}(\mathrm{t}) > \mathrm{P}(0)e^{-\alpha \mathrm{t}} > 0$$
$$\mathrm{G}(\mathrm{t}) > \mathrm{G}(0)e^{-\beta \mathrm{t}} > 0$$
$$\mathrm{D}(\mathrm{t}) > \mathrm{D}(0)e^{-\theta \mathrm{t}} > 0.$$

Consequently, the mathematical model is epidemiologically and mathematically well posed in $\mathrm{E} = \left\{(\mathrm{P}, \mathrm{G}, \mathrm{D}) \in \mathbb{R}^{3+}\right\}$. □

Theorem 2 *The system (1) admits the following equilibrium points:*

$$\begin{cases} P^* &= \dfrac{n(\alpha\gamma - g_2)}{\alpha^2\gamma} = \dfrac{n(\alpha\gamma - g_2)}{\alpha(\alpha\beta - g_1 g_2)} \\ G^* &= \dfrac{-n}{\alpha\gamma} = \dfrac{n g_1}{\alpha\beta - g_1 g_2} \\ D^* &= \dfrac{n d_1}{\theta\alpha\gamma}(\gamma - \delta) = \dfrac{n d_1(\gamma - \delta)}{\theta(\alpha\beta - g_1 g_2)} \end{cases} \quad \textit{with:} \quad \begin{cases} \alpha &= g_1 + d_1 + \mu \\ \beta &= g_2 + g_3 + \mu \\ \theta &= \omega + \mu \\ \gamma &= \frac{g_2}{\alpha} - \frac{\beta}{g_1} \\ \delta &= \frac{g_2}{\alpha} + \frac{g_3}{d_1} \end{cases}$$

with the following condition:

$$\alpha\beta \neq g_1 g_2.$$

Theorem 3 *The equilibrium point* $(\mathrm{P}^*, \mathrm{G}^*, \mathrm{D}^*)$ *is locally asymptotically stable.*

Proof The Jacobian matrix is given by

$$\begin{bmatrix} -\alpha & g_2 & 0 \\ g_1 & -\beta & 0 \\ d_1 & g_3 & -\theta \end{bmatrix}$$

$$\Rightarrow Q(\lambda) = \begin{vmatrix} -\alpha - \lambda & g_2 & 0 \\ g_1 & -\beta - \lambda & 0 \\ d_1 & g_3 & -0 - \lambda \end{vmatrix}.$$

From which it follows that the eigenvalues are roots of the following characteristic polynomial in lambda:

$$Q(\lambda) = (-\theta - \lambda)(\lambda^2 + (\alpha + \beta)\lambda + (\alpha\beta - g_1 g_2)).$$

Then applying the Routh–Hurwitz stability criterion, we deduce that the Jacobian matrix evaluated at the equilibrium point (P^*, G^*, D^*) has three negative eigenvalues and hence the equilibrium state is locally asymptotically stable. □

Theorem 4 *The system (1) admits a unique solution:* $\begin{pmatrix} G(t) \\ P(t) \\ D(t) \end{pmatrix}$ *under the condition:*

$$\alpha\beta \neq g_1 g_2.$$

Proof The system can be described by the differential equation:

$$X'(t) = AX(t) + B,$$

where $X(t) = \begin{pmatrix} G(t) \\ P(t) \\ D(t) \end{pmatrix}$, $A = \begin{pmatrix} -\alpha & g2 & 0 \\ gl & -\beta & 0 \\ dl & g3 & -\theta \end{pmatrix}$ and $B = \begin{pmatrix} n \\ 0. \\ 0 \end{pmatrix}$

The matrix A is diagonalizable, and its eigenvalues are

$$\begin{pmatrix} \lambda_1 \\ \lambda_2 \\ \lambda_3 \end{pmatrix} = \begin{pmatrix} -\theta \\ -\beta/2 - \alpha/2 + 1/2\sqrt{\alpha^2 - 2\,\beta\,\alpha + \beta^2 + 4\,gl\,g2}\,. \\ -\beta/2 - \alpha/2 - 1/2\sqrt{\alpha^2 - 2\,\beta\,\alpha + \beta^2 + 4\,gl\,g2} \end{pmatrix}$$

Then A can be factored as $A = PDP^{-1}$ where

$$D = \begin{pmatrix} \lambda_1 & 0 & 0 \\ 0 & \lambda_2 & 0 \\ 0 & 0 & \lambda_3 \end{pmatrix}$$

and

$$P = \begin{pmatrix} 2\,\frac{g2\,(\beta+\alpha-\sqrt{\Delta}-2\,\theta)\,gl}{(dl\,\sqrt{\Delta}+(\beta-\alpha)dl+2\,g3\,gl)(\beta-\alpha-\sqrt{\Delta})} & 2\,\frac{g2\,(\beta+\alpha+\sqrt{\Delta}-2\,\theta)\,gl}{(-dl\,\sqrt{\Delta}+(\beta-\alpha)dl+2\,g3\,gl)(\beta-\alpha+\sqrt{\Delta})} & 0 \\ -\frac{(\beta+\alpha-\sqrt{\Delta}-2\,\theta)\,gl}{dl\,\sqrt{\Delta}+(\beta-\alpha)dl+2\,g3\,gl} & -\frac{(\beta+\alpha+\sqrt{\Delta}-2\,\theta)\,gl}{-dl\,\sqrt{\Delta}+(\beta-\alpha)dl+2\,g3\,gl} & 0 \\ 1 & 1 & 1 \end{pmatrix}$$

with $\Delta = \alpha^2 - 2\beta\alpha + \beta^2 + 4glg2$.

We define

$$Y(t) = P^{-1}X(t).$$

Then

$$Y'(t) = DY(t) + P^{-1}B$$

$$Y(t) = \begin{pmatrix} x(t) \\ y(t) \\ z(t) \end{pmatrix}, \; P^{-1}B = \begin{pmatrix} B1 \\ B2 \\ B3 \end{pmatrix}.$$

The solution of this system can be expressed as

$$x(t) = -\frac{B1}{\lambda_1} + \frac{e^{\lambda_1 t}(\lambda_1 x0 + B1)}{\lambda_1}, \tag{2}$$

$$y(t) = -\frac{B2}{\lambda_2} + \frac{e^{\lambda_2 t}(\lambda_2 y0 + B2)}{\lambda_2}, \tag{3}$$

$$z(t) = -\frac{B3}{\lambda_3} + \frac{e^{\lambda_3 t}(\lambda_3 z0 + B3)}{\lambda_3}, \tag{4}$$

where

$$\begin{pmatrix} x0 \\ y0 \\ z0 \end{pmatrix} = P^{-1}X(0).$$

So the general solution can be written as

$$X(t) = PY(t).$$

□

4 Sensitivity Analysis

The population dynamics engendered by the mathematical model depends on the values of the parameters used. Sensitivity analysis allows to determine how a relative change in a parameter p induces a relative change in a variable V. One of the tools used is the normal forward sensitivity index which is given by the ratio of the relative change in a variable V to the relative change in a parameter p $\left(\mathrm{I}_\mathrm{p}^\mathrm{V} = \frac{\frac{dV}{V}}{\frac{dp}{p}} = \frac{\frac{dV}{dp}}{\frac{V}{p}} \right)$. For example, a value of $\mathrm{I}_\mathrm{p}^\mathrm{V} = 1$ means that increasing p by 10% will induce an increase of V by 10% while $\mathrm{I}_\mathrm{p}^\mathrm{V} = -0.5$ will indicate that increasing p by 10% decreases V by 5%.

The calculus of sensitivity index is given by the following equations:

$$\begin{cases} \dfrac{\partial G^*}{\partial g_1} \times \dfrac{g_1}{G^*} &= \dfrac{\beta(\alpha - g_1)}{\alpha\beta - g_1 g_2} > 0 \\ \dfrac{\partial G^*}{\partial g_2} \times \dfrac{g_2}{G^*} &= \dfrac{-g_2(\alpha - g_1)}{\alpha\beta - g_1 g_2} < 0 \\ \dfrac{\partial G^*}{\partial g_3} \times \dfrac{g_3}{G^*} &= \dfrac{-\alpha g_3}{\alpha\beta - g_1 g_2} < 0. \end{cases}$$

We can see that, an increase of the parameter g_1 will induce an increase of G. While the increase of g_2 or g_3 will decrease G.

$$\begin{cases} \dfrac{\partial D^*}{\partial d_1} \times \dfrac{d_1}{D^*} &= -\dfrac{g_1 g_2 - \beta(g_1 + \mu)}{(\alpha\beta - g_1 g_2)^2} = \dfrac{g_1 g_3 + g_1\mu + g_2\mu + g_3\mu + \mu^2}{(\alpha\beta - g_1 g_2)^2} > 0 \\ \dfrac{\partial D^*}{\partial \omega} \times \dfrac{\omega}{G^*} &= \dfrac{-\omega}{(\mu + \omega)(\alpha\beta - g_1 g_2)} < 0. \end{cases}$$

An increase of the parameter d_1 will induce an increase of D. While the increase of ω will decrease G.

The value of each relative index can be obtained from Table 2, using the values of different parameters given in Table 3.

Table 2 Normal forward sensitivity index

	g_1	g_2	g_3	d_1	ω
I^{G^*}	$\dfrac{\beta(\alpha - g_1)}{\alpha\beta - g_1 g_2}$	$-\dfrac{g_2(\alpha - g_1)}{\alpha\beta - g_1 g_2}$	$-\dfrac{\alpha g_3}{\alpha\beta - g_1 g_2}$	$\dfrac{-n g_1 \beta}{(\alpha\beta - g_1 g_2)^2}$	0
I^{D^*}				$\dfrac{(\beta(g_1 + \mu) - g_1 g_2)}{\alpha\beta - g_1 g_2}$	$-\dfrac{(\omega/\mu + \omega)}{\alpha\beta - g_1 g_2}$

Table 3 Values of different parameters used for sensitivity analysis and simulation

Parameter	Value
n	1936460
g_1	14.7×10^{-2}
g_2	0.04
g_3	0.022
d_1	0.0045
μ	0.001
ω	0.007

Sensitivity analysis is important for decision-makers looking for an optimal strategy that reduces the prevalence of gestational diabetes and type 2 diabetes.

5 Numerical Simulation

Values of the different parameters used in the mathematical model were determined from the following sources: High Commission for Planning (HCP), International Diabetes Federation(IDF), Perspective Monde, Moroccan Ministry of Health (MoH), American Diabetes Association (ADA), World Population Prospects (WPP).

A mathematical simulation is carried out to evaluate the progression of women of childbearing age through the three different stages: healthy, gestational diabetes and diabetes.

Figure 3 indicates that, with time, a significant increase in diabetes is seen among women of childbearing age. In opposition to this, an important decrease is observed among healthy women.

To illustrate the effect of action on parameters, an implementation is made by decreasing value of the parameter g_1 (from 0.0147 to 0.01) and increasing value of the parameter g_2 (from 0.04 to 0.05).

Figure 4 shows clearly that this strategy has a positive impact on the evolution of women of childbearing age in each stage. Precisely, a decrease of parameter g_1 and an increase of parameter g_2 ensure a rise of the number of healthy women of childbearing age by 11% and a decrease by 9% and 30% in the corresponding numbers of diabetic women and women with gestational diabetes, respectively.

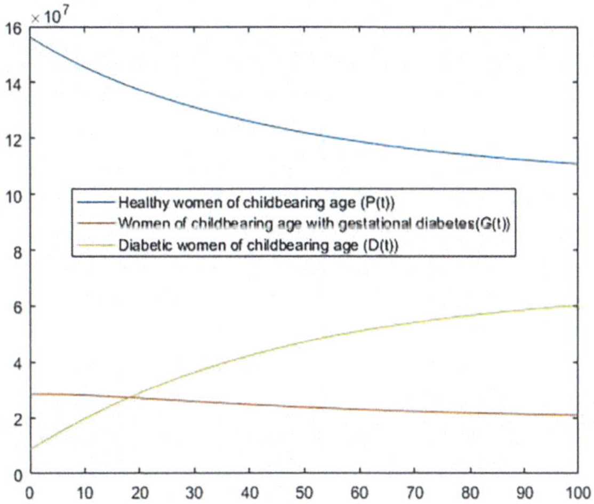

Fig. 3 Evolution of women of reproductive age through the stages P, G and D

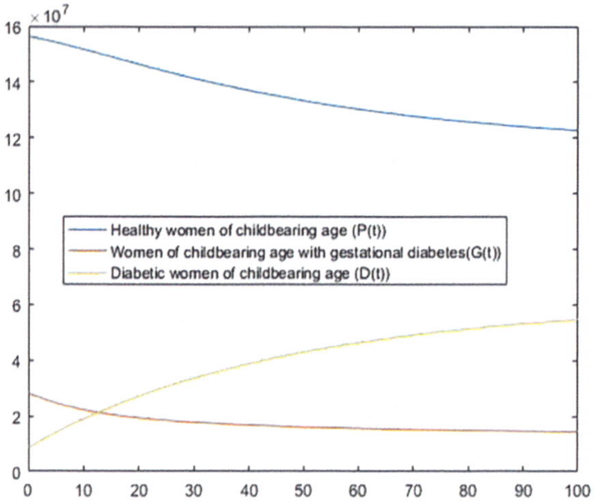

Fig. 4 Variation of parameters g_1 and g_2

6 Discussion

Stability analysis showed that our mathematical model has an equilibrium point which is locally asymptotically stable. Using the results given by sensitivity analysis and numerical simulations, some pragmatic recommendations can be formulated to help decision-makers to reduce the gestational diabetes. Indeed, theoretically, the normal forward sensitivity index IG_1(0.92) indicates that decreasing g_1 by 10% will reduce the population with gestational diabetes with nearly 10%. Pragmatically, this reduction can be achieved by increasing screening for gestational diabetes and sensitizing for healthy lifestyle that prevents both from gestational diabetes and type 2 diabetes, especially for women who had gestational diabetes in previous pregnancies or those with risk factors like family history, pre-pregnancy overweight/obesity, excessive gestational weight gain and excessive central body fat deposition. Such a strategy will also limit as far as possible the number of gestational diabetes women who are susceptible to become definitely diabetic during pregnancy or after giving birth. As indicated by the American Diabetes Association (ADA), although it may be a scary diagnosis, gestational diabetes is treatable and manageable but pregnant women at risk should act quickly.

The normal forward sensitivity index IG_2(0.73) shows also that the population of women with gestational diabetes can be controlled by increasing the flow g_2 from G to P but the strategy for this is similar to that proposed for g_1.

To control the size of the population of women with type 2 diabetes, the model indicates that two kind of actions should be considered by: (1) obviously reducing the rate d_2 at which adult women become diabetic. This can mainly be achieved by adopting the well-known strategies based on healthy diet, regular physical effort and overweight/obesity control. (2) Limiting the number of women with gestational diabetes who become diabetic. However, it should be noted that decreasing mortality will also increase the number of women with diabetes. In fact, although diabetes is not curable, people who have diabetes can have a "normal" life (like people without diabetes) provided they avoid complications of diabetes such as kidney failure (nephropathy), eye damage (retinopathy), foot damage and cardiovascular disease.

7 Conclusion

In this paper, a simple mathematical model was presented, focussing on the gestational diabetes issue. The main purpose is to provide health decision-makers with pragmatic recommendations that should lead to efficient cost-effective strategies allowing pregnant women to avoid gestational diabetes and its unfortunate consequences on the health of women and that of their children.

Sensitivity analysis followed by numerical simulations confirms that prevention in terms of gestational diabetes and type 2 diabetes will promote a healthy life for women in general and those of childbearing age in particular.

References

1. American Diabetes Association: Classification and diagnosis of diabetes: Standards of medical care in diabetes-2021. Diabetes Care **44**(Supplement 1), S15–S33 (2021)
2. Boutayeb, A., Lamlili, M.E., Boutayeb, W.: A review of compartmental mathematical models used in diabetology. In: Disease Prevention and Health Promotion in Developing Countries, pp. 217–250 (2020)
3. Danyliv, A., Gillespie, P., O'Neill, C., Tierney, M., O'Dea, A., McGuire, B.E., Glynn, L.G., Dunne, F.P.: The cost-effectiveness of screening for gestational diabetes mellitus in primary and secondary care in the republic of ireland. Diabetologia **59**(3), 436–444 (2016)
4. Daud, A.A.M., Toh, C.Q., Saidun, S.: Development and analysis of a mathematical model for the population dynamics of diabetes mellitus during pregnancy. Math. Models Comput. Simul. **12**(4), 620–630 (2020)
5. Diabetes Canada: Gestational diabetes. https://www.diabetes.ca/about-diabetes/gestational (2021)
6. Donaldson, S., McIntyre, D.: Suggested consensus for the diagnosis of gestational diabetes. Diabetes Manag. **4**(1), 11 (2014)
7. IDF: Gestational diabetes. https://www.idf.org/our-activities/care-prevention/gdm (2019)
8. IDF: Gestational prevalence. idf diabetes atlas 10th edition 2021. https://diabetesatlas.org/data/en/indicators/14/ (2021)
9. International Association of Diabetes and Pregnancy Study Groups Consensus Panel and others: International association of diabetes and pregnancy study groups recommendations on the diagnosis and classification of hyperglycemia in pregnancy. Diabetes Care **33**(3), 676–682 (2010)
10. Li, G., Wei, T., Ni, W., Zhang, A., Zhang, J., Xing, Y., Xing, Q.: Incidence and risk factors of gestational diabetes mellitus: a prospective cohort study in qingdao, china. Front. Endocrinol. **11** (2020)
11. Lohse, N., Marseille, E., Kahn, J.G.: Development of a model to assess the cost-effectiveness of gestational diabetes mellitus screening and lifestyle change for the prevention of type 2 diabetes mellitus. Int. J. Gynecol. Obstetr. **115**, S20–S25 (2011)

12. López Stewart, G.: Diagnostic criteria and classification of hyperglycaemia first detected in pregnancy: a world health organization guideline (2014)
13. The women's royal hospital Victoria Australia: Gestational diabetes. https://www.thewomens.org.au/health-information/pregnancy-and-birth/pregnancy-problems/pregnancy-problems-in-later-pregnancy/gestational-diabetes (2021)
14. World Health Organization: Diagnostic criteria and classification of hyperglycaemia first detected in pregnancy. https://apps.who.int/iris/bitstream/handle/10665/85975/WHO_NMH_MND_13.2_eng.pdf (2013)

Predictive Modeling of Diabetes Risk Factors and Early Complications Using Machine Learning and Deep Learning Approaches

Abdelaadim Khriss, Mohammed Badaoui, Wiam Boutayeb, Hannah Al Ali, and Aissa Kerkour Elmiad

Abstract Diabetes, an ever-growing chronic disease, poses significant challenges when it is not detected early and diagnosed promptly. Recent advances in medical science exploit various machine learning techniques, including ontology-based ML approaches, to develop automated systems capable of efficiently detecting diabetic patients. This paper delves into an in-depth evaluation of various machine learning algorithms, with a particular focus on model performance and the influence of feature selection methods, in the context of diabetes prediction. The study highlights the important role of these techniques in optimizing automated systems for early and accurate diabetes detection, thus contributing to more effective healthcare interventions.

1 Introduction

Diabetes is a global health problem, often linked to factors such as obesity or high blood glucose levels. Diabetes is a persistent condition characterized by high blood

A. Khriss
Faculty of Sciences, University Mohammed First, Oujda, Morocco
e-mail: abdelaadim.khriss20@ump.ac.ma

M. Badaoui
Higher School of Technologie, University Mohammed First, Oujda, Morocco

W. Boutayeb (✉)
Higher Scool of Education and Training, Oujda, Morocco
e-mail: wiam.boutayeb@ump.ac.ma

H. Al Ali
Emirates Aviation University, Dubai, United Arab Emirates
e-mail: hannah.alali@emirates.com

A. Kerkour Elmiad
Faculty of Sciences, University Mohammed First, Oujda, Morocco
e-mail: a.kerkour@ump.ac.ma

A. Boutayeb and M. E. N. Lamlili (eds.), *Computational Mathematics and Modelling for Diabetes*, Industrial and Applied Mathematics,
https://doi.org/10.1007/978-981-96-1925-2_9

glucose levels, which in the long term can lead to serious damage to the heart, blood vessels, eyes, kidneys, and nerves. The most common type of diabetes is type 2, which generally occurs in adults when the body is resistant to insulin or does not produce enough of it. Over the past 30 years, the prevalence of type 2 diabetes has increased dramatically in countries of all income levels with an impact on [2]. Type 1 diabetes, formerly known as juvenile diabetes or insulin-dependent diabetes, is a chronic disease in which the pancreas produces little or no insulin. A global target has been set to halt the rise in diabetes and obesity by 2025. An estimated 422 million people worldwide have diabetes, the majority living in low- and middle-income countries, and 1.5 million deaths are directly attributed to diabetes each year. The number of cases and the prevalence of diabetes have risen steadily over the last few decades [2]. Diabetes, therefore represents a major cause of death worldwide, highlighting the importance of early detection of signs through the use of emerging technologies such as machine learning and ontologies.

Machine learning (ML) is positioning itself as one of the fastest-growing areas of computing, offering multiple applications. It encompasses the process of extrapolating meaningful information from large data sets. ML techniques are used in a variety of sectors, including medical diagnostics, marketing, industry, and other scientific fields [9, 31, 38]. ML algorithms are frequently used in medical data analysis due to their efficiency [13]. Different forms of ML, such as classification, regression, and clustering, are implemented according to the specific problems to be solved, each one produces unique results and impacts.

The aim of this study is to leverage advanced machine learning and deep learning techniques to predict the onset of diabetes. Our approach involves the use of a various set of methodologies, including decision trees, naive Bayes, Support Vector Machine (SVM), K-Nearest Neighbor (KNN), Logistic Regression, and Artificial Neural Networks (ANN). All these algorithms are renowned for their effectiveness in predictive modeling. These methods will be used to assess and predict diabetes risk and associated complications.

A key aspect of our strategy is the incorporation of sophisticated feature selection techniques. We use methods such as Least Absolute Shrinkage and Selection Operator (LASSO), Recursive Feature Elimination (RFE), Random Forest, and genetic algorithm. Each one of these methods aims to identify the most influential factors contributing to diabetes risk. This step is essential for refining our predictive models by eliminating the least relevant variables, thereby improving the accuracy and efficiency of our predictions.

The model with the highest predictive accuracy is further examined using a probabilistic model. This additional analysis aims to disentangle the complex relationships and dependencies between the identified risk factors. To do this, we use a graphical probabilistic decision tree model, a technique that uncovers probabilistic dependencies. This approach offers a more detailed insight into how different factors interact to influence diabetes and its early complications.

Our methodology combines a range of sophisticated techniques, from conventional machine learning to the application of probabilistic models and advanced feature selection methods. This comprehensive approach is designed to produce more accurate and interpretable diabetes prediction models that effectively capture the complexity of interactions between different risk factors.

2 Related Work

Recently, a number of researchers have published studies focused on identifying diabetic patients through machine learning methods. In [20], the authors present a model aimed at anticipating the presence of diabetes in a patient. This approach relies on the effectiveness of powerful machine learning algorithms, exploiting measures such as precision, recall, and F1-measurement. Their study uses the Pima Indian Diabetes (PIDD) dataset to predict the development of diabetes based on a diagnosis [11]. The performance of the logistic regression (LR), Naive Bayes (NB), and K-nearest neighbor (KNN) algorithms was 94%, 79%, and 69%, respectively. In [21], the authors compare seven machine learning algorithms on the dataset for predicting diabetes. They conclude that the logistic regression model and SVM are most effective for predicting diabetes. A neural network model with two hidden layers achieved an accuracy of 88.6%. The study described in [10] explores various machine-learning classification algorithms on the PIID [1] dataset, where logistic regression displays the best accuracy. Sarwar et al. The paper [30] discusses predictive analytics in healthcare, using several machine learning algorithms on a medical dataset. In [25], the authors propose a predictive model for diabetes incorporating external factors and regular parameters such as glucose, BMI, age, and insulin. Classification accuracy is improved with a new dataset compared to the existing one. Paper [29] presents the application of eight ML algorithms on 521 instances, with logistic regression, linear and non-linear kernel SVM, random forest, decision tree, adaptive reinforcement classifier, K-nearest neighbor, and naive Bayes. The Random Forest classifier achieves 98% accuracy. In [12], researchers use various machine learning algorithms, with the Random Forest classifier standing out with 98% accuracy. Paper [3] proposes a new approach to diabetes prediction using meaningful attributes, with ANN, Random Forest, and K-means, with ANN offering the best accuracy. The paper [37] presents a method using machine learning algorithms in Hadoop-based clusters, demonstrating high accuracy in diabetes prediction. In [33], four machine learning algorithms are used to predict diabetes at an early stage, with the Random Forest classifier achieving 87.66% accuracy. The work in [17] explores the prediction and classification of diabetes complications using various supervised classification algorithms.

Two machine learning approaches, in paper [6] are proposed. The authors used the Random Forest algorithm and XGBoost and the results showed that the latter outperforms with an accuracy rate of 74.10% In paper [6], the authors propose two machine learning approaches, using the Random Forest algorithm and XGBoost, the latter outperforming the others with an accuracy rate of 74.10%. Authors of the paper [7, 35] analyze significant risk factors for type 2 diabetes mellitus (T2DM) in Gulf countries using the Gulf COAST dataset. Using principal component analysis (PCA) and decision tree algorithms, the study identifies hypertension and dyslipidemia as important predictors of T2DM. It also highlights age and obesity as important risk factors. The research emphasizes the need for lifestyle changes, including regular exercise, a balanced diet, and routine medical check-ups, to prevent the onset of T2DM. The study highlights the importance of data mining techniques in understanding and managing diabetes risk factors. Finally, [28] assesses various machine learning algorithms, where the Naive Bayes and Random Forest classifiers achieve 80% accuracy.

Despite the advancements, these studies have some limitations. Many are based on the PIDD dataset, which may not generalize well to other populations due to demographic and genetic differences. This study addresses this limitation by using the Gulf COAST dataset, encompassing various demographic from the Gulf region, aiming for more generalizable results. Additionally, prior research often compared a limited number of machine learning algorithms and did not consistently employ comprehensive feature selection techniques, leading to potential overfitting and less interpretable models. By employing a thorough evaluation of model performance and the influence of feature selection methods, this study aims to enhance the accuracy and interpretability of diabetes prediction models, ultimately contributing to more effective healthcare interventions.

In the following section, we will present our methodology in detail. This important phase of our study will shed full light on the approaches, methods and techniques we have selected to achieve our research objectives. By setting out our methodology in a transparent and detailed manner, we aim to establish a solid basis for understanding and analyzing our results.

3 Methodology

3.1 Overview

This study aims to predict diabetes and its early complications using a comprehensive and multi-faceted methodology. The approach integrates a variety of machine learning and deep learning techniques combined with sophisticated feature selection methods to create accurate predictive models.

To implement these predictive models, it is important to understand the dataset used in this study. The following subsection provides a detailed description of the dataset.

3.2 Dataset Description

Data from the Gulf COAST database was used to conduct a case-control epidemiological investigation. The data was collected between 2012 and 2013 and was obtained through surveys of admitted patients in this time period who had acute coronary syndrome (ACS). The study encompassed a total of 29 hospitals across four Gulf countries (UAE, Bahrain, Kuwait, and Qatar). Out of the 4,061 patients registered in the database and having consented to the use of their information, a sample of 3,372 patients was randomly selected. This selection was based on the criterion of attendance at the corresponding hospitals. The main objective of our study is to anticipate risk factors for type 2 diabetes (T2DM). Only non-diabetic patients and individuals with T2DM were included in the sample, with a requirement for a substantial portion of their information to be available.

Having established the dataset, we now delve into the specific variables included in our study and their significance in building a robust predictive model.

3.2.1 Variables

Each individual variable plays an important role in building a general of an individual's health profile, offering valuable insights into potential risk factors associated with diabetes. The following exploration presents and categorizes these variables, highlighting their respective contributions to the overall goal of developing an effective diabetes prediction model.

- **Demographic Information**
  - `Country`—The country of the individual.
  - `Gender`—The gender of the individual.
  - `Age`—The age of the individual.
  - `Marital_Status`—The marital status of the individual.
  - `Work`—The occupational status of the individual.
  - `Education`—The educational background of the individual.
- **Clinical Parameters**
  - `Cardiac_Arrest_Admission`—History of cardiac arrest upon admission.
  - `Non_Cardiac_Condition`—Presence of non-cardiac conditions.
  - `Hypertension`—Presence of hypertension.
  - `Dyslipidemia`—Presence of dyslipidemia.
  - `Smoking_History`—History of smoking.
  - `Waist`—Waist circumference.
  - `BMI`—Body Mass Index.
  - `Admission_Blood_Glucose`—Blood glucose level upon admission.
  - `Fasting_Blood_Glucose`—Fasting blood glucose level.
  - `HbA1C_Admission`—HbA1C level upon admission.

- **Laboratory Values**
  - `Lipid_24_Collected`—Collection of lipid profile data.
  - `Cholesterol`—Cholesterol level.
  - `Triglycerides`—Triglyceride level.
  - `LDL`—Low-Density Lipoprotein level.
  - `HDL`—High-Density Lipoprotein level.
- **Physiological and Lifestyle Factors**
  - `Stress`—Level of stress.
  - `Creatinine_Clearance`—Creatinine clearance, indicating renal function.
  - `Sleep_Apnea`—Presence of sleep apnea.
  - `Heart_Rate`—Heart rate.
  - `Systolic_BP`—Systolic blood pressure.
  - `Diastolic_BP`—Diastolic blood pressure.

The selection of these variables was guided by their availability in the Gulf COAST dataset. Although other variables could potentially contribute to the prediction of diabetes, we were limited by the data provided. Therefore, these were the only available variables included in our study.

3.2.2 Attributes of the Target Population

The examined cohort, comprising 3,372 people residing in Bahrain, Kuwait, Oman, or the United Arab Emirates, includes 1,907 people diagnosed with type 2 diabetes. This various group, aged between 18 and 99, has a mean age of 60 ± 12 years. This means that most participants are between around 48 and 72 years of age, with a few outside this range. In addition, the population includes a balanced representation of gender, socio-economic backgrounds, and lifestyle factors, ensuring a comprehensive and multifaceted exploration of type 2 diabetes risk factors within this various demographic.

Examining the data presented in Fig. 1a, it functions as a visual depiction of diabetes distribution within a specific population. The figure is divided into *"Diabetes"* and *"No Diabetes"* segments, indicating the presence or absence of diabetes among study participants. The predominant red segment, encompassing 56% of the population, represents individuals with diabetes, suggesting that slightly over half of the studied population is affected by the condition.

Conversely, the smaller blue segment, comprising 43% of the population, signifies individuals without diabetes. This observation implies that less than half of the studied population is free from the condition.

Shifting the focus to Fig. 1b, it reveals a notable disparity in diabetes prevalence between genders. Female exhibit a higher percentage of individuals with diabetes, coupled with a lower percentage of those without the condition. This points to a potential gender-based discrepancy in the occurrence of diabetes.

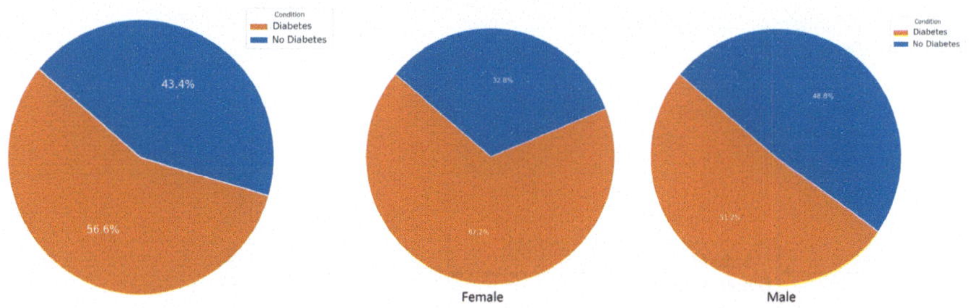

(a) Percentage Distribution of Diabetes (b) Distribution of Diabetes Across Genders

Fig. 1 Exploring diabetes: a gender-based distribution analysis and overall percentage insights

With a clear understanding of the dataset and the target population, we move on to the next critical phase of our methodology: feature selection. This process is essential to refine our dataset and enhance the predictive power of our models.

3.3 Feature Selection

Selecting features is an important stage in developing a model, requiring the identification and preservation of the most informative attributes in a dataset while discarding irrelevant or redundant ones. Not only does this enhance the predictive performance of the model but also aids in understanding the underlying patterns within the data [18].

In the specialized realm of predicting diabetes, the intricate nature of the disease requires a careful selection of features that markedly contribute to the model's predictive accuracy. We explore various feature selection methods, with a particular emphasis on embedded and wrapper methods.

Fig. 2 represents the feature selection process in machine learning. The *"original set"* is presented as the initial dataset or problem set, serving as the groundwork for subsequent steps. This is followed by *"subset generation"* phase, involving the creation of subsets from the original set through the application of various techniques. These generated sub-assemblies are then subjected to in-depth evaluation in the subsequent *"Sub-Assembly Evaluation"* stage, leading to a conclusive examination of the results in the *"Results Evaluation"* phase. The iterative nature of the process persists until a predetermined stopping criterion is reached, as indicated by the *"Stopping Criteria"*. This subsection provides a comprehensive overview of the feature selection methodology used, highlighting its importance in refining machine learning models.

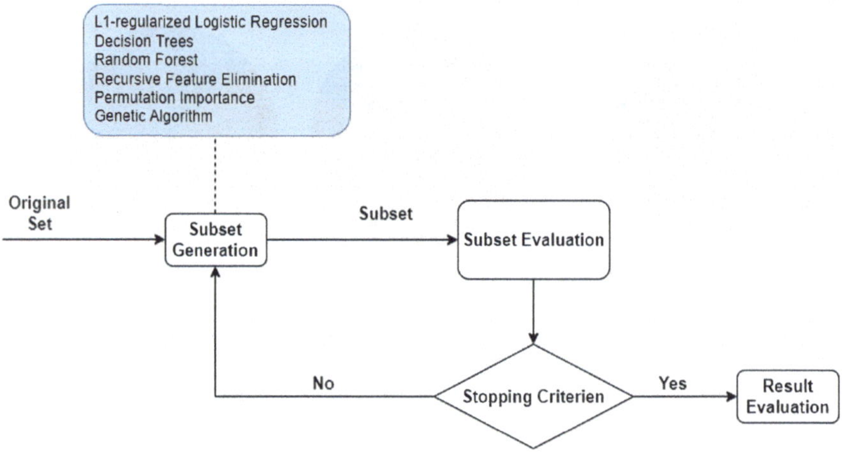

Fig. 2 Feature selection process

3.3.1 L1-Regularized Logistic Regression

L1-Regularized Logistic Regression, commonly known as LASSO [26] (Least Absolute Shrinkage and Selection Operator), is an extension of logistic regression that introduces a penalty term to the standard cost function. The cost function, denoted as $J(\theta)$, is formulated as follows:

$$J(\theta) = -\frac{1}{m}\sum_{i=1}^{m}[y^{(i)}\log(h_\theta(x^i)) + (1-y^{(i)})\log(1-h_\theta(x^{(i)}))] + \lambda\sum_{j=1}^{n}|\theta_j|. \quad (1)$$

In this equation, m represents the number of training samples, n is the number of features, θ corresponds to the coefficients of the logistic regression model, $h_\theta(x)$ denotes the logistic function, and λ governs the regularization strength. The term $|\theta_j|$ specifically signifies the absolute value of the j-th coefficient.

The inclusion of the penalty term ($\lambda \sum_{j=1}^{n} |\theta_j|$) in the cost function is a distinctive characteristic of L1 regularization. This term is important in promoting sparsity within the solution space, as it induces a tendency for certain coefficients to become exactly zero. Consequently, LASSO not only performs logistic regression but also serves as a feature selection mechanism by favoring models with fewer non-zero coefficients. The parameter λ allows practitioners to adjust the level of regularization, influencing the trade-off between model complexity and the goodness of fit to the training data.

3.3.2 Decision Tree

Let D be the dataset at a particular node, and $I(D)$ be the impurity measure (Gini impurity or entropy) [14]. For a binary split on a feature X_i at a threshold θ, the impurity of the split is calculated as

$$I_{\text{split}}(D, X_i, \theta) = \frac{N_{\text{left}}}{N} \cdot I(D_{\text{left}}) + \frac{N_{\text{right}}}{N} \cdot I(D_{\text{right}}), \tag{2}$$

where N is the total number of samples at the node, N_{left} and N_{right} are the number of samples in the left and right child nodes after the split, and D_{left} and D_{right} are the datasets in the left and right child nodes.

The best split is determined by finding the one that minimizes the impurity across all possible splits for all features.

3.3.3 Random Forest

Random Forest [16] is a powerful ensemble learning method widely used for feature selection. It incorporates two main sources of randomness:

1. **Bootstrapped Sampling:** Each tree in the forest is trained on a random subset of the original dataset sampled with replacement (bootstrap sample).
2. **Random Feature Selection:** When deciding the split at each node, only a random subset of features is considered, adding an additional layer of randomness to the model.

The prediction of a Random Forest $F(x)$ for a given input x is calculated as the average prediction of all trees in the forest:

$$F(x) = \frac{1}{N_{\text{trees}}} \sum_{i=1}^{N_{\text{trees}}} f_i(x), \tag{3}$$

where N_{trees} is the total number of trees in the forest, and $f_i(x)$ represents the prediction of the i-th tree.

In addition to its predictive power, Random Forest is valuable for feature selection. The feature importance in Random Forest is often quantified based on the average decrease in impurity across all trees when a particular feature is used for splitting. Features that consistently lead to a greater reduction in impurity are considered more important in the modeling process.

3.3.4 Recursive Feature Elimination

Recursive Feature Elimination (RFE) is a feature selection technique that aims to improve the performance of a model by iteratively removing the least important features [36]. The process involves the following steps:

1. Train the model on the entire set of features.
2. Rank the features based on their importance or contribution to the model.
3. Eliminate a predefined number or a percentage of the least important features.
4. Repeat the process until the desired number of features is reached or until the model performance stops improving.

The selection criterion for determining feature importance depends on the specific machine learning algorithm used. Common metrics include coefficients in linear models or feature importances in tree-based models.

RFE is particularly useful when dealing with high-dimensional datasets where the number of features is large compared to the number of samples. By iteratively pruning the least important features, RFE helps prevent overfitting and improves the interpretability of the model.

The effectiveness of RFE depends on finding the right balance between feature elimination and model performance. It provides a systematic approach to selecting relevant features and is widely used in applications where feature interpretability and model simplicity are important.

3.3.5 Permutation Importance

Permutation Importance is a technique used for assessing the importance of individual features in a predictive model [4]. Unlike traditional feature importance methods that rely on model-specific metrics, permutation importance provides a model-agnostic approach.

The process of calculating permutation importance involves the following steps:

1. Train the model on the original dataset and record its performance metric (e.g., accuracy, F1-score), denoted as M_{original}.
2. For each feature X_i of interest:
   a. Randomly permute the values of X_i in the dataset.
   b. Reevaluate the model on the modified dataset and record the new performance metric, denoted as $M_{\text{permuted}}(X_i)$.
   c. Calculate the permutation importance for feature X_i as

$$\text{Permutation Importance}(X_i) = M_{\text{original}} - M_{\text{permuted}}(X_i). \quad (4)$$

Permutation importance provides a robust measure of feature importance by considering how the model's performance degrades when the relationship between a feature and the target variable is disrupted.

This technique is particularly valuable for models with complex interactions and non-linear relationships, as it captures the overall impact of each feature on the model's predictive ability. Additionally, permutation importance is model-agnostic, making it applicable to a wide range of machine learning algorithms.

Researchers and practitioners often use permutation importance as a complementary tool for feature selection and interpretation, providing insights into the relative contributions of different features in the model [8, 19, 24].

3.3.6 Genetic Algorithm

Genetic Algorithm (GA) is a powerful optimization technique inspired by the principles of natural selection [5]. When applied to feature selection, GA aims to identify subsets of features that maximize the performance of a machine learning model. The process involves the following steps:

1. **Initialization:** Creates an initial population of potential feature subsets represented as binary strings, where each bit corresponds to the inclusion or exclusion of a feature. Let $X_{\text{pop}}^{(i)}$ denote the i-th individual in the population.
2. **Evaluation:** Assesses the fitness of each individual in the population using a predefined fitness function. In feature selection, the fitness function $f(X_{\text{pop}}^{(i)})$ measures the performance of a model based on the selected subset of features.
3. **Selection:** Chooses individuals from the population for reproduction based on their fitness. The probability of selecting an individual is proportional to its fitness. Let $X_{\text{selected}}^{(i)}$ denote the i-th selected individual.
4. **Crossover:** Applies crossover (recombination) to pairs of selected individuals to create offspring. The crossover operation combines features from two parents to produce new potential solutions. Let $X_{\text{offspring}}^{(i)}$ denote the i-th offspring.
5. **Mutation:** Introduces random changes to the offspring, such as flipping bits in the binary representation. Mutation introduces diversity into the population. Let $X_{\text{mutated}}^{(i)}$ denote the i-th mutated offspring.
6. **Replacement:** Replaces the old generation with the new generation of individuals, which includes both parents and offspring. Let $X_{\text{new}}^{(i)}$ denote the i-th individual in the new generation.
7. **Termination:** Repeats the process for a specified number of generations or until convergence criteria are met.

After carefully selecting the relevant features, the next step involves applying various machine learning algorithms to predict diabetes. The following section outlines the machine learning predicting algorithms employed in this study (Fig. 3).

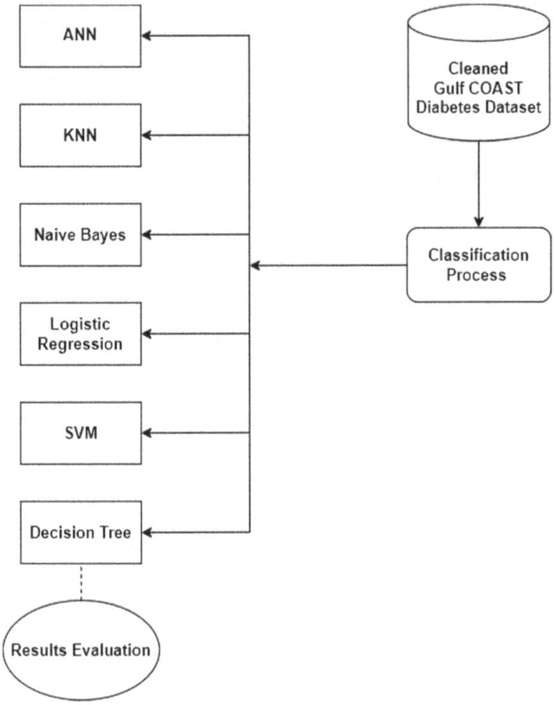

Fig. 3 Workflow of prediction

3.4 Machine Learning Predicting Algorithms

The task of predicting diabetes is a complex challenge that requires the utilization of advanced machine learning techniques to discern intricate patterns within datasets. We explore a various set of classification algorithms employed in the development of predictive models for diabetes. Each algorithm brings unique strengths and characteristics, contributing to the nuanced landscape of predictive analytics in the medical domain [32].

Figure 4 presents an overview of the workflow, depicting the sequential steps involved in the analysis of the cleaned dataset. The process encompasses the application of various machine learning classification algorithms, such as Artificial Neural Networks (ANN), K-Nearest Neighbors (KNN), Naive Bayes, Logistic Regression, Support Vector Machines (SVM), and Decision Tree. Subsequently, a Results Evaluation is conducted to assess the performance of these algorithms.

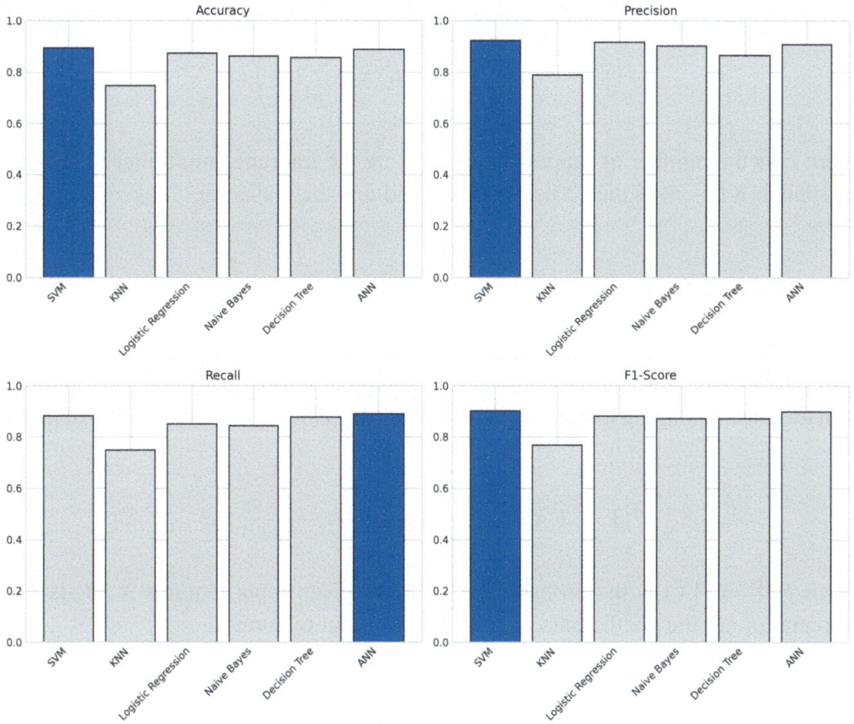

Fig. 4 Performance evaluation of machine learning models: a comparative analysis of accuracy, precision, recall, and F1-score

3.4.1 K-Nearest Neighbors (KNN)

KNN [15] operates on the principle of proximity, classifying data points based on the characteristics of their nearest neighbors. For a given instance x_i, its class is determined by the majority class among its k-nearest neighbors. Mathematically, the prediction function $f(x_i)$ for KNN is defined as

$$f(x_i) = \text{majority class}\left(\text{neighbors}(x_i)\right),$$

where neighbors(x_i) denotes the k nearest neighbors of x_i.

3.4.2 Support Vector Machines (SVM)

SVM [27] find optimal hyperplanes to separate different classes in high-dimensional spaces. For a binary classification problem, SVM seeks to find the hyperplane that maximizes the margin between the two classes. The decision function for SVM can be expressed as

$$f(x) = \text{sign}\left(\sum_{i=1}^{N} \alpha_i y_i K(x, x_i) + b\right),$$

where N is the number of support vectors, α_i are the Lagrange multipliers, y_i is the class label, $K(x, x_i)$ is the kernel function, and b is the bias term.

3.4.3 Logistic Regression

Logistic Regression [23] is a classic yet robust algorithm for binary classification tasks. The logistic function, or sigmoid function, is employed to model the probability of an instance belonging to the positive class. The logistic regression model is defined as

$$P(Y = 1|X) = \frac{1}{1 + e^{-(\beta_0 + \beta_1 X_1 + \ldots + \beta_n X_n)}},$$

where $P(Y = 1|X)$ is the probability of class 1 given input features X, β_0 is the intercept, β_i are the coefficients, and X_i are the input features.

3.4.4 Naive Bayes

Naive Bayes classifiers [34] are based on Bayesian probability principles, assuming independence between features. The probability of an instance belonging to a particular class is computed using Bayes' theorem:

$$P(Y|X) = \frac{P(X|Y)P(Y)}{P(X)},$$

where $P(Y|X)$ is the posterior probability of class Y given features X, $P(X|Y)$ is the likelihood, $P(Y)$ is the prior probability, and $P(X)$ is the evidence.

3.4.5 Artificial Neural Networks (ANN)

ANN [22] are inspired by the human brain's architecture and are capable of capturing intricate patterns in complex datasets. For a simple feedforward neural network, the output y is computed as

$$y = \sigma\left(\sum_{j=1}^{M} w_j \sigma\left(\sum_{i=1}^{N} x_i w_{ij} + b_j\right) + b\right),$$

where N is the number of input features, M is the number of hidden units, x_i are the input features, w_{ij} are the weights, b_j are the biases, σ is the activation function, and b is the output layer bias.

In the next section, we will examine the prediction results using these different models and feature selection methods, in order to identify important factors in the detection of early complications associated with diabetes.

4 Evaluation

This section presents the examination of machine learning models used for predicting the complications associated with diabetes. These models were evaluated to determine their predictive capabilities, interpretability and computational efficiency. The evaluation focuses not only on comparing the models but also on understanding the impact of different feature selection strategies on their performance. Specific measures are introduced and applied to discover the strengths and limitations of these approaches in the context of the machine learning task in point. This evaluation aims not only to assess the accuracy and reliability of the models but also to inform the selection of the most effective feature selection technique for the early identification of diabetes-related complications. By analyzing results and interpreting parameters, this evaluation guides decision-making towards the adoption of optimal strategies for improving diagnostic accuracy. The probabilistic decision tree method then is applied in the best model. The evaluation aims to select the most appropriate approach and features to effectively address early detection of diabetes-related complications in the set of models considered.

4.1 Evaluation Metrics

The evaluation of machine learning models is important in gauging their efficacy in predictive tasks. In the realm of predicting diabetes, various metrics are employed to systematically assess and quantify the performance of various models.

4.1.1 Accuracy

Accuracy serves as a fundamental metric, providing a comprehensive perspective on a model's correctness by quantifying the ratio of accurately predicted instances to the total number of instances in the dataset.

$$\text{Accuracy} = \frac{\text{Number of Correct Predictions}}{\text{Total Number of Predictions}}. \tag{5}$$

4.1.2 Precision

Precision delves into the model's ability to make accurate positive predictions, computed as the ratio of true positive predictions to the sum of true positives and false positives.

$$\text{Precision} = \frac{\text{True Positives}}{\text{True Positives} + \text{False Positives}}. \quad (6)$$

4.1.3 Recall

Recall, synonymous with sensitivity or true positive rate, assesses the model's capacity to capture all pertinent instances, calculated by the ratio of true positive predictions to the sum of true positives and false negatives.

$$\text{Recall} = \frac{\text{True Positives}}{\text{True Positives} + \text{False Negatives}}. \quad (7)$$

4.1.4 F1-Score

The F1-Score, which combines precision and recall, provides a balanced measure considering both false positives and false negatives. It is particularly valuable in scenarios with imbalanced class distributions.

$$\text{F1-Score} = 2 \times \frac{\text{Precision} \times \text{Recall}}{\text{Precision} + \text{Recall}}. \quad (8)$$

As we progress through the evaluation process, these metrics collectively provide nuanced insights into the predictive capabilities of machine learning models. Examining accuracy, precision, recall, and F1-score provides an in-depth understanding of their strengths and limitations, enabling informed decisions to be made when selecting models that contribute optimally to diabetes prediction.

4.2 Results

Table 1 presents a comprehensive overview of the performance evaluation for machine learning algorithms applied to the prediction of diabetes.

The results presented in Table 1 provide insightful observations regarding the effectiveness of various machine learning algorithms for diabetes prediction. Notably, SVM emerges as the predominant algorithm, achieving a remarkable accuracy of 0.8933, signifying its ability to accurately classify instances. SVM also demonstrated a remarkable precision of 0.9224 and an F1-score of 0.9024, underlining its effec-

Table 1 Performances evaluation

Algorithm	Accuracy	Precision	Recall	F1-Score
SVM	**0.8933**	**0.9224**	0.8833	**0.9024**
KNN	0.7467	0.7877	0.7480	0.7673
Logistic Regression	0.8726	0.9145	0.8515	0.8819
Naive Bayes	0.8607	0.9008	0.8435	0.8712
Decision Tree	0.8548	0.8642	0.8780	0.8711
ANN	0.8874	0.9057	**0.8912**	0.8984

tiveness in minimizing false positives while striking a judicious balance between precision and recall.

Conversely, KNN showed diminished performance across all measures, recording a precision of 0.7467 and an F1-score of 0.7673. This suggests that the KNN proximity-based approach may not match the requirements of the diabetes prediction task as effectively as other algorithms under evaluation.

Logistic regression, Naive Bayes, decision tree, and ANN show competitive performance, with accuracies ranging from 0.8548 to 0.8874. These algorithms exhibit various strengths in precision, recall, and F1-Score, offering a nuanced perspective on their performance attributes.

4.2.1 Analysis of Feature Selection

The presented Table 2 offers a summary of features chosen by various feature selection methods, encompassing LASSO, Decision Tree, Random Forest, Recursive Feature Elimination, Permutation Importance, and Genetic Algorithm. These methods reveal distinct inclinations in feature selection.

Notably, certain features such as Age, Hypertension, Dyslipidemia, and Cholesterol consistently emerge as selections across multiple methods, implying their potential significance in the predictive modeling context. The disparities in feature selection outcomes stem from the inherent characteristics of each method and the specific criteria guiding their selection process.

Understanding these methodological variations holds importance in the pursuit of selecting the most pertinent features to construct robust models. As different algorithms prioritize distinct facets of the dataset, a comprehensive grasp of these nuances facilitates informed decision-making. The outcomes provide valuable insights into how each method influences feature selection within the context of the analyzed dataset, thereby contributing to a more nuanced understanding of the modeling process.

Table 2 Feature selection methods comparison

Features	LASSO	DT	RF	RFE	PI	GA
Country	X		X	X		
Gender				X		X
Age	X	X	X	X	X	
Marital_Status	X	X				X
Work	X			X		
Cardiac_Arrest_Admission						X
Non_Cardiac_Condition						X
Hypertension	X	X	X		X	X
Dyslipidemia	X	X	X	X	X	X
Smoking_History	X		X	X		
Waist		X	X	X	X	
BMI	X	X	X	X	X	
Admission_Blood_Glucose	X	X	X		X	X
Fasting_Blood_Glucose	X	X	X		X	X
HbA1C_Admission	X	X	X		X	X
Lipid_24_Collected						
Cholesterol	X	X	X	X	X	X
Triglycerides		X	X	X	X	
LDL	X	X	X	X	X	X
HDL	X	X	X	X	X	X
Stress		X			X	
Creatinine_Clearance		X	X	X	X	
Education	X		X	X		X
Sleep_Apnea						
Heart_Rate		X	X		X	
Systolic_BP		X		X	X	X
Diastolic_BP		X	X	X	X	X

LASSO

The visual portrayal of the LASSO feature selection process Fig. 5 clearly shows the significance of *Dyslipidemia*, *LDL*, and *Work*, as evidenced by their notably higher values. These features play a substantial role in influencing outcomes related to diabetes, suggesting that variations or patterns within them carry a more pronounced impact on the model's predictive capabilities than other features.

Conversely, *Creatinine_Clearance*, *BMI*, and *Age* appear to exert a lesser influence, as indicated by their lower values. This suggests that these particular features contribute to the model's prediction to a comparatively lesser extent or hold a reduced impact on the overall predictive performance of the model.

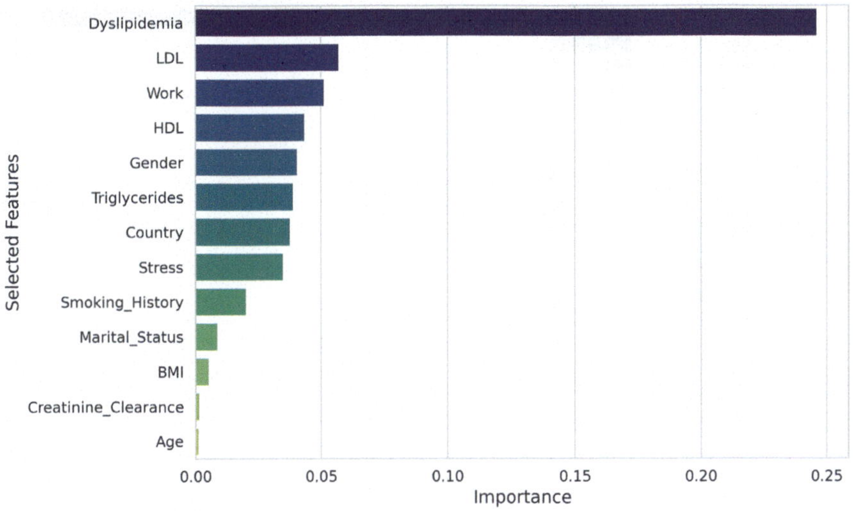

Fig. 5 Features importance using LASSO

Decision Tree

Utilizing decision tree feature selection, as depicted in the provided Fig. 6, the most important feature identified is *HbA1c_Admission*, likely indicating the level of hemoglobin A1c, a form chemically linked to sugar, at the time of admission. Subsequently, significant features include *Fasting_Blood_Glucose*, *Admission_Blood_Glucose*, *Heart_Rate*, *BMI*, *HDL*, *Creatinine_Clearance*, *Dyslipidemia*, *Cholesterol*, *Triglycerides*, *Systolic_BP*, *Diastolic_BP*, *Waist*, *Age*, and *"Hypertension"*. Conversely, *Marital_Status* and *Stress* emerge as the least influential features.

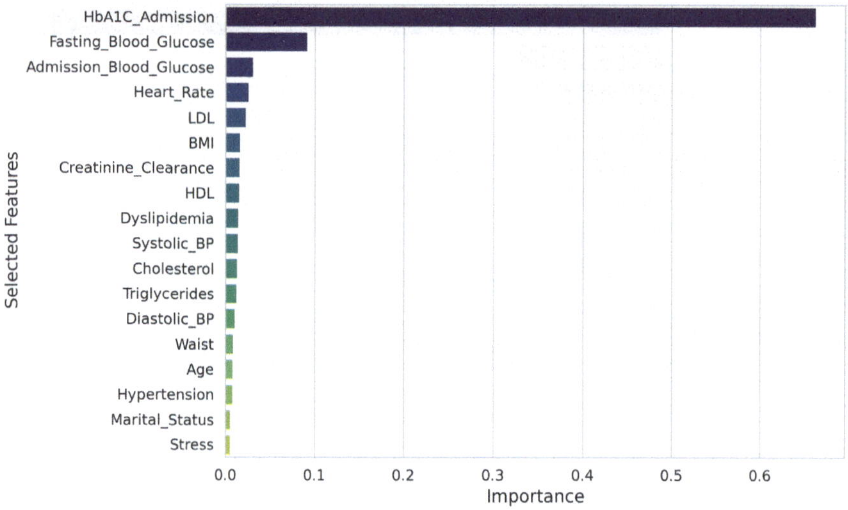

Fig. 6 Features importance using decision tree model

Random Forest

Considering the insights provided by Fig. 7, the important feature identified as the most influential is *HbA1c_Admission*. This particular feature is believed to signify the level of hemoglobin A1c-a variant of hemoglobin chemically associated with sugar.[1] Subsequently, the features that exhibit noteworthy importance in descending order include *Fasting_Blood_Glucose*, *Admission_Creatinine*, *LDL*, *Dyslipidemia*, *Cholesterol*, *Hypertension*, *Age*, *BMI*, *Heart_Rate*, *Triglycerides*, *HDL*, *Systolic_BP*, *Diastolic_BP*, *Education*, *Waist*, *Country*, and *Smoking_History*. This ranking implies the relative significance of these features in contributing to the predictive performance of the random forest model.

[1] At the time of admission.

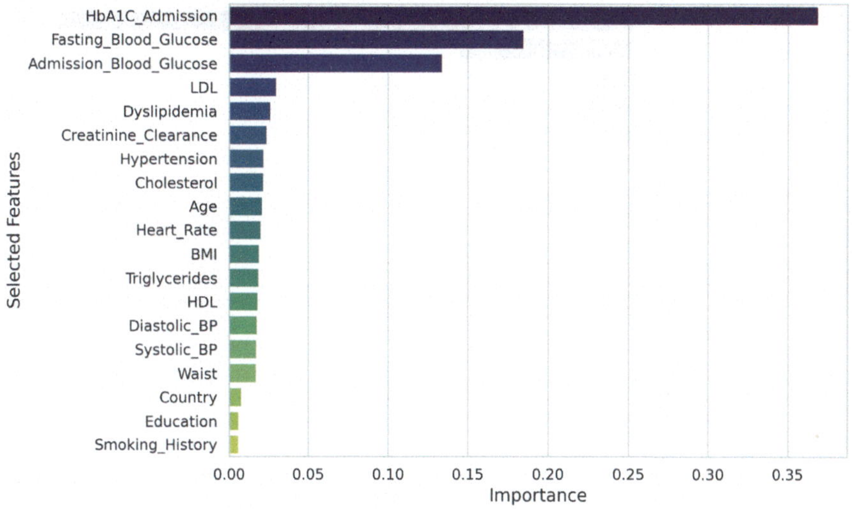

Fig. 7 Features importance using random forest model

Permutation Importance

Considering the insights provided by Fig. 8, the most important feature identified is *HbA1c_Admission*. This feature likely indicates the level of hemoglobin A1c-a form of hemoglobin chemically linked to sugar-at the time of admission. Subsequently, the features that exhibit noteworthy importance in descending order include *Fasting_Blood_Glucose*, *Admission_Blood_Glucose*, *Dyslipidemia*, *LDL*, *Creatinine_Clearance*, *Heart_Rate*, *Age*, *Triglycerides*, *BMI*, *Diastolic_BP*, *HDL*, *Waist*, and *Systolic_BP*. The least important feature is *Cholesterol*.

For Recursive Feature Elimination (RFE) and Genetic Algorithm (GA), the approach to feature selection does not involve assigning explicit importance scores to individual features. Instead, both techniques operate through iterative elimination processes.

In the case of RFE, features are systematically removed in successive iterations, with the algorithm assessing the impact on model performance at each step. The elimination decisions are based on the features contributions to overall performance metrics, resulting in a refined subset optimized for the chosen evaluation criteria.

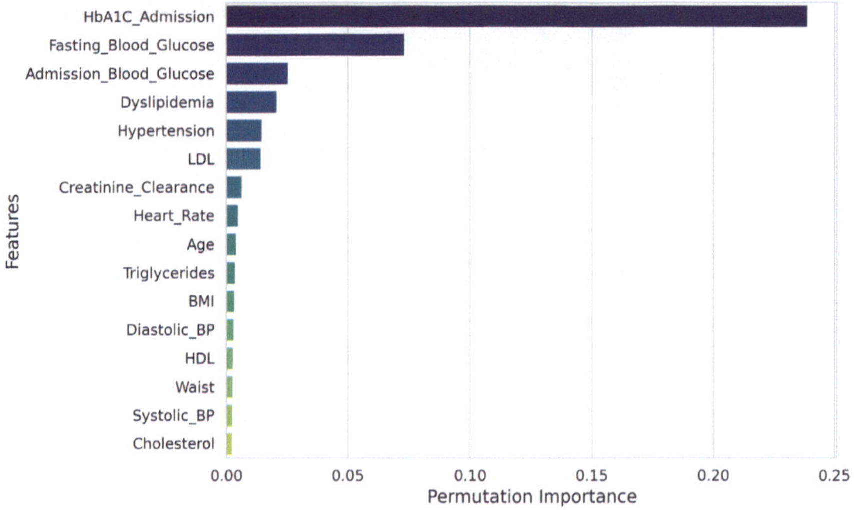

Fig. 8 Features importance using permutation importance

Similarly, genetic algorithm adopts an evolutionary methodology, treating features as genes within a population. Over successive generations, feature subsets undergo combination and modification, and their performance is evaluated using a fitness function. Features are selectively eliminated or retained based on their contribution to the model's fitness, leading to the evolution of feature subsets that exhibit improved predictive capabilities. both RFE and genetic algorithm achieve feature selection by iteratively refining feature subsets, utilizing elimination strategies to enhance the predictive efficiency and effectiveness of the model.

4.2.2 Predictive Modeling with Selected Features

Our selected strategy involves utilizing the SVM for predictive modeling, driven by its exceptional performance in our specific case. The meticulous process of feature selection, wherein features with significant contributions to predictive accuracy are identified, establishes SVM as the preferred algorithm. The decision to employ SVM is based on its capacity to effectively manage intricate relationships within the selected features, resulting in robust and accurate predictions. The use of SVM highlights a strategic alignment with our dataset's characteristics, emphasizing its effectiveness in achieving optimal classification outcomes based on the chosen subset of features. This section seeks to illustrate how the collaborative synergy between feature selection and the SVM algorithm enhances the overall predictive modeling capabilities tailored to our unique context.

Table 3 presents the performance evaluation with feature selection. Notably, the genetic algorithm stands out with the highest accuracy of 0.9037, precision of 0.9358,

Table 3 Performance evaluation with feature selection

Algorithm	Accuracy	Precision	Recall	F1-Score
LASSO	0.8993	0.9304	0.8859	0.9076
Decision Tree	0.8963	0.9276	0.8833	0.9049
Random Forest	0.8978	0.9278	0.8859	0.9064
Recursive Feature Elimination	0.6756	0.6975	0.7401	0.7181
Permutation Importance	0.8489	0.8706	0.8568	0.8636
Genetic Algorithm	**0.9037**	**0.9358**	**0.8886**	**0.9116**

recall of 0.8886, and F1-Score of 0.9116, marking it as the top-performing algorithm among the evaluated methods. LASSO, Decision Tree, and Random Forest also demonstrate commendable performance, with accuracies above 0.89 and well-balanced precision, recall, and F1-Score values. However, Recursive Feature Elimination lags behind in accuracy and other metrics, indicating potential limitations in its feature selection approach. Permutation Importance shows strong overall performance, particularly in precision and F1-Score. The results emphasize the effectiveness of feature selection in enhancing the predictive capabilities of these algorithms, with the genetic algorithm leading in achieving optimal classification outcomes.

It is noteworthy that the utilization of the genetic algorithm for feature selection yields results that outperform the performance achieved without employing feature selection, as shown in Fig. 1. This significant observation highlights the impactful role of feature selection strategies, particularly the genetic algorithm, in enhancing the overall predictive capabilities of the machine learning model for diabetes prediction. The notable enhancement in performance serves as evidence of the effectiveness of feature selection, highlighting its role in optimizing classification outcomes within the context of our study.

4.2.3 Predictive Insights: Diabetes Complications Analysis

The examination of the decision tree, as shown in Fig. 9, yields valuable insights regarding the initial indicators of complications associated with diabetes. A factor identified is the level of *HbA1C_Admission*, a critical gauge of long-term blood sugar control. Individuals maintaining *HbA1C_Admission* levels below 6.328729 exhibit a diminished likelihood of encountering diabetes-related complications. Additionally, early signs manifest with lower *Fasting_Blood_Glucose* levels, specifically below 8.013898, contributing to a reduced risk of complications. In cases where *HbA1C_Admission* equals or surpasses 6.328729, the model further refines early signs by incorporating additional factors. Notably, individuals with elevated *HbA1C_Admission* levels, notably exceeding 7.11, face a heightened susceptibility to complications, underscoring the critical role of glycemic control in promptly

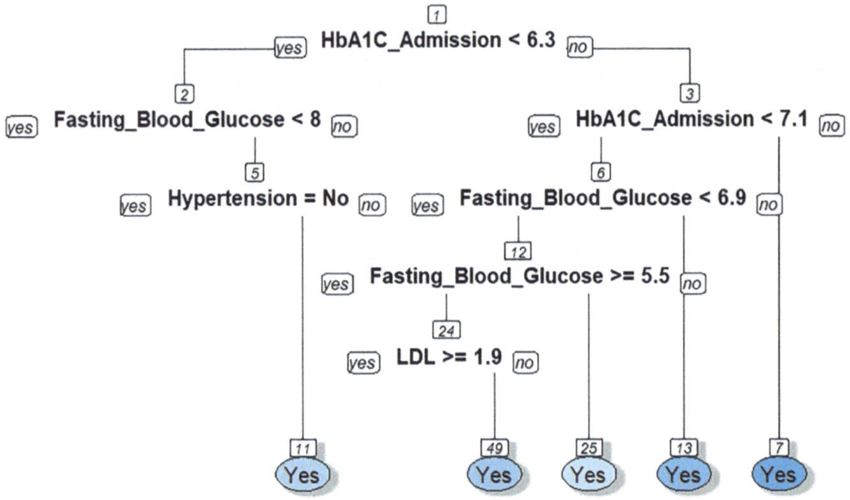

Fig. 9 Decision tree predictive insights

identifying diabetes-related issues. Moreover, the decision tree highlights specific conditions such as *LDL* levels and *Fasting_Blood_Glucose* concentrations as contributors to predicting complications. This provides healthcare professionals with invaluable insights into potential early markers for intervention. Overall, this analysis emphasizes the significance of vigilant monitoring of these key parameters for the early identification and effective management of complications associated with diabetes.

5 Discussion

In our analysis, we assessed the effectiveness of various machine learning algorithms in predicting diabetes, scrutinizing both model performance and the influence of feature selection methods. The outcomes underscored the superior performance of the SVM, which exhibited the highest accuracy, precision, and F1-score, establishing it as a robust choice for diabetes prediction. Conversely, the KNN algorithm displayed comparatively diminished performance, implying limitations in its suitability for this specific task. Logistic regression, Naive Bayes, decision tree, and ANN yielded competitive results, showcasing unique strengths within their respective frameworks.

An in-depth analysis of feature selection methods revealed discernible preferences, consistently identifying specific features like age, hypertension, dyslipidemia, and cholesterol as noteworthy predictors. Notably, the genetic algorithm emerged as the most efficient method for feature selection, contributing to enhanced classification outcomes.

Our investigation delved further into predictive insights through decision tree analysis. Important factors, including *HbA1C_Admission* and *LDL* levels, emerged as indicators of diabetes-related complications. This intricate understanding empowers healthcare professionals to tailor interventions based on individual risk profiles, highlighting the important significance of early monitoring and glycemic control.

6 Conclusion

The study aimed to tackle the ever-growing challenge of diabetes, emphasizing the importance of early detection and prediction to enhance healthcare interventions. By leveraging advanced machine learning and deep learning techniques, the research evaluated various algorithms, focusing on model performance and feature selection methods to predict diabetes risk and its early complications. The integration of sophisticated feature selection techniques such as LASSO, RFE, Random Forest, and genetic algorithms played a role in refining the predictive models by identifying the most influential factors. The methodology involved the application of various machine learning algorithms including decision trees, naive Bayes, SVM, KNN, logistic regression, and artificial neural networks, which were assessed for their effectiveness in predictive modeling.

Future perspectives include further refining these models by incorporating larger and more various datasets to improve generalizability. In addition, exploring hybrid models that combine the strengths of different machine learning algorithms could lead to even more robust predictions. Implementing real-time predictive analytics in clinical settings will be critical for timely interventions. Continued advances in machine learning techniques and their applications in healthcare have the potential to revolutionize diabetes management and ultimately reduce the global burden of this chronic disease.

Acknowledgements We gratefully thank the Principal Investigator of Gulf Coast Professor Mohammad Zubaid for providing the Gulf Coast dataset. In addition, we would like to express our sincere gratitude to "The Mohammed Bin Rashed University" and "Gulf Coast registry" for giving the permission to use Gulf Coast dataset.

References

1. Piid—dataset. https://catalog.data.gov/dataset/piid. Accessed: 2024-06-29
2. World health organization — diabetes. https://www.who.int/health-topics/diabetes#tab=tab_1. Accessed on 2024-06-29
3. Alam, T.M., Iqbal, M.A., Ali, Y., Wahab, A., Ijaz, S., Baig, T.I., Hussain, A., Malik, M.A., Raza, M.M., Ibrar, S., et al.: A model for early prediction of diabetes. Inform. Med. Unlocked **16**, 100204 (2019)
4. Altmann, A., Toloşi, L., Sander, O., Lengauer, T.: Permutation importance: a corrected feature importance measure. Bioinformatics **26**(10), 1340–1347 (2010)

5. Babatunde, O.H., Armstrong, L., Leng, J., Diepeveen, D.: A genetic algorithm-based feature selection (2014)
6. Barik, S., Mohanty, S., Mohanty, S., Singh, D.: Analysis of prediction accuracy of diabetes using classifier and hybrid machine learning techniques. In: Intelligent and Cloud Computing: Proceedings of ICICC 2019, vol. 2, pp. 399–409. Springer (2021)
7. Boutayeb, W., Badaoui, M., Al Ali, H., Boutayeb, A., Lamlili, M.: Use of data mining in the prediction of risk factors of type 2 diabetes mellitus in gulf countries. Math. Model. Comput. **4**(8), 638–645 (2021)
8. Brenning, A.: Transforming feature space to interpret machine learning models. arXiv preprint arXiv:2104.04295 (2021)
9. Castiglioni, I., Rundo, L., Codari, M., Di Leo, G., Salvatore, C., Interlenghi, M., Gallivanone, F., Cozzi, A., D'Amico, N.C., Sardanelli, F.: Ai applications to medical images: from machine learning to deep learning. Physica Medica **83**, 9–24 (2021)
10. Cıhan, P., Coşkun, H.: Performance comparison of machine learning models for diabetes prediction. In: 2021 29th Signal Processing and Communications Applications Conference (SIU), pp. 1–4. IEEE (2021)
11. DeMouy, J., Chamberlain, J., Harris, M., Marchand, L.: The pima Indians: pathfinders of health. Nat. Inst. Diabetes Digestive Kidney Diseases, Bethesda, MD (1995)
12. Emon, M.U., Keya, M.S., Kaiser, M.S., Tanha, T., Zulfiker, M.S., et al.: Primary stage of diabetes prediction using machine learning approaches. In: 2021 International Conference on Artificial Intelligence and Smart Systems (ICAIS), pp. 364–367. IEEE (2021)
13. Garg, A., Mago, V.: Role of machine learning in medical research: a survey. Comput. Sci. Rev. **40**, 100370 (2021)
14. Grabczewski, K., Jankowski, N.: Feature selection with decision tree criterion. In: Fifth International Conference on Hybrid Intelligent Systems (HIS'05), pp. 6–pp. IEEE (2005)
15. Guo, G., Wang, H., Bell, D., Bi, Y., Greer, K.: Knn model-based approach in classification. In: On The Move to Meaningful Internet Systems 2003: CoopIS, DOA, and ODBASE: OTM Confederated International Conferences, CoopIS, DOA, and ODBASE 2003, Catania, Sicily, Italy, November 3–7, 2003. Proceedings, pp. 986–996. Springer (2003)
16. Hasan, M.A.M., Nasser, M., Ahmad, S., Molla, K.I.: Feature selection for intrusion detection using random forest. J. Inf. Secur. **7**(3), 129–140 (2016)
17. Jian, Y., Pasquier, M., Sagahyroon, A., Aloul, F.: A machine learning approach to predicting diabetes complications. In: Healthcare, vol. 9, p. 1712. MDPI (2021)
18. Jović, A., Brkić, K., Bogunović, N.: A review of feature selection methods with applications. In: 2015 38th International Convention on Information and Communication Technology, Electronics and Microelectronics (MIPRO), pp. 1200–1205. IEEE (2015)
19. Kaneko, H.: Interpretation of machine learning models for data sets with many features using feature importance. ACS Omega **8**(25), 23218–23225 (2023)
20. Khaleel, F.A., Al-Bakry, A.M.: Diagnosis of diabetes using machine learning algorithms. Mater. Today: Proc. **80**, 3200–3203 (2023)
21. Khanam, J.J., Foo, S.Y.: A comparison of machine learning algorithms for diabetes prediction. Ict Express **7**(4), 432–439 (2021)
22. Krogh, A.: What are artificial neural networks? Nat. Biotechnol. **26**(2), 195–197 (2008)
23. LaValley, M.P.: Logistic regression. Circulation **117**(18), 2395–2399 (2008)
24. Mi, X., Zou, B., Zou, F., Hu, J.: Permutation-based identification of important biomarkers for complex diseases via machine learning models. Nat. Commun. **12**(1), 3008 (2021)
25. Mujumdar, A., Vaidehi, V.: Diabetes prediction using machine learning algorithms. Procedia Comput. Sci. **165**, 292–299 (2019)
26. Ng, A.Y.: Feature selection, l 1 vs. l 2 regularization, and rotational invariance. In: Proceedings of the Twenty-First International Conference on Machine Learning, p. 78 (2004)
27. Noble, W.S.: What is a support vector machine? Nat. Biotechnol. **24**(12), 1565–1567 (2006)
28. Pavani, K., Anjaiah, P., Krishna Rao, N., Deepthi, Y., Noel, D., Lokesh, V.: Diabetes prediction using machine learning techniques: a comparative analysis. In: Energy Systems, Drives and Automations: Proceedings of ESDA 2019, pp. 419–428. Springer (2020)

29. Rady, M., Moussa, K., Mostafa, M., Elbasry, A., Ezzat, Z., Medhat, W.: Diabetes prediction using machine learning: a comparative study. In: 2021 3rd Novel Intelligent and Leading Emerging Sciences Conference (NILES), pp. 279–282. IEEE (2021)
30. Sarwar, M.A., Kamal, N., Hamid, W., Shah, M.A.: Prediction of diabetes using machine learning algorithms in healthcare. In: 2018 24th International Conference on Automation and Computing (ICAC), pp. 1–6. IEEE (2018)
31. Shinde, P.P., Shah, S.: A review of machine learning and deep learning applications. In: 2018 Fourth International Conference on Computing Communication Control and Automation (ICCUBEA), pp. 1–6. IEEE (2018)
32. Sidey-Gibbons, J.A., Sidey-Gibbons, C.J.: Machine learning in medicine: a practical introduction. BMC Med. Res. Methodol. **19**, 1–18 (2019)
33. Tripathi, G., Kumar, R.: Early prediction of diabetes mellitus using machine learning. In: 2020 8th International Conference on Reliability, Infocom Technologies and Optimization (Trends and Future Directions) (ICRITO), pp. 1009–1014. IEEE (2020)
34. Webb, G.I., Keogh, E., Miikkulainen, R.: Naïve bayes. Encycl. Mach. Learn. **15**(1), 713–714 (2010)
35. Wiam, B., Mohammed, B., Hannah, A.A., Abdesslam, B., Mouhssine, D.: Machine learning methods in the detection of type 2 diabetes mellitus risk factors. In: International Conference of Machine Learning and Computer Science Applications, pp. 81–91. Springer (2022)
36. Yan, K., Zhang, D.: Feature selection and analysis on correlated gas sensor data with recursive feature elimination. Sens. Actuators B: Chem. **212**, 353–363 (2015)
37. Yuvaraj, N., SriPreethaa, K.: Diabetes prediction in healthcare systems using machine learning algorithms on hadoop cluster. Cluster Comput. **22**(Suppl 1), 1–9 (2019)
38. Zhang, L., Tan, J., Han, D., Zhu, H.: From machine learning to deep learning: progress in machine intelligence for rational drug discovery. Drug Discov. Today **22**(11), 1680–1685 (2017)

An Assessment of the Influence of Lifestyle Factors on the Risk of T2D in Adults Over the Age of 50

Abdelaadim Khriss, Mohammed Badaoui, Wiam Boutayeb, Hannah Al Ali, and Aissa Kerkour Elmiad

Abstract In this study, we investigate the impact of various lifestyle factors, such as marital status, employment status, and education level, on the risk of Type 2 Diabetes (T2D) in individuals over the age of 50. Using a large sample size and machine learning techniques, we further analyze these parameters. The results indicate significant associations between marital status, employment status, and education level, and the prevalence of T2D in this population. Notably, being single or never married is associated with an increased risk of T2D. In terms of occupational status, different categories have varying influences on the risk of T2D. The study highlights a statistically significant correlation between lower levels of education and an increased likelihood of developing T2D. These findings offer an in-depth view of the relationship between lifestyle factors and the risk of T2D in adults over 50, providing a basis for future research and interventions.

A. Khriss
Faculty of Sciences, University Mohammed First, Oujda, Morocco
e-mail: abdelaadim.khriss20@ump.ac.ma

M. Badaoui
Higher School of Technologie, University Mohammed First, Oujda, Morocco

W. Boutayeb (✉)
Higher School of Education and Training, Oujda, Morocco
e-mail: wiam.boutayeb@ump.ac.ma; wiam.boutayeb@gmail.com

University Mohamed Premier, Oujda, Morocco

H. Al Ali
Emirates Aviation University, Dubai, United Arab Emirates
e-mail: hannah.alali@emirates.com

A. Kerkour Elmiad
Faculty of Sciences, University Mohammed First, Oujda, Morocco
e-mail: a.kerkour@ump.ac.ma

A. Boutayeb and M. E. N. Lamlili (eds.), *Computational Mathematics and Modelling for Diabetes*, Industrial and Applied Mathematics,
https://doi.org/10.1007/978-981-96-1925-2_10

1 Introduction

In recent years, the global incidence of Type 2 Diabetes (T2D) has seen a rapid and widespread increase [3, 21]. This increase is largely due to a combination of factors such as environmental changes, lifestyle modifications and an aging population [6, 11, 13]. This article aims to explore in depth the complex interconnections between various lifestyle elements and their influence on the likelihood of developing T2D, particularly in people aged over 50. A closer look at the impact of diet and physical activity on the risk of T2D shows that a diet high in calorie-dense foods, typically linked to a Western-style dietary model, combined with a lack of physical activity, are singled out as the main culprits. This combination not only contributes significantly to rising rates of T2D, but is also at the root of the global obesity crisis [20]. Interestingly, studies suggest that elements such as increased abdominal obesity and misplaced fat accumulation, rather than traditional measures such as Body Mass Index (BMI), play a greater role in increasing the risk of T2D. Socio-demographic factors also appear to be crucial influencers of T2D risk. Factors such as age, ethnicity, family medical history and low socio-economic status are all recognized as impacting variables, particularly in regions such as Europe, known for its aging population, diverse immigrant groups and considerable socio-economic variation [14]. The subtle effect of marital status is also examined, with research indicating that people who remain married, despite significant weight gain, are less likely to develop diabetes than those who divorce [19]. Although the number of studies looking specifically at the role of work in the risk of T2D is limited, it is accepted that certain work-related factors can contribute to the disease [1]. For example, long periods of sedentary work are considered a potential risk factor for obesity and increased vulnerability to T2D [12]. Education level, another lifestyle factor, is examined for its link to risk T2D [17, 22, 23]. A low level of education often goes hand in hand with a lower socio-economic status, increasing the risk of pre-diabetes and T2D [8].

To understand the complexity of T2D risk, lifestyle factors, including diet, physical activity, habits and socio-demographic components, need to be studied in depth. Emphasis is placed on the need for further research to improve our understanding of these complex relationships, and to develop targeted approaches for effective T2D prevention.

2 Aim of the Current Study

The main objective of this study is to explore in depth the complexity of interactions between various lifestyle factors and their subsequent impact on the likelihood of developing T2D, with a particular focus on adults aged 50 and over.

Targeting this age group is of crucial importance for several reasons. Firstly, diabetes, particularly type 2, is a widespread health problem in the aging population [4, 9]. In addition, older people with T2D have an increased propensity for geriatric syndromes compared with their non-diabetic peers of the same age [10]. In addition, these individuals are at greater risk of diabetes-related complications, such as heart attacks and strokes [24]. Finally, diabetes can lead to serious health problems such as heart disease, stroke, kidney disease, eye problems and nerve damage that can lead to amputation [16].

Our research will involve a comprehensive examination of various lifestyle determinants. These include marital status, which may influence health behaviors and healthcare accessibility. Additionally, we will study occupational status, which may affect physical activity levels and dietary habits. Finally, we will consider education level, which may influence health knowledge and access to healthcare resources.

To conduct our survey in a reliable and valid manner, we will use a large sample size. In addition, we will apply advanced machine learning methods to analyze our data. The chi-square test of independence, a statistical tool, will be used to assess the significant relationship between the aforementioned categorical lifestyle variables and T2D in our research context.

Complementing the chi-square test, we will employ the gradient boosting machine method. This model will be deployed to anticipate the impact of lifestyle factors on the risk of developing T2D.

Through this comprehensive research, we aim to deepen our understanding of the complex relationships between lifestyle factors and the risk of T2D. These findings could contribute to the development of targeted strategies for more effective prevention of T2D.

In addition, our research could provide valuable information for public health policies and interventions. By identifying lifestyle factors that significantly amplify the risk of T2D in the elderly, we could help develop policies and interventions that promote healthier lifestyles, thereby reducing the risk of T2D in this demographic group. These findings could have significant implications for the health and well-being of older people, contributing to the overall goal of reducing the global burden of T2D.

3 Method

3.1 Sample and Procedure

Data for the case-control study were extracted from the "Gulf COAST" database. These data, collected by the Gulf COAST registry between 2012 and 2013, were obtained through surveys of patients admitted for acute coronary syndrome (ACS) during this period. The study encompassed 29 hospitals in four Gulf countries (UAE, Bahrain, Kuwait and Qatar). Of the 4,061 patients registered in the database and having consented to the use of their information, a sample of 2,621 patients aged over 50 was selected, based on the criterion of attendance at the corresponding hospitals. The main aim of the study was to anticipate risk factors for T2D T2D. Only non-diabetic patients and those with T2D were included in the sample, on condition that they provided a substantial proportion of their own information.

3.2 Measures

3.2.1 Outcome Variable

In the Gulf COAST registry dataset, the *DM* variable plays a crucial role in determining the incidence and effects of T2D within the study group. Defined as a binary variable, it effectively differentiates between those who have diabetes and those who do not. This binary categorization offers a simple but essential division, enabling researchers and medical practitioners to segregate the study group based on whether they have this metabolic condition or not. The presence of such a binary marker is critical for a detailed examination of the dataset, facilitating the discovery of trends, associations, and risk factors linked to diabetes.

Figure 1a illustrates two separate categories, labeled as "Yes" and "No" which signify the presence or absence of diabetes. The segment colored in blue, symbolizing the "Yes" category, escalates to nearly 60%, implying that about 60% of people within this age bracket are diagnosed with diabetes. Conversely, the segment in orange, which represents the "No" category, reaches close to 40%, indicating that nearly 40% of people in the same age bracket are not diagnosed with diabetes.

Changing the focus to Fig. 1b offers an in-depth illustration of the incidence of diabetes in people over 50, broken down by gender. It shows that about 70% of women over 50 are diabetic, leaving around 30% without the disease. On the other hand, about 55% of men in this age bracket have been diagnosed with diabetes, with the remaining 45% being non-diabetic. This information highlights a greater occurrence of diabetes in women over 50 compared to men of the same age group.

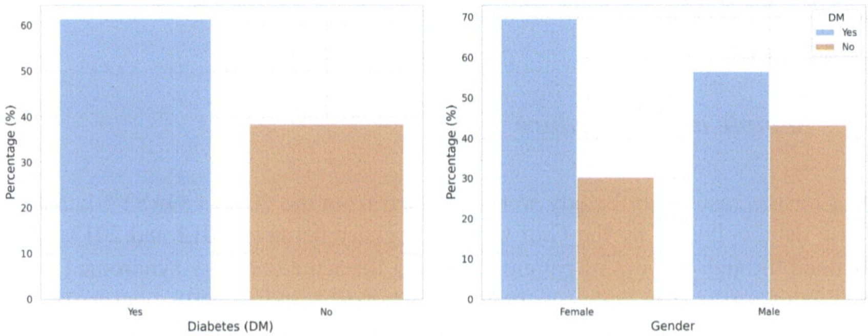

(a) Diabetes in Over-50s'

(b) Over-50s' diabetes by Gender

Fig. 1 Exploring diabetes: a gender-based distribution analysis and overall percentage insights

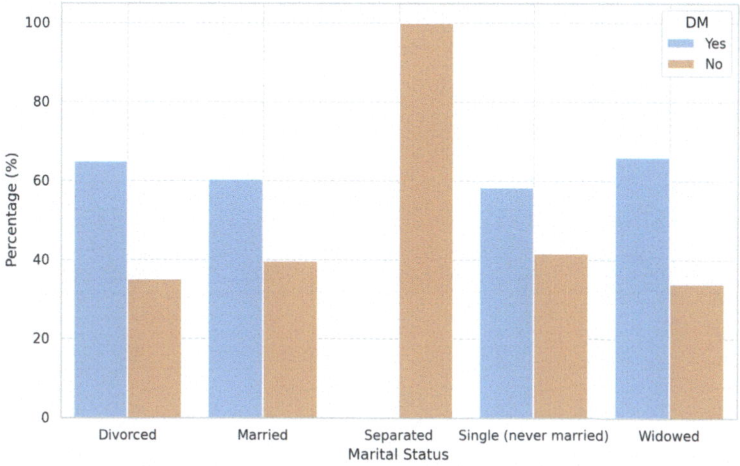

Fig. 2 Distribution of T2D based on marital status

3.2.2 Marital Status

The *Marital Status* is a key attribute that captures the varied marital conditions of individuals over 50 who took part in the research conducted in the Gulf countries between 2012 and 2013. This attribute provides a detailed insight into the social and relational facets of the participants, particularly in relation to diabetes, which is the main focus of the study.

The five statuses—*Married, Widowed, Divorced, Single (never married)*, and *Separated*—offer a comprehensive classification that is crucial in investigating possible correlations between marital status and outcomes related to diabetes. Acknowledging that people in different marital status categories may exhibit unique health behaviors, the *Marital Status* attribute proves to be an essential resource for deciphering the complex links between relational patterns and the incidence and management of diabetes.

Figure 2 illustrates the prevalence of diabetes across marital statuses. It is noteworthy that divorced and widowed individuals have a higher incidence of diabetes, at 64% and 65%, respectively. Conversely, those who are married or have never been married show slightly lower rates, at 60% and 58%, respectively. The separated group shows no diabetes cases, but this could be due to insufficient data. These observations hint at a possible link between marital status and the occurrence of diabetes.

3.2.3 Work

The variable *Work* represents the employment status of individuals, and is a non-numeric variable that includes three specific categories: *No*, *"Yes, Part-time"* and *"Yes, Full-time"*.

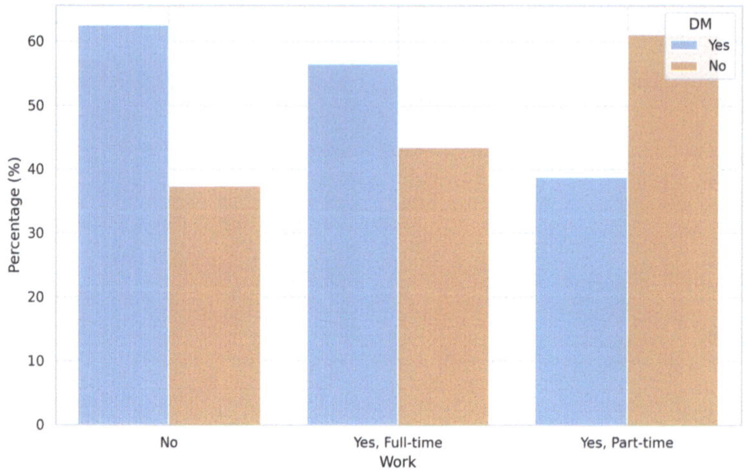

Fig. 3 Distribution of T2D based on employment status

This categorical division plays a crucial role in investigating possible correlations between employment status and diabetes-related outcomes in the study population. The various employment statuses could act as indicators for different lifestyle factors, such as stress levels, physical activity, and access to healthcare resources, which can all affect the incidence and management of diabetes.

The *Work* variable provides a detailed analysis of how employment status could influence the risk factors and progression of diabetes. For instance, full-time employees may face different health management challenges and opportunities compared to those who work part-time or not at all.

The results provided in Fig. 3 shed light on the correlation between employment status and the incidence of diabetes. Remarkably, individuals classified as *"No" (unemployed)* display the highest diabetes percentage at 62%, succeeded by those in full-time employment *"Yes, Full-time"* with a marginally lower yet significant percentage of 56.46%. On the other hand, individuals involved in part-time work *"Yes, Part-time"* demonstrate a considerably lower diabetes prevalence at 38.81%, coupled with a higher percentage of individuals without diabetes at 61.19%. This information implies a possible link between employment status and diabetes, with a trend showing higher diabetes prevalence among those unemployed or in full-time employment.

3.2.4 Education

The *Education* variable is a categorical feature comprises three distinct categories: *High School*, *No School*, and *College*.

This categorical breakdown provides a valuable lens through which researchers can explore the potential relationships between education and diabetes-related outcomes within the study population. Educational attainment is often linked to socio-economic status, health literacy, and access to resources, all of which can influence the prevalence and management of diabetes.

The *Education* variable enables a nuanced examination of how different levels of education might impact the risk factors and progression of diabetes. Individuals with varying educational backgrounds may have different health behaviors, levels of health literacy, and approaches to disease management.

The results in Fig. 4 illuminate the connection between the level of education and the incidence of diabetes. The data notably uncovers a noticeable trend where an increase in education levels corresponds to a decrease in diabetes prevalence. For instance, college graduates show a diabetes prevalence of 58%, which is lower than those with a high school education at 61% or those without any formal education also at 61%. In contrast, the proportion of individuals without diabetes exhibits a reverse pattern, implying that higher education levels might be linked to a higher probability of not having diabetes.

It should be noted that these results should be approached with caution, especially given that a correlation does not automatically imply a cause-and-effect relationship. The observed links may be influenced by various factors such as age, lifestyle, genetic factors and socio-economic conditions. To fully understand these trends, a more in-depth analysis is required, taking into account these complex influences and using a broader data set. In addition, it is essential to take into account the representativity of the sample and any possible bias in the data collection process to achieve a more thorough interpretation.

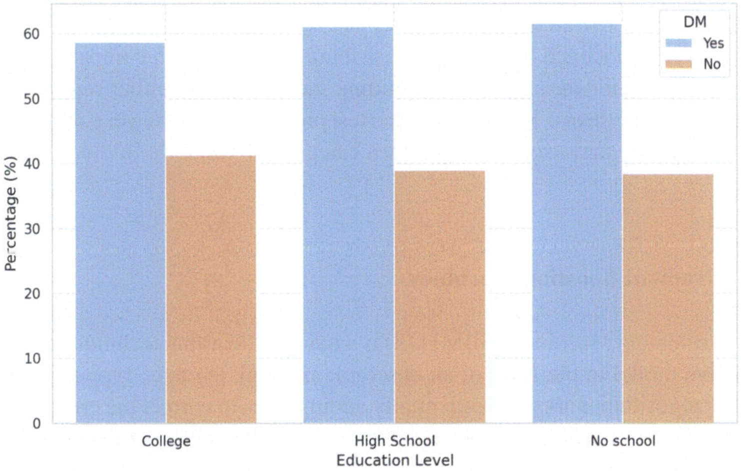

Fig. 4 Distribution of T2D based on education level

3.3 Analytic Approach

In this section, we outline the analytical methods used for examining the data and extracting important findings. Our approach combines both the Chi-Squared (χ^2) test and gradient boosting machines to gain a comprehensive understanding of the influence of lifestyle factors on T2D.

The algorithms were selected for their exclusive ability to (1) provide information on variable influence and (2) efficiently handle categorical and ordinal input variables, in addition to continuous variables.

3.3.1 Chi-Squared (χ^2) Test

The statistical method [15] is used to determine the association or independence between two categorical variables. It assesses whether observed frequencies in a contingency table significantly deviate from the expected frequencies under the assumption of independence.

It is expressed mathematically through the formula:

$$\chi^2 = \sum \frac{(O_i - E_i)^2}{E_i}. \tag{1}$$

Here, O_i represents the observed frequency in each cell of the contingency table, E_i denotes the expected frequency for the corresponding cell, and the summation extends over all cells. The test statistic, χ^2, quantifies the squared differences between observed and expected frequencies normalized by the expected frequencies. Degrees of freedom (df) are determined by the product of the number of rows (r) minus one and the number of columns (c) minus one, i.e., $df = (r - 1) \times (c - 1)$. The critical value or p-value obtained from the χ^2 distribution table is then compared to assess the statistical significance of the test, guiding the decision to either reject or fail to reject the null hypothesis. The Chi-Squared test provides a concise yet comprehensive means to evaluate the association between categorical variables in diverse fields of study.

3.3.2 Gradient Boosting Machines

Gradient Boosting Machines (GBM) [18] is a machine learning technique that builds a predictive model in the form of an ensemble of weak learners, typically decision trees. The algorithm works by sequentially adding trees to correct the errors made by the preceding ones, with each tree fitting to the residuals of the combined ensemble. The final prediction is a weighted sum of the individual tree predictions.

The mathematical formula for GBM involves the construction of an additive model composed of weak learners (usually decision trees). Let's denote the predictive model as $F(x)$ and the individual weak learners as $h_i(x)$, where i represents the iteration. The final model is built as follows:

$$F(x) = \sum_{i=1}^{N} \beta_i h_i(x).$$

Here, N is the number of weak learners, β_i is the weight assigned to the i-th weak learner, and $h_i(x)$ represents the output of the i-th weak learner for the input x. The algorithm iteratively adds weak learners to minimize a specified loss function by adjusting the weights and creating new weak learners that correct the errors of the combined ensemble up to that point.

The choice of GBM as the method for this study is due to its numerous advantages and proven effectiveness in a wide range of applications. Gradient Boosting has demonstrated superior accuracy compared to other ensemble methods and single model techniques. Its ability to reduce bias and variance by iteratively improving the model makes it highly effective for complex datasets [5]. GBM can handle various types of data (numeric, categorical) and can be customized with different loss functions to suit specific problems, enhancing its versatility in different contexts [2]. The method is robust to overfitting, especially with proper regularization techniques like shrinkage and subsampling, which helps in building generalizable models [7]. Despite being an ensemble method, the interpretability of the model remains relatively high due to the use of decision trees, which are inherently interpretable and can provide insights into feature importance. By citing these papers and highlighting their contributions, the robustness and versatility of GBM are well-supported, justifying its application in this study.

4 Results

The results of this study point to the importance of understanding the complex interaction between lifestyle choices and diabetes risk. Detailed research findings are presented in the following sections.

4.1 The Impact of Marital Status

The Chi-Squared test results in Table 1 reveal a significant association between Marital Status and the presence of T2D in the studied population. Both the Chi-Squared statistic ($X^2 = 20.775$, df$=4$, $p < 0.001$) and the likelihood ratio test (value = 21.090, df = 4, $p < 0.001$) consistently indicate a strong statistical relationship. The sample size of 3372 further strengthens the reliability of the findings. These results suggest that individuals' marital status is not independent of their likelihood of having T2D. While the statistical significance is established, it is crucial to recognize that these findings imply association and not causation.

Table 1 Chi-squared test results—marital status and T2D

	Value	df	p
X^2	20.775	4	<0.001
N	3372		

Analysis of the Table 2 results on the influence of marital status on the T2D reveals significant variations between the different categories. The data indicate that *(Single, never married)* status is strongly associated with changes in the T2D variable, with a notable influence of 0.125027. *Separated* individuals also show a marked influence 0.079464, exceeding that of *Divorced* individuals 0.029422 and *Married* individuals 0.057320. *Widowed* and *Married* individuals show similar influences on the T2D variable, with values of 0. 057579 and 0.057320, respectively. These results suggest that marital status may play a significant role in variations in the T2D, highlighting the importance of further exploring the dynamics between marital status and T2D in the context of this study.

Table 2 Influence of marital status on T2D

Marital Status	Influence
Marital_Status_Divorced	0.029422
Marital_Status_Married	0.057320
Marital_Status_Separated	0.079464
Marital_Status_Single (never married)	0.125027
Marital_Status_Widowed	0.057579

4.2 The Impact of Employment Status

The Chi-Squared test results in Table 3 for the association between *Work* and the presence of T2D reveal a highly significant relationship. The Chi-Squared statistic ($X^2 = 105.589$) with 2 degrees of freedom and a p-value less than 0.001 indicates a substantial departure from the null hypothesis of independence. This suggests that the distribution of individuals with and without T2D significantly differs across different work categories. The large Chi-Squared value implies a strong association, reinforcing the conclusion that there is a statistically significant link between work status and the likelihood of having T2D in the studied population of 3372 individuals.

Table 3 Chi-squared test results—employment status and T2D

	Value	df	p
X^2	105.589	2	<0.001
N	3372		

Table 4 sheds light on the influence of various work statuses on the likelihood of individuals having T2D. According to the provided influence values, individuals who are not employed (*Work_No*) exhibit a moderate impact with a value of 0.064801, implying that being unemployed is associated with a discernible influence on the likelihood of having T2D. On the other hand, individuals engaged in full-time employment (*Work_Yes, Full-time*) demonstrate a relatively lower influence with a value of 0.039654, suggesting a milder association between full-time employment and the occurrence of T2D. Notably, individuals working part-time (*Work_Yes, Part-time*) show the highest influence with a value of 0.165700, indicating a substantial impact on the likelihood of having T2D for those in part-time employment. This nuanced quantification provides valuable insights into the varying degrees of association between work statuses and the prevalence of T2D in the studied population, offering a more detailed understanding of the role of employment in influencing health outcomes.

Table 4 Influence of work on T2D

Employment status	Influence
Work_No	0.064801
Work_Yes, Full-time	0.039654
Work_Yes, Part-time	0.165700

4.3 *The Impact of Education Level*

The Chi-Squared test results in Table 5 for the impact of education level on the presence of T2D in the studied population indicate a statistically significant association. The Chi-Squared statistic ($X^2 = 25.339$) with 2 degrees of freedom and a p-value less than 0.001 suggests a significant departure from the null hypothesis of independence. This implies that there is a meaningful relationship between individuals' education levels and the likelihood of having T2D.

Shifting focus to Table 6 which illustrates the quantified influence of different education levels on the likelihood of individuals having T2D. Notably, individuals with a *Education_College* exhibit the highest influence with a substantial value of

Table 5 Chi-squared test results—education level and T2D

	Value	df	p
X^2	25.339	2	<0.001
N	3372		

0.287289, suggesting a pronounced impact on the occurrence of T2D. In contrast, completing *Education_High School* is associated with a comparatively lower influence, as indicated by the influence value of 0.026521. Individuals with no formal education, represented by the category *Education_No school*, demonstrate a moderate influence of 0.067222 on the likelihood of having T2D. These findings imply that higher educational attainment, specifically a college education, is more strongly associated with the occurrence of T2D in the studied population. The table thus contributes valuable insights into the nuanced relationship between education levels and the prevalence of T2D, highlighting the varying degrees of influence associated with different educational backgrounds.

Table 6 Influence of education on T2D

Education level	Influence
Education_College	0.287289
Education_High School	0.026521
Education_No school	0.067222

It's essential to note that while statistical significance is established, these findings do not imply causation. Further investigation and consideration of potential confounding factors are necessary to understand the nature and implications of the relationship between lifestyle factors and the likelihood of T2D.

5 Strengths and Limitations

This study has significant strengths that reinforce its credibility. Firstly, the sample of 2,621 patients aged over the age of 50, from 29 hospitals in four Gulf countries, offers a diverse representation of the target population. Data were collected from in-depth surveys of patients admitted for acute coronary syndrome (ACS), suggesting the availability of detailed information on lifestyle factors such as marital status, work and education. In addition, participants informed consent demonstrates an ethical approach to data collection, reinforcing the rigor of the study. The clear objective of anticipating risk factors for T2D also lends a precise direction to the research. However, certain limitations need to be taken into consideration. The geographical limitation to Gulf countries could restrict the generalizability of the results to other populations. In addition, the selection of participants based on the criterion of hospital attendance may introduce a selection bias, while the retroactivity of data between 2012 and 2013 may lead to limitations related to the availability of information. Dependencies on participants responses and the possible presence of unmeasured variables may influence the study's internal validity.

6 Conclusion

In conclusion, our study thoroughly investigated the impact of lifestyle factors, such as marital status, occupational status and education level, on the risk of T2D in adults over 50. Results obtained from a large sample size and using machine learning techniques reveal significant associations between these variables and the prevalence of T2D.

In particular, marital status emerged as a determining factor, with a notable association between being single or never married and an increased risk of T2D. In addition, occupational status showed variable influences on diabetes risk, underlining the importance of considering the various job categories in understanding this relationship. In addition, a statistically significant correlation was found between higher levels of education and an increased likelihood of developing T2D.

These results underline the importance of considering lifestyle aspects in the prevention and management of T2D in adults over 50. They provide a solid basis to guide future research and interventions aimed at better understanding and reducing the risks associated with this disease. By adopting a holistic lifestyle approach, it is possible to formulate more effective strategies for the health promotion and prevention of T2D in this specific population.

Future research should focus on conducting longitudinal studies to observe how changes in marital status, employment, and education levels over time affect the risk of developing T2D. This would help in understanding causality and the long-term impact of these lifestyle factors. Interventional studies targeting marital status could also be developed and evaluated, aiming to mitigate the increased risk of T2D in single or never-married individuals through social support programs or targeted health promotion activities. Additionally, investigating the effectiveness of workplace health programs tailored to different job categories, especially for those in part-time or full-time employment, would be beneficial. Programs focusing on reducing sedentary behavior and promoting healthy lifestyles could further enhance our understanding of these influences.

Examining the impact of educational campaigns on diabetes awareness and prevention, particularly targeting populations with lower educational attainment, is another crucial area for future research. Assessing whether increasing health literacy can reduce the prevalence of T2D would provide valuable insights. Exploring the role of socio-economic and cultural factors in the relationship between lifestyle variables and T2D risk can help tailor interventions to be culturally sensitive and more effective. Conducting gender-specific analyses to understand if and how the associations between lifestyle factors and T2D risk differ between men and women could lead to more personalized prevention strategies.

Investigating the interplay between genetic predispositions and lifestyle factors in the development of T2D is also essential. Identifying high-risk individuals through genetic screening combined with lifestyle interventions could be a future direction. Evaluating the impact of public health policies that promote healthy lifestyles, such as policies to increase physical activity or improve access to healthy foods, particularly among older adults, could further our understanding. Finally, utilizing technology, such as wearable devices and health apps, to monitor lifestyle behaviors and provide real-time interventions for individuals at risk of T2D would be a promising area for future research. By pursuing these research directions, future studies can build on the findings of this study to develop more targeted and effective strategies for the prevention and management of T2D in adults over 50.

Acknowledgements We gratefully thank the Principal Investigator of Gulf Coast Professor Mohammad Zubaid for providing the Gulf Coast dataset. In addition, we would like to express our sincere gratitude to "The Mohammed Bin Rashed University" and "Gulf Coast registry" for giving the permission to use Gulf Coast dataset.

References

1. Beulens, J.W., Pinho, M.G., Abreu, T.C., den Braver, N.R., Lam, T.M., Huss, A., Vlaanderen, J., Sonnenschein, T., Siddiqui, N.Z., Yuan, Z., et al.: Environmental risk factors of type 2 diabetes—an exposome approach. Diabetologia **65**(2), 263–274 (2022)
2. Chen, T., Guestrin, C.: Xgboost: a scalable tree boosting system. In: Proceedings of the 22nd ACM SIGKDD International Conference on Knowledge Discovery and Data Mining, pp. 785–794 (2016)
3. Collaboration, E.R.F., et al.: Diabetes mellitus, fasting blood glucose concentration, and risk of vascular disease: a collaborative meta-analysis of 102 prospective studies. The Lancet **375**(9733), 2215–2222 (2010)
4. ElSayed, N.A., Aleppo, G., Aroda, V.R., Bannuru, R.R., Brown, F.M., Bruemmer, D., Collins, B.S., Hilliard, M.E., Isaacs, D., Johnson, E.L., et al.: 13. older adults: standards of care in diabetes–2023. Diab. Care **46**(Supplement_1), S216–S229 (2023)
5. Friedman, J.H.: Greedy function approximation: a gradient boosting machine. Ann. Stat. 1189–1232 (2001)
6. Geng, T., Zhu, K., Lu, Q., Wan, Z., Chen, X., Liu, L., Pan, A., Liu, G.: Healthy lifestyle behaviors, mediating biomarkers, and risk of microvascular complications among individuals with type 2 diabetes: a cohort study. PLoS Med. **20**(1), e1004135 (2023)
7. Hastie, T., Tibshirani, R., Friedman, J.H., Friedman, J.H.: The elements of statistical learning: data mining, inference, and prediction, vol. 2. Springer (2009)
8. Hill-Briggs, F., Adler, N.E., Berkowitz, S.A., Chin, M.H., Gary-Webb, T.L., Navas-Acien, A., Thornton, P.L., Haire-Joshu, D.: Social determinants of health and diabetes: a scientific review. Diab. Care **44**(1), 258 (2021)
9. Kalyani, R.R., Golden, S.H., Cefalu, W.T.: Diabetes and aging: unique considerations and goals of care. Diab. Care **40**(4), 440 (2017)

10. Kirkman, M.S., Briscoe, V.J., Clark, N., Florez, H., Haas, L.B., Halter, J.B., Huang, E.S., Korytkowski, M.T., Munshi, M.N., Odegard, P.S., et al.: Diabetes in older adults. Diab. Care **35**(12), 2650 (2012)
11. Kolb, H., Martin, S.: Environmental/lifestyle factors in the pathogenesis and prevention of type 2 diabetes. BMC Med. **15**(1), 1–11 (2017)
12. Kumar, L., Mohammadnezhad, M.: Health care workers' perceptions on factors affecting diabetes self-management among type 2 diabetes mellitus patients in fiji: a qualitative study. Front. Public Health **10**, 779266 (2022)
13. Kyrou, I., Tsigos, C., Mavrogianni, C., Cardon, G., Van Stappen, V., Latomme, J., Kivelä, J., Wikström, K., Tsochev, K., Nanasi, A., et al.: Sociodemographic and lifestyle-related risk factors for identifying vulnerable groups for type 2 diabetes: a narrative review with emphasis on data from Europe. BMC Endocr. Disord. **20**, 1–13 (2020)
14. Manios, Y., Androutsos, O., Lambrinou, C.P., Cardon, G., Lindstrom, J., Annemans, L., Mateo-Gallego, R., de Sabata, M.S., Iotova, V., Kivela, J., et al.: A school-and community-based intervention to promote healthy lifestyle and prevent type 2 diabetes in vulnerable families across Europe: design and implementation of the feel4diabetes-study. Public Health Nutr. **21**(17), 3281–3290 (2018)
15. McHugh, M.L.: The chi-square test of independence. Biochemia Medica **23**(2), 143–149 (2013)
16. Meneilly, G.S., Knip, A., Miller, D.B., Sherifali, D., Tessier, D., Zahedi, A., Committee, D.C.C.P.G.E., et al.: Diabetes in older people. Can. J. Diab. **42**, S283–S295 (2018)
17. Na-Ek, N., Srithong, J., Aonkhum, A., Boonsom, S., Charoen, P., Demakakos, P.: Educational level as a cause of type 2 diabetes mellitus: caution from triangulation of observational and genetic evidence. Acta Diabetol. 1–9 (2022)
18. Natekin, A., Knoll, A.: Gradient boosting machines, a tutorial. Front. Neurorobotics **7**, 21 (2013)
19. de Oliveira, C.M., Viater Tureck, L., Alvares, D., Liu, C., Horimoto, A.R.V.R., Balcells, M., de Oliveira Alvim, R., Krieger, J.E., Pereira, A.C.: Relationship between marital status and incidence of type 2 diabetes mellitus in a Brazilian rural population: the Baependi heart study. PloS One **15**(8), e0236869 (2020)
20. Smith, A.D., Crippa, A., Woodcock, J., Brage, S.: Physical activity and incident type 2 diabetes mellitus: a systematic review and dose-response meta-analysis of prospective cohort studies. Diabetologia **59**(12), 2527–2545 (2016)
21. System, U.S.R.D.: 2018 USRDS annual data report: epidemiology of kidney disease in the United States (2018)
22. Yuan, S., Larsson, S.C.: An atlas on risk factors for type 2 diabetes: a wide-angled mendelian randomisation study. Diabetologia **63**, 2359–2371 (2020)
23. Zhang, J., Chen, Z., Pärna, K., van Zon, S.K., Snieder, H., Thio, C.H.: Mediators of the association between educational attainment and type 2 diabetes mellitus: a two-step multivariable mendelian randomisation study. Diabetologia **65**(8), 1364–1374 (2022)
24. Zheng, Y., Ley, S.H., Hu, F.B.: Global aetiology and epidemiology of type 2 diabetes mellitus and its complications. Nat. Rev. Endocrinol. **14**(2), 88–98 (2018)

The manufacturer's authorised representative in the EU is Springer Nature Customer Service Centre GmbH, Europaplatz 3, 69115 Heidelberg, Germany. If you have any concerns regarding our products, please contact ProductSafety@springernature.com

Printed and bound by CPI Group (UK) Ltd, Croydon, CR0 4YY
01/10/2026
02225932-0008